Probabilistic Risk Analysis: Foundations and Methods

Probabilistic Risk Analysis: Foundations and Methods

Tim Bedford
Delft University of Technology
and
University of Strathclyde

Roger Cooke
Delft University of Technology

CAMBRIDGE
UNIVERSITY PRESS

University Printing House, Cambridge CB2 8BS, United Kingdom

Cambridge University Press is part of the University of Cambridge.

It furthers the University's mission by disseminating knowledge in the pursuit of education, learning and research at the highest international levels of excellence.

www.cambridge.org
Information on this title: www.cambridge.org/9780521773201

© Cambridge University Press 2001

This publication is in copyright. Subject to statutory exception and to the provisions of relevant collective licensing agreements, no reproduction of any part may take place without the written permission of Cambridge University Press.

First published 2001
7th printing 2011

A catalogue record for this publication is available from the British Library

Library of Congress Cataloguing in Publication data
Bedford T.
Mathematical tools for probabilistic risk analysis / Tim Bedford and Roger M. Cooke.
    p.   cm.
Includes bibliographical references and index.
ISBN 0 521 77320 2
    1. Reliability (Engineering) Mathematics. 2. Risk assessment. 3. Probabilities. I. Cooke, Roger M., 1946- II. Title.

TA169.B44 2001
620'.00452'0151–dc21   00-046762

ISBN 978-0-521-77320-1 Hardback

Cambridge University Press has no responsibility for the persistence or accuracy of URLs for external or third-party internet websites referred to in this publication, and does not guarantee that any content on such websites is, or will remain, accurate or appropriate.

Contents

Illustrations	*page*	xiii
Tables		xvi
Preface		xix

	Part I: Introduction	1
1	**Probabilistic risk analysis**	3
1.1	Historical overview	4
	1.1.1 The aerospace sector	4
	1.1.2 The nuclear sector	5
	1.1.3 The chemical process sector	8
	1.1.4 The less recent past	9
1.2	What is the definition of risk?	9
1.3	Scope of probabilistic risk analyses	11
1.4	Risk analysis resources	12
	1.4.1 Important journals	12
	1.4.2 Handbooks	12
	1.4.3 Professional organizations	12
	1.4.4 Internet	13
	Part II: Theoretical issues and background	15
2	**What is uncertainty?**	17
2.1	The meaning of meaning	17
2.2	The meaning of uncertainty	19
2.3	Probability axioms	21
	2.3.1 Interpretations	22
2.4	Savage's theory of rational decision	24
	2.4.1 Savage's axioms	26
	2.4.2 Quantitative probability	28

	2.4.3	Utility	28
	2.4.4	Observation	28
2.5	Measurement of subjective probabilities		30
2.6	Different types of uncertainty		33
2.7	Uncertainty about probabilities		35

3 Probabilistic methods — 39

3.1	Review of elementary probability theory		39
3.2	Random variables		41
	3.2.1	Moments	42
	3.2.2	Several random variables	43
	3.2.3	Correlations	44
	3.2.4	Failure rates	45
3.3	The exponential life distribution		47
	3.3.1	Constant test intervals	48
	3.3.2	Exponential failure and repair	50
3.4	The Poisson distribution		51
3.5	The gamma distribution		52
3.6	The beta distribution		53
3.7	The lognormal distribution		54
3.8	Stochastic processes		55
3.9	Approximating distributions		58

4 Statistical inference — 61

4.1	Foundations		61
4.2	Bayesian inference		63
	4.2.1	Bayes' Theorem	64
	4.2.2	An example with the exponential distribution	67
	4.2.3	Conjugate distributions	69
	4.2.4	First find your prior	70
	4.2.5	Point estimators from the parameter distribution	74
	4.2.6	Asymptotic behaviour of the posterior	74
4.3	Classical statistical inference		75
	4.3.1	Estimation of parameters	75
	4.3.2	Non-parametric estimation	77
	4.3.3	Confidence intervals	78
	4.3.4	Hypothesis testing	79

5 Weibull Analysis — 83

5.1	Definitions		85
5.2	Graphical methods for parameter fitting		85
	5.2.1	Rank order methods	86
	5.2.2	Suspended or censored items	88

	5.2.3 The Kaplan–Meier estimator	91
5.3	Maximum likelihood methods for parameter estimation	92
5.4	Bayesian estimation	94
5.5	Extreme value theory	94

Part III: System analysis and quantification — 97

6	**Fault and event trees**	**99**
6.1	Fault and event trees	99
6.2	The aim of a fault-tree analysis	100
6.3	The definition of a system and of a top event	103
	6.3.1 External boundaries	103
	6.3.2 Internal boundaries	104
	6.3.3 Temporal boundaries	104
6.4	What classes of faults can occur?	104
	6.4.1 Active and passive components	105
	6.4.2 Primary, secondary and command faults	105
	6.4.3 Failure modes, effects and mechanisms	105
6.5	Symbols for fault trees	106
6.6	Fault tree construction	106
6.7	Examples	108
	6.7.1 Reactor vessel	108
	6.7.2 New Waterway barrier	109
6.8	Minimal path and cut sets for coherent systems	110
	6.8.1 Cut sets	110
	6.8.2 Path sets	112
6.9	Set theoretic description of cut and path sets	112
	6.9.1 Boolean algebra	112
	6.9.2 Cut set representation	114
	6.9.3 Path set representation	115
	6.9.4 Minimal cut set/path set duality	115
	6.9.5 Parallel and series systems	117
6.10	Estimating the probability of the top event	117
	6.10.1 Common cause	118
7	**Fault trees – analysis**	**121**
7.1	The MOCUS algorithm for finding minimal cut sets	121
	7.1.1 Top down substitution	121
	7.1.2 Bottom up substitution	122
	7.1.3 Tree pruning	122
7.2	Binary decision diagrams and new algorithms	123
	7.2.1 Prime implicants calculation	129

	7.2.2	Minimal p-cuts	130
	7.2.3	Probability calculations	132
	7.2.4	Examples	132
	7.2.5	The size of the BDD	134
7.3	Importance		135

8 Dependent failures — 140

8.1	Introduction	140
8.2	Component failure data versus incident reporting	140
8.3	Preliminary analysis	141
8.4	Inter-system dependencies	143
8.5	Inter-component dependencies – common cause failure	143
8.6	The square root bounding model	143
8.7	The Marshall–Olkin model	143
8.8	The beta-factor model	146
	8.8.1 Parameter estimation	147
8.9	The binomial failure rate model	148
8.10	The α-factor model	151
8.11	Other models	151

9 Reliability data bases — 153

9.1	Introduction		153
9.2	Maintenance and failure taxonomies		156
	9.2.1	Maintenance taxonomy	156
	9.2.2	Failure taxonomy	157
	9.2.3	Operating modes; failure causes; failure mechanisms and failure modes	158
9.3	Data structure		160
	9.3.1	Operations on data	161
9.4	Data analysis without competing risks		163
	9.4.1	Demand related failures: non-degradable components	163
	9.4.2	Demand related failures: degradable components	164
	9.4.3	Time related failures; no competing risks	165
9.5	Competing risk concepts and methods		166
	9.5.1	Subsurvivor functions and identifiability	168
	9.5.2	Colored Poisson representation of competing risks	170
9.6	Competing risk models		172
	9.6.1	Independent exponential competing risk	172
	9.6.2	Random clipping	175
	9.6.3	Random signs	175
	9.6.4	Conditionally independent competing risks	177
	9.6.5	Time window censoring	179

9.7	Uncertainty	179
	9.7.1 Uncertainty due to non-identifiability: bounds in the absence of sampling fluctuations	180
	9.7.2 Accounting for sampling fluctuations	182
	9.7.3 Sampling fluctuations of Peterson bounds	182
9.8	Examples of dependent competing risk models	184
	9.8.1 Failure effect	185
	9.8.2 Action taken	186
	9.8.3 Method of detection	188
	9.8.4 Subcomponent	189
	9.8.5 Conclusions	189
10	**Expert opinion**	**191**
10.1	Introduction	191
10.2	Generic issues in the use of expert opinion	192
10.3	Bayesian combinations of expert assessments	192
10.4	Non-Bayesian combinations of expert distributions	194
10.5	Linear opinion pools	199
10.6	Performance based weighting – the classical model	199
	10.6.1 Calibration	200
	10.6.2 Information	202
	10.6.3 Determining the weights	203
	10.6.4 Approximation of expert distributions	206
10.7	Case study – uncertainty in dispersion modeling	208
11	**Human reliability**	**218**
11.1	Introduction	218
11.2	Generic aspects of a human reliability analysis	220
	11.2.1 Human error probabilities	220
	11.2.2 Task analysis	220
	11.2.3 Performance and error taxonomy	221
	11.2.4 Performance shaping factors	223
11.3	THERP – technique for human error rate prediction	224
	11.3.1 Human error event trees	226
	11.3.2 Performance shaping factors	227
	11.3.3 Dependence	227
	11.3.4 Time dependence and recovery	228
	11.3.5 Distributions for HEPs	228
11.4	The Success Likelihood Index Methodology	230
11.5	Time reliability correlations	232
11.6	Absolute Probability Judgement	235
11.7	Influence diagrams	236
11.8	Conclusions	238

12	**Software reliability**	**240**
12.1	Qualitative assessment – ways to find errors	240
	12.1.1 FMECAs of software-based systems	240
	12.1.2 Formal design and analysis methods	241
	12.1.3 Software sneak analysis	241
	12.1.4 Software testing	241
	12.1.5 Error reporting	242
12.2	Software quality assurance	242
	12.2.1 Software safety life-cycles	242
	12.2.2 Development phases and reliability techniques	243
	12.2.3 Software quality	245
	12.2.4 Software quality characteristics	245
	12.2.5 Software quality metrics	245
12.3	Software reliability prediction	245
	12.3.1 Error seeding	247
	12.3.2 The Jelinski–Moranda model	247
	12.3.3 Littlewood's model	248
	12.3.4 The Littlewood–Verral model	249
	12.3.5 The Goel–Okumoto model	250
12.4	Calibration and weighting	251
	12.4.1 Calibration	251
	12.4.2 Weighted mixtures of predictors	253
12.5	Integration errors	253
12.6	Example	255
	Part IV: Uncertainty modeling and risk measurement	**257**
13	**Decision theory**	**259**
13.1	Preferences over actions	261
13.2	Decision tree example	262
13.3	The value of information	264
	13.3.1 When do observations help?	267
13.4	Utility	268
13.5	Multi-attribute decision theory and value models	269
	13.5.1 Attribute hierarchies	270
	13.5.2 The weighting factors model	271
	13.5.3 Mutual preferential independence	271
	13.5.4 Conditional preferential independence	274
	13.5.5 Multi-attribute utility theory	277
	13.5.6 When do we model the risk attitude?	280
	13.5.7 Trade-offs through time	281

13.6	Other popular models	281
	13.6.1 Cost–benefit analysis	281
	13.6.2 The analytic hierarchy process	283
13.7	Conclusions	283

14 Influence diagrams and belief nets — 286

14.1	Belief networks	286
14.2	Conditional independence	288
14.3	Directed acyclic graphs	289
14.4	Construction of influence diagrams	290
	14.4.1 Model verification	292
14.5	Operations on influence diagrams	294
	14.5.1 Arrow reversal	294
	14.5.2 Chance node removal	294
14.6	Evaluation of influence diagrams	295
14.7	The relation with decision trees	295
14.8	An example of a Bayesian net application	296

15 Project risk management — 299

15.1	Risk management methods	300
	15.1.1 Identification of uncertainties	300
	15.1.2 Quantification of uncertainties	302
	15.1.3 Calculation of project risk	302
15.2	The Critical Path Method (CPM)	302
15.3	Expert judgement for quantifying uncertainties	304
15.4	Building in correlations	305
15.5	Simulation of completion times	305
15.6	Value of money	306
15.7	Case study	307

16 Probabilistic inversion techniques for uncertainty analysis — 316

16.1	Elicitation variables and target variables	318
16.2	Mathematical formulation of probabilistic inversion	319
16.3	PREJUDICE	320
	16.3.1 Heuristics	320
	16.3.2 Solving for minimum information	321
16.4	Infeasibility problems and PARFUM	322
16.5	Example	323

17 Uncertainty analysis — 326

17.1	Introduction	326
	17.1.1 Mathematical formulation of uncertainty analysis	326
17.2	Monte Carlo simulation	327

	17.2.1 Univariate distributions	327
	17.2.2 Multivariate distributions	328
	17.2.3 Transforms of joint normals	329
	17.2.4 Rank correlation trees	330
	17.2.5 Vines	334
17.3	Examples: uncertainty analysis for system failure	339
	17.3.1 The reactor example	339
	17.3.2 Series and parallel systems	341
	17.3.3 Dispersion model	342
17.5	Appendix: bivariate minimally informative distributions	346
	17.5.1 Minimal information distributions	346
18	**Risk measurement and regulation**	**350**
18.1	Single statistics representing risk	350
	18.1.1 Deaths per million	350
	18.1.2 Loss of life expectancy	351
	18.1.3 Delta yearly probability of death	353
	18.1.4 Activity specific hourly mortality rate	354
	18.1.5 Death per unit activity	355
18.2	Frequency *vs* consequence lines	355
	18.2.1 Group risk comparisons; ccdf method	356
	18.2.2 Total risk	359
	18.2.3 Expected disutility	360
	18.2.4 Uncertainty about the fC curve	361
	18.2.5 Benefits	362
18.3	Risk regulation	362
	18.3.1 ALARP	362
	18.3.2 The value of human life	363
	18.3.3 Limits of risk regulation	365
18.4	Perceiving and accepting risks	365
	18.4.1 Risk perception	367
	18.4.2 Acceptability of risks	368
18.5	Beyond risk regulation: compensation, trading and ethics	369

Bibliography	373
Index	390

Illustrations

1.1	Risk curve	10
3.1	A schematic representation of a bathtub curve	46
3.2	Availability of a component under constant test intervals	49
3.3	Exponential failure and repair	50
3.4	A lognormal density	54
3.5	A lognormal failure rate	55
4.1		65
4.2	Prior and posterior density and distribution functions	68
4.3	Prior and posterior density and distribution functions (100 observations)	70
5.1	Densities for the Weibull distribution	85
5.2	A Weibull plot of the data in Table 5.4	89
5.3	Revised Weibull plot for Table 5.6	91
6.1	An event tree	100
6.2	The Cassini event tree	101
6.3	Security system	102
6.4	AND and OR gates	103
6.5	Common gates and states	107
6.6	Schematic diagram for the reactor protection system	108
6.7	Fault tree for the reactor protection system	109
6.8	Schematic diagram for the New Waterway water-level measurement system	110
6.9	Fault tree for the New Waterway water-level measurement system	111
6.10	Cut set fault tree representation for the reactor protection example	114
6.11	Path set fault tree representation for the reactor protection example	116
6.12	Dual tree for the reactor protection example	116
6.13	Very simple fault tree	119
7.1	Cut set calculation for the reactor protection system	122
7.2	A power system	124
7.3	The simple coherent fault tree from Example 7.2	125
7.4	The simple non-coherent fault tree from Example 7.3	126
7.5	A binary decision tree for Example 7.2	127

7.6	Application of the simplification rules	128
7.7	BDD for Example 7.3	128
7.8	BDD representation for the reactor protection example	133
7.9	BDD representation for the electrical power system example	133
7.10	Probability calculations for the electrical power system example	134
7.11	Fault tree example	137
8.1	Auxiliary Feedwater System	141
9.1	Maintenance and failure in time	159
9.2	Hierarchical categories	160
9.3	Superposed and pooled time histories	163
9.4	Calendar time picture of censored data	170
9.5	Censored failure data from four plants	173
9.6	Data fields for coloring	185
9.7	Coloring of 'failure effect'	186
9.8	Coloring of 'action taken'	187
9.9	Coloring of 'method of detection'	188
9.10	Coloring of 'subcomponent'	189
10.1	The expert's interpolated density	203
10.2	Expert ranking	204
10.3	Interpolation of expert quantiles	206
10.4	Combination of expert distributions	207
10.5	Range graphs	214
10.6	Range graphs	215
11.1	Classification of expected cognitive performance	222
11.2	The dynamics of GEMS	223
11.3	Example human error event tree	226
11.4	Example human time reliability correlation	229
11.5	Hypothetical lognormal density of HEPs	230
11.6	An ID for human error probabilities	236
12.1	Classification of software reliability models	244
12.2	A u-plot	252
12.3	Cumulative times until failure	254
12.4	Expected number of failures as function of time using LV model	255
12.5	The u-plot	256
13.1	The decision tree for the research project	263
13.2	The decision tree for the extended research proposal	265
13.3	Simple attribute hierarchy	270
13.4	The construction of a marginal value function	273
13.5	Indifference curves for cost and delay	276
13.6	Indifference curves for cost and performance given delay $= 5$	276
13.7	The trade-off between cost and performance	278
14.1	A simple belief net	286
14.2	Alarm influence diagram	292
14.3	A Bayesian belief net and the corresponding moral graph	294
14.4	An influence diagram for fire risk	298
15.1	A simple network	303

15.2	Determining the critical path	303
15.3	Form of the loss function for overspending	306
15.4	Program standard offer	308
15.5	Program alternative offer	309
15.6	Triangular distribution	312
15.7	Distribution of project duration	313
15.8	Distribution of project duration, with new option	314
16.1	Box model for soil migration	319
16.2	Marginal distributions for the target variables	324
17.1	The diagonal band distribution with parameter α	331
17.2	A dependence tree	333
17.3	Partial correlation vine	338
17.4	Rank correlation tree for power lines, Case 2	339
17.5	Rank correlation tree for power lines, Case 3	339
17.6	Distributions for powerline system failure probability	340
17.7	Distribution for series system lifetime	341
17.8	Mean series system lifetime, depending on rank correlation	341
17.9	Mean parallel system lifetime, depending on rank correlation	342
17.10	Unconditional cobweb plot	343
17.11	Conditional cobweb plot: high concentration at 0.5 km downwind	344
17.12	Conditional cobweb plot: mid-range concentration at 0.5 km downwind	345
17.13	Conditional cobweb plot: low concentration at 0.5 km downwind	346
18.1	Frequency consequence lines	356
18.2	The Dutch group risk criterion	357
18.3	Unacceptable risks	358
18.4	Equally risky activities	360
18.5	Changes in numbers of deaths of motor cyclists in different US states	366

Tables

5.1	Swedish nuclear power plant data failure fields	84
5.2	A small mortality table for failure data	87
5.3	Median ranks	88
5.4	A small mortality table with median rank estimates	88
5.5	Failure times and suspensions	89
5.6	Revised median rank estimates, with suspensions	90
5.7	Kaplan–Meier estimates	92
5.8	Extreme value distributions	96
6.1	Laws of Boolean algebra	113
7.1	Truth table for Example 7.2	125
7.2	Truth table for Example 7.3	126
7.3	The p-cut calculation for the electrical power system	135
7.4	Comparison of fault tree size and BDD size for various test cases	135
7.5	FV importance of basic events	137
7.6	Approximate conditional probability of basic events given failure of Component 2 and T	138
8.1	Generic causes of dependent component failures	144
8.2	Some β-factors	148
9.1	Maintenance jobs and schedules	157
9.2	Maintenance and failure type	158
9.3	Component socket time histories, 314 pressure relief valves	160
10.1	Expert 1 data	212
10.2	Expert 3 data	212
10.3	Expert 4 data	213
10.4	Expert 5 data	213
10.5	Expert 8 data	214
10.6	Global weights for the five experts	216
10.7	Equal weights for the five experts	216
11.1	Some generic HEPs used in WASH-1400	219
11.2	Stages in a human reliability analysis	221
11.3	Failure modes for the three performance levels	224

11.4	HEP modification factors, Table 20-16 from [Swain and Guttmann, 1983]	227
11.5	Levels of dependency	228
11.6	Typical values of the distribution	235
11.7	Marginal and conditional probabilities for the ID in Figure 11.6	237
14.1	Conditional probability tables	297
15.1	An example Project Uncertainty Matrix	301
15.2	Criticality indices for activities	313
16.1	Quantile information for elicitation variables at various downwind distances	323
16.2	Product moment correlation matrix for target variables	325
16.3	Rank correlation matrix for target variables	325
17.1	Moments and quantiles for uncertainty distribution	340
18.1	Annual risk of death in the United States	352
18.2	Loss of life expectancy	352
18.3	Hourly specific mortality rates	354
18.4	Deaths per 10^9 km traveled	355
18.5	Risk goals for various technologies	363
18.6	Median costs of life-saving measures per sector	364

Preface

We have written this book for numerate readers who have taken a first university course in probability and statistics, and who are interested in mastering the conceptual and mathematical foundations of probabilistic risk analysis. It has been developed from course notes used at Delft University of Technology. An MSc course on risk analysis is given there to mathematicians and students from various engineering faculties. A selection of topics, depending on the specific interests of the students, is made from the chapters in the book. The mathematical background required varies from topic to topic, but all relevant probability and statistics are contained in Chapters 3 and 4.

Probabilistic risk analysis differs from other areas of applied science because it attempts to model events that (almost) never occur. When such an event does occur then the underlying systems and organizations are often changed so that the event cannot occur in the same way again. Because of this, the probabilistic risk analyst must have a strong conceptual and mathematical background.

The first chapter surveys the history of risk analysis applications. Chapter 2 explains why probability is used to model uncertainty and why we adopt a subjective definition of probability in spite of its limitations. Chapters 3 and 4 provide the technical background in probability and statistics that is used in the rest of the book. The remaining chapters are more-or-less technically independent of each other, except that Chapter 7 must follow Chapter 6, and 14 should follow 13. The final chapter gives a broad overview of risk measurement problems and looks into the future of risk analysis.

Almost all the chapters are concluded with exercises. The answers to these exercises are not given in the book, but bona fide teachers who are using the book in conjunction with their courses may contact David Tranah (dtranah@cambridge.org) and ask for a PDF file with the solutions to the problems.

Acknowledgements

A large number of people, including many students of our course 'Risk Analysis', have given us comments on this book. Special thanks are due to Frank Phillipson, Mart Janssen, Frank Rabouw, Linda van Merrienboer, Eeke Mast, Gilbert Pothoff, Erwin van Iperen, Lucie Aarts, Mike Frank, Antoine Rauzy, Christian Pressyl, Joe Fragola, Floor Koornneef and co-authors of various chapters, Bernd Kraan, Jan Norstrøm and Lonneke Holierhoek.

Special thanks go of course to our families and friends for putting up with us during the preparation of this manuscript.

Part I

Introduction

1

Probabilistic risk analysis

Probabilistic risk analysis (PRA), also called quantitative risk analysis (QRA) or probabilistic safety analysis (PSA), is currently being widely applied to many sectors, including transport, construction, energy, chemical processing, aerospace, the military, and even to project planning and financial management. In many of these areas PRA techniques have been adopted as part of the regulatory framework by relevant authorities. In other areas the analytic PRA methodology is increasingly applied to validate claims for safety or to demonstrate the need for further improvement. The trend in all areas is for PRA to support tools for management decision making, forming the new area of *risk management*.

Since PRA tools are becoming ever more widely applied, and are growing in sophistication, one of the aims of this book is to introduce the reader to the main tools used in PRA, and in particular to some of the more recent developments in PRA modeling. Another important aim, though, is to give the reader a good understanding of uncertainty and the extent to which it can be modeled mathematically by using probability. We believe that it is of critical importance not just to understand the mechanics of the techniques involved in PRA, but also to understand the foundations of the subject in order to judge the limitations of the various techniques available. The most important part of the foundations is the study of uncertainty. What do we mean by uncertainty? How might we quantify it?

After the current introductory chapter, in Part two we discuss theoretical issues such as the notion of uncertainty and the basic tools of probability and statistics that are widely used in PRA. Part three presents basic modeling tools for engineering systems, and discusses some of the techniques available to quantify uncertainties (both on the basis of reliability data, and using expert judgement). In Part four we discuss uncertainty modeling and risk measurement. The aim is to show how dependent uncertainties are important and can be modeled, how value judgements can be combined with uncertainties to make optimal decisions under uncertainty, and how uncertainties and risks can be presented and measured.

1.1 Historical overview

This introductory section reviews the recent history of these developments, focusing in particular on the aerospace, nuclear and chemical process sectors.

1.1.1 The aerospace sector

A systematic concern with risk assessment methodology began in the aerospace sector following the fire of the Apollo test AS-204 on January 27, 1967, in which three astronauts were killed. This one event set the National Aeronautics and Space Administration (NASA) back 18 months, involved considerable loss of public support, cost NASA salaries and expenses for 1500 people involved in the subsequent investigation, and ran up $410 million in additional costs [Wiggins, 1985]. Prior to the Apollo accident, NASA relied on its contractors to apply 'good engineering practices' to provide quality assurance and quality control.

On April 5, 1969 the Space Shuttle Task Group was formed in the Office of Manned Space Flight of NASA. The task group developed 'suggested criteria' for evaluating the safety policy of the shuttle program which contained quantitative safety goals. The probability of mission completion was to be at least 95% and the probability of death or injury per mission was not to exceed 1%. These numerical safety goals were not adopted in the subsequent shuttle program.

The reason for rejecting quantitative safety goals given at the time was that low numerical assessments of accident probability could not guarantee safety: '... the problem with quantifying risk assessment is that when managers are given numbers, the numbers are treated as absolute judgments, regardless of warnings against doing so. These numbers are then taken as fact, instead of what they really are: subjective evaluations of hazard level and probability.' ([Wiggins, 1985], p. 85).

An extensive review of the NASA safety policy following the Challenger accident of January 28, 1986 brought many interesting facts to light. A quantitative risk study commissioned by the US Air Force in 1983 estimated the Challenger's solid rocket booster failure probability per launch as 1 in 35. NASA management rejected this estimate and elected to rely on

1.1 Historical overview

their own engineering judgment, which led to a figure of 1 in 100.000 [Colglazier and Weatherwas, 1986]

It has also become clear that distrust of reassuring risk numbers was not the reason that quantitative risk assessment was abandoned. Rather, initial estimates of catastrophic failure probabilities were so high that their publication would have threatened the political viability of the entire space program. For example, a General Electric 'full numerical probabilistic risk assessment' on the likelihood of successfully landing a man on the moon indicated that the chance of *success* was 'less than 5%'. When the NASA administrator was shown these results, he 'felt that the numbers could do irreparable harm, and disbanded the effort' [Bell and Esch, 1989].

By contrast, a congressional report on the causes of the Shuttle accident (quoted in [Garrick, 1989]) concluded that the qualitative method of simply identifying failures leading to loss of vehicle accidents (the so-called *critical items*) was limited because not all elements posed an equal threat. Without a means of identifying the *probability* of failure of the various elements NASA could not focus its attention and resources effectively.

Since the shuttle accident, NASA has instituted programs of quantitative risk analysis to support safety during the design and operations phases of manned space travel. The NASA risk assessment effort reached a high point with the publication of the SAIC Shuttle Risk Assessment [Fragola, 1995]. With this assessment in hand, NASA was able to convince the US Congress that the money spent on shuttle development since the Challenger accident had been well used, even though no failure paths had been eliminated. The report showed that the *probability* of the most likely failure causes had been significantly reduced.

The European space program received a setback with the failure of the maiden flight of Ariane 5. A board of inquiry [ESA, 1997] revealed that the disaster was caused by software errors and the management of the software design. The accident demonstrates the problem of integrating working technologies from different environments into a new reliable system.

1.1.2 The nuclear sector

Throughout the 1950s, following Eisenhower's 'Atoms for Peace' program, the American Atomic Energy Commission (AEC) pursued a philosophy of risk assessment based on the 'maximum credible accident'. Because 'credible accidents' were covered by plant design, residual risk was estimated by studying the hypothetical consequences of 'incredible accidents'. An early study released in 1957 focused on three scenarios of radioactive releases from a 200 megawatt nuclear power plant operating 30 miles from a large population center. Regarding the probability of such releases the study concluded that 'no one knows now or will ever know the exact magnitude of this low probability'.

Successive design improvements were intended to reduce the probability of a catastrophic release of the reactor core inventory. Such improvements could have no visible impact on the risk as studied with the above methods. On the other hand, plans were being drawn for reactors in the 1000 megawatt range located close to population centers, and these developments would certainly have a negative impact on the consequences of the 'incredible accident'.

The desire to quantify and evaluate the effects of these improvements led to the introduction of *probabilistic* risk analysis. As mentioned above, the basic methods of probabilistic risk assessment originated in the aerospace program in the 1960s. The first full scale application of these methods, including an extensive analysis of the accident consequences, was undertaken in the *Reactor Safety Study* WASH-1400 [NRC, 1975] published by the US Nuclear Regulatory Commission (NRC). This study is rightly considered to be the first modern PRA.

The reception of the *Reactor Safety Study* in the scientific community may best be described as turbulent. The American Physical Society [APS, 1975] conducted an extensive review of the first draft of the *Reactor Safety Study*. In the accompanying letter attached to their report, physicists Wolfgang Panofsky, Victor Weisskopf and Hans Bethe concluded, among other things, that the calculation methods were 'fairly unsatisfactory', that the emergency core cooling system is unpredictable and that relevant physical processes 'which could interfere with its functioning have not been adequately analyzed', and that 'the consequences of an accident involving major radioactive release have been underestimated as to casualties by an order of magnitude'. The final draft of the *Reactor Safety Study* was extensively reviewed by, among others, the Environmental Protection Agency [EPA, 1976] and the Union of Concerned Scientists [Union of Concerned Scientists, 1977].

In 1977 the United States Congress passed a bill creating a special 'review panel' of external reactor safety experts to review the 'achievements and limitations' of the *Reactor Safety Study*. The panel was led by Prof. Harold Lewis, and their report is known as the 'Lewis Report' [Lewis et al., 1979]. While the Lewis Report recognized the basic validity of the PRA methodology and expressed appreciation for the pioneering effort put into the *Reactor Safety Study*, they also uncovered many deficiencies in the treatment of probabilities. They were led to conclude that the uncertainty bands claimed for the conclusions in the *Reactor Safety Study* were 'greatly understated'.

1.1 Historical overview

Significantly, the Lewis Report specifically endorsed the use of subjective probabilities in the *Reactor Safety Study*.

In January 1979 the NRC distanced itself from the results of the *Reactor Safety Study*: 'In particular, in light of the Review Group conclusions on accident probabilities, the Commission does not regard as reliable the *Reactor Safety Study*'s numerical estimate of the overall risk of a reactor accident.' [NRC, 1979]

The future of PRA after the NRC's announcement of 1979 did not look bright. However, the dramatic events of March 1979 served to change that. In March 1979 the Three Mile Island – 2 (TMI) Nuclear Generating Unit suffered a severe core damage accident. Subsequent study of the accident revealed that the accident sequence had been predicted by the *Reactor Safety Study*. The probabilities associated with that sequence, particularly those concerning human error, do not appear realistic in hindsight.

Two influential independent analyses of the TMI accident, the Report of the President's Commission on the Accident at Three Mile Island [Kemeny et al, 1979] and the Rogovin Report [Rogovin and Frampton, 1980], credited the *Reactor Safety Study* with identifying the small loss-of-coolant accidents as the major threat to safety, and recommended that greater use should be made of probabilistic analyses in assessing nuclear plant risks. They also questioned whether the NRC was capable of regulating the risks of nuclear energy, and recommended that the regulatory body be massively overhauled (which recommendation was not carried out).

Shortly thereafter a new generation of PRAs appeared in which some of the methodological defects of the *Reactor Safety Study* were avoided. The US NRC released *The Fault Tree Handbook* [Vesely et al., 1981] in 1981 and the PRA *Procedures Guide* [NRC, 1983] in 1983 which shored up and standardized much of the risk assessment methodology. Garrick's review [Garrick, 1984] of PRAs conducted in the aftermath of the Lewis report discussed the main contributors to accident probability identified at the plants. He also noted the necessity to model uncertainties properly in order to use PRA as a management tool, and suggested the use of on-line computer PRA models to guide plant management (a process now called a 'living PRA').

The accident at the Chernobyl Nuclear Power Plant occurred on April 26, 1986. A test was carried out in order to determine how long the reactor coolant pumps could be operated using electrical power from the reactors' own turbine generator under certain abnormal conditions. At the beginning of the test some of the main coolant pumps slowed down, causing a reduction of coolant in the core. The coolant left began to boil, adding reactivity to the core (due to the so-called positive void coefficient of the RBMK reactor type). This caused a sudden increase in power which could not be controlled because the control systems worked too slowly. The power surge caused the fuel to overheat and disintegrate. Pieces of fuel ejected into the coolant then caused a steam explosion whose force blew the cover off the reactor. There were 31 early deaths and (amongst other radiological effects) a 'real and significant' increase in childhood carcinoma of the thyroid [OECD, 1996]. Thousands of people have been displaced and blame the reactor accident for all sorts of health problems.

The management of western nuclear power corporations moved quickly to assure the public that this type of accident could not occur in the US and western Europe because of the difference in reactor design. The Three Mile Island and Chernobyl accidents, in addition to regular publicity about minor leaks of radioactive material from other power stations and processing plans, however, have fostered a climate of distrust in nuclear power and in the capacity of management to run power stations properly.

Besides technical advances in the methodology of risk analyses, the 1980s and 1990s have seen the further development of numerical safety goals. Examples are the USNRC policy statement of 1986 [NRC, 1986] and the UK *Tolerability of risk* document [HSE, 1987]. These documents seek to place the ALARP principle ('as low as reasonably possible') into a numerical framework by defining upper levels of intolerable risk and lower levels of broadly tolerable risk.

1.1.3 The chemical process sector

In the process sector, government authorities became interested in the use of probabilistic risk analysis as a tool for estimating public exposure to risk in the context of licensing and citing decisions. Important European efforts in this direction include two studies of refineries on Canvey Island ([HSE, 1978], [HSE, 1981]) in the UK, a German sulphuric acid plant study [Jäger, 1983], and the Dutch LPG and COVO studies ([TNO, 1983], [COVO, 1982]). The COVO study [COVO, 1982] was a risk analysis of six potentially hazardous objects in the Rijnmond area. The group which performed the study later formed the consulting firm Technica, which has since played a leading role in risk analysis.

The impetus for much of this work was the Post-Seveso Directive [EEC, 1982] adopted by the European Community following the accidental release of dioxin by a chemical plant near Seveso, Italy. The directive institutes a policy of risk management for chemical industries handling hazardous materials in which each member state has the responsibility for developing its own risk management methodology. The Dutch government took the lead in requiring quantitative risk analyses of potentially hazardous objects, and has invested heavily in developing tools and methods for implementing this policy. This lead has been followed by several other member countries.

Dutch legislation requires the operator of a facility dealing with hazardous substances to submit an 'External Safety Report' (Externe Veiligheid Rapport, EVR). About 70 companies in the Netherlands fall under this reporting requirement. EVRs are to be updated every 5 years.

The quantitative risk analysis required in EVRs may be broken down into four parts. The first part identifies the undesirable events which may lead to a threat to the general public. The second part consists of an effect and damage assessment of the undesired events. The third part calculates the probability of damage, consisting of the probability of an undesired release of hazardous substances, and the probability of propagation through the environment causing death. The fourth part determines the individual and group risk associated with the installation.

For new installations, the risk of death of an 'average individual' exposed at any point outside the installation perimeter for an entire year must not exceed 10^{-6}. The group risk requirement stipulates that the probability of 10^n or more fatalities ($n > 1$) must not exceed 10^{-3-2n} per year. If this probability exceeds 10^{-5-2n} per year then further reduction is required.

Risk based regulation is now common in many different sectors. An overview of some of the risk goals currently set is given in Table 18.5.

1.1.4 The less recent past

We have concentrated on the developments in a few important sectors since the end of the second world war. The reader interested in looking further into the past is referred to [Covello and Mumpower, 1985] which starts about 3200 BC, and [Bernstein, 1996].

1.2 What is the definition of risk?

The literature on the subject of risk has grown rapidly in the last few years, and the word 'risk' is used in many different ways. The purpose of this section is to discuss briefly what we mean by risk, and in what way the concept can be described in a mathematical setting. The limitations of the mathematical approach to measuring risk will be highlighted in Chapter 18. Our discussion here is largely drawn from [Kaplan and Garrick, 1981].

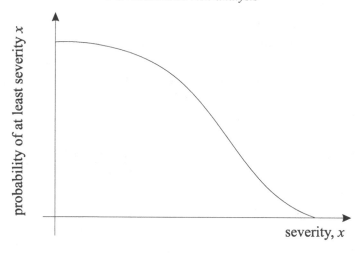

Fig. 1.1. Risk curve

A *hazard* is considered as a source of danger but the concept does not contain any notion of the likelihood with which that danger will actually impact on people or on the environment. We are often *uncertain* about whether or not a hazard will actually lead to negative consequences (that is, whether the potentiality will be converted into actuality). As will be argued in Chapter 2, that uncertainty can – in principle – be quantified by *probability*. The definition of *risk* combines both of the above elements. A risk analysis tries to answer the questions:

(i) What can happen?
(ii) How likely is it to happen?
(iii) Given that it occurs, what are the consequences?

Kaplan and Garrick [Kaplan and Garrick, 1981] define risk to be a set of *scenarios* s_i, each of which has a *probability* p_i and a *consequence* x_i. If the scenarios are ordered in terms of increasing severity of the consequences then a risk curve can be plotted, for example as shown in Figure 1.1. The risk curve illustrates what is the probability of *at least* a certain number of casualties in a given year.

Kaplan and Garrick [Kaplan and Garrick, 1981] further refine the notion of risk in the following way. First, instead of talking about the probability of an event, they talk about the *frequency* with which such an event might take place. They then introduce the notion of uncertainty about the frequency (the 'probability of a frequency'). This more sophisticated notion of risk will be discussed further in Chapter 2.

For the moment the reader should bear in mind that the basic method of risk analysis is the identification and quantification of scenarios, probabilities and consequences. The tools that will be described in this book are dedicated to these tasks.

1.3 Scope of probabilistic risk analyses

The goals and sizes of probabilistic risk analyses vary widely. The nuclear sector has made the largest commitments of resources in this area. Some of the larger chemical process studies have been of comparable magnitude, although the studies performed in connection with the Post-Seveso Directive tend to be much smaller. In the aerospace sector, the methods have not yet been fully integrated into the existing design and operations management structures, although the United Space Alliance (the consortium operating the space shuttle for NASA) is probably currently the furthest in an integrated risk management approach.

The *Procedures Guide* [NRC, 1983], gives a detailed description of the various levels of commitment of resources in nuclear PRAs, and we summarize this below, as the general format is applicable to other sectors as well. Three levels are distinguished:

level 1, systems analysis;
level 2, systems plus containment analysis;
level 3, systems, containment and consequence analysis.

These are explained further below.

Level 1: systems analysis The systems analysis focuses on the potential release of hazardous substances and/or energy from the design envelope of the facility in question. In the case of a nuclear reactor, this concerns primarily the release of radioactive material from the reactor core. Phases within level 1 include event tree modeling, analysis of human reliability impacts, database development, accident sequence quantification, external event analysis and uncertainty analysis.

Level 2: containment analysis The containment analysis includes an analysis of pathways into the biosphere, and a transport analysis to determine important parameters such as composition of release, quantity of release, time profile of release, physics of release event (e.g. elevated or ground source, pressurized or ambient, etc.).

Level 3: consequence analysis Here the pathways by which the hazardous material can reach and affect man are identified, using an analysis of the dispersion through the atmosphere and groundwater, propagation through the food chain, and the toxicological effects in the human body, in both the short and long term. In addition the economic impact of land interdiction, clean-up, relocation etc. should be included.

1.4 Risk analysis resources

1.4.1 Important journals

The journal that best covers the area discussed in this book is *Reliability Engineering and System Safety*. There are several journals covering the mathematical theory of reliability such as *IEEE Transactions on Reliability* and *Microelectronics and Reliability*, and many (applied) statistics and OR journals that frequently publish papers on reliability: *Technometrics, Applied Statistics, Operations Research, Lifetime Data Analysis, Biometrika, The Scandinavian Journal of Statistics* etc. Papers on risk analysis are to be found in engineering journals and also in a number of interdisciplinary journals. Amongst the engineering journals are *Journal of Hazardous Materials* and *Safety Science*. Interdisciplinary journals covering risk analysis are *Risk Analysis, Risk: Health, Safety and Environment, The Journal of Risk Research* and *Risk, Decision and Policy*.

1.4.2 Handbooks

Important textbooks and handbooks in this area are: [Vesely *et al.*, 1981], [NRC, 1983], [Swain and Guttmann, 1983], [Kumamoto and Henley, 1996], [CCPS, 1989], [O'Connor, 1994].

1.4.3 Professional organizations

The Society of Reliability Engineers (SRE) has branches in many countries, as does the IEEE Reliability Society. The International Association for Probabilistic Safety Assessment and Management (IAPSAM) is an international organization which organizes the PSAM conferences. The European Safety and Reliability Association, ESRA, is the European umbrella organization of national associations in the area of risk and reliability analysis and organizes the yearly ESREL conferences. The European Safety Reliability and Data Association, ESReDA, is another European organization with industrial members that organizes regular seminars. The Society for Risk Analysis and the Risk Analysis and Policy Association are two US based interdisciplinary organizations.

1.4.4 Internet

The IEEE Reliability Society maintains an excellent website, with references to other sites, information about standards and much more. The URL is **http://engine.ieee.org/society/rs/**. Riskworld is a good source of information about risk analysis and can be found at **http://www.riskworld.com/**. The European Safety and Reliability Society maintains an up-to-date list of links at its site **http://www.esra.be**.

Part II

Theoretical issues and background

2

What is uncertainty?

2.1 The meaning of meaning

Our focus in this book is practical. Unfortunately, in order to be practical, one must occasionally dabble in philosophy. An opening chapter on uncertainty is one of those occasions. Formulating a clear, consistent and workable point of view at the beginning prevents confusion and ambiguity from cumulating and erupting at inopportune moments. We will never straighten out the meaning of uncertainty if we don't know what 'meaning' means. Let's just step through some examples, starting with something unproblematic:

(I) It will rain in Cincinnati on Jan. 1 2050.

Of course we have to specify what we mean by 'Cincinnati on Jan. 1 2050', and exactly how much precipitation is needed to qualify as 'rain'. Having done that, we can be uncertain about whether the precipitation count in what we agree to call Cincinnati on Jan. 1 2050 is sufficient for 'rain'. Less straightforward is

(II) The number of atoms in the universe is 10^{80}.

Even though we will never be able in practice to count the atoms in the universe, we can imagine *in principle* that they could be counted, even though we cannot say offhand how that might be done. In this sense we can be uncertain about (II). More precisely, we are uncertain about how exactly (II) would be resolved, *and* uncertain about what the resolution would be; but we are certain that it can *in principle* be resolved.

What about the following?

(III) Bread, sugar, milk, orange juice, coffee.

This is a string of meaningful words but it does not have the syntactic form of a declarative sentence – it's a shopping list. This string cannot be true or false, and we cannot be uncertain about it.

According to the logical positivist theory of meaning – which we take to be adequate for practical scientific contexts – the meaning of a declarative sentence is its set of *truth conditions*; i.e. the rules for deciding whether the sentence is true or false [Schlick, 1936]. Thus, the meaning of a declarative sentence is that which enables it to acquire a truth value. It follows inexorably that no truth conditions = no meaning. So what about the following?

(IV) Colorless green ideas sleep furiously.

Noam Chomsky gave this as an example of a string of words which satisfies the syntactic conditions for 'sentencehood' but still does not have declarative meaning. More precisely, the meaning of the individual words rather suggests that truth conditions do not exist. Colorless green ideas do not sleep furiously, nor do they sleep un-furiously. They don't sleep at all; in fact there aren't any green ideas, colorless or otherwise. We could arbitrarily assign truth conditions to (IV), and then be uncertain whether the conditions for 'true' actually hold; but the normal road to truth conditions, based on the meaning of the constituent words, leads nowhere. As it stands, (IV) has no declarative meaning; we cannot be uncertain about its truth because there are no rules for adjudicating its truth. It does have meaning in another sense, of course, which is why it became a popular bumper sticker in the US.

What about the following?

(V) God exists.

The unvarnished logical positivist looks for truth conditions and, finding none, declares this as meaningless as the previous sentence. Being practical, we refer this one to the theologians with a request to specify the conditions under which this would be graded 'true' or 'false'. Of course, if (V) is stipulated to be true in every possible world, then its declarative meaning is the same as that of every other proposition which is true in every possible world, like 'A rose is a rose' or 'A sad saint is a sad saint'.

Moving on:

(VI) Next weekend I will clean my cellar.

The truth conditions for (VI) are disturbingly clear; hence I can be uncertain whether the conditions for 'true' will really hold. However, my uncertainty is different than in (I). (I) concerns 'what will be the case' whereas (VI) concerns 'what will I decide to do'. Although we can speak meaningfully about uncertainty for (VI), we *cannot* measure this uncertainty in any accepted sense of the word [Cooke, 1986]. Uncertainty about what the subject him/herself will decide is termed *volitional uncertainty*. We return to this example in Section 2.5.

Closer to home:

(VII) My flashlight has failure rate 0.001 per hour.

A constant failure rate means that the flashight's probability of failing at time t, given that it hasn't failed up to t, is constant in t. Can we be uncertain about that? Some people find this similar to rain in Cincinnati, and others see more resemblance to green ideas. In any case, the constant failure rate is a *model*, and uncertainty in (VII) is uncertainty in the truth of the model.

Of course all models, including this one, are false. The failure rate will never be *really* constant. Strictly speaking, our probability for (VII) must therefore be zero; but that somehow misses the point. There are many ways we could associate truth conditions with (VII). For example we could assign 'my flashlight' to a very large population of 'similar' flashlights, measure the failure rate (as a function of time) in this population, and stipulate that (VII) is true if the failure rate is constant and equal to 0.001. It might not be equally clear to everyone that two given flashlights are 'similar', and this must be clarified beforehand. The failure rate of a flashlight is not a property of the flashlight like weight and color, it is rather a property of a population to which the given flashlight is assigned.

The point is, we cannot be uncertain about a sentence until we have assigned it a meaning, i.e. truth conditions. *Model uncertainty* is currently a hot topic; can we make sense of the uncertainty in the constant failure rate model for my flashlight? Of course we can, once we have stipulated truth conditions for the model. Such discussions heat up when we try to talk about uncertainty *without* first clarifying the truth conditions.

2.2 The meaning of uncertainty

Uncertainty is that which disappears when we become certain.

We become certain of a declarative sentence when (a) truth conditions exist and (b) the conditions for the value 'true' hold. That's what we mean by becoming certain. In practical scientific and engineering contexts, certainty is achieved through observation, and uncertainty is that which is removed by observation. Hence in these contexts uncertainty is concerned with the results of possible observations.

Uncertainty must be distinguished from ambiguity. Ambiguity is removed on the level of words by linguistic conventions. As we saw above, a syntactically well formed string of meaningful words does not always constitute a meaningful declarative sentence. Hence, there may also be ambiguity with regard to truth conditions of propositions. This ambiguity must be removed before we can meaningfully discuss uncertainty.

The two notions of uncertainty and ambiguity become contaminated when observations are described in an ambiguous language. We assume that it is always possible to reduce any given ambiguity to any desired level. It is impossible to remove all ambiguity, and the work of disambiguation goes on until the residual ambiguities are not worth the effort required to remove them.

To be studied quantitatively, uncertainty must be provided with a mathematical 'representation'. A full mathematical representation comprises three things:

- *axioms* specifying the formal properties of uncertainty;
- *interpretations* connecting the primitive terms in the axioms with observable phenomena (these have been variously called *operational definitions, correspondence rules, semantic rules, epistemic correlations*, see [Nagel, 1961]);
- *measurement procedures* providing, together with supplementary assumptions, practical methods for interpreting the axiom system.

The historical origins of uncertainty make highly entertaining reading [Hacking, 1975, Bernstein, 1996]. In the beginning of the twentieth century there was a great deal of discussion about the formal properties of 'probability'. J.M. Keynes [Keynes, 1973] argued that probability was a partial ordering on a set of propositions. Many people believed that conditional probability was the appropriate concept to take as primitive [Popper, 1959]. Conditional probability is a mapping from pairs of propositions onto the [0,1] interval.

Eventually the view of A.N. Kolmogorov [Kolmogorov, 1933] prevailed, and the formal properties of probability are laid down in Kolmogorov's axioms. These basically say that probability is a positive normalized measure over a field of 'possible worlds' or 'possible states of nature'. This will be explained below, but for the present, it suffices to say that probability has the formal properties of a measure of area on a surface with total area 1. This currently has the status of a 'core theory'; it is simple, intuitive, and adequate for the vast majority of applications. Every feature of this core theory has been challenged, generalized, strengthened, and relaxed, but no other serious contender has emerged, and Kolmogorov's axioms remain the point of departure for foundational discussions.

In saying this, we have not said how probability is to be interpreted or measured. Nonetheless, we have said quite a lot. To illustrate, we consider briefly a set of 'representations of uncertainty' which do not behave as measures. We term this the class of 'fuzzy representations', as the myriad fuzzy-based representations of uncertainty are the salient members of this class.

Let A and B be objects for which an uncertainty operator U is defined:

$$U(A) \in [0,1]; \quad U(B) \in [0,1].$$

Further, let the Boolean operations 'AND' and 'OR' be defined on A and B, and let U be defined on 'A AND B' and on 'A OR B'. Then U is called a fuzzy representation if for some functions G, H with

$$G, H : [0,1] \times [0,1] \to [0,1]$$

we have

$$U(A \text{ AND } B) = G(U(A), U(B))$$

and

$$U(A \text{ OR } B) = H(U(A), U(B)).$$

Thus, for fuzzy representations, the uncertainty in 'A AND B' depends only on the *uncertainty* of A and the *uncertainty* of B, and not on A and B themselves.

For example, one of the more popular interpretations (see [Zadeh, 1965]) is:

$$U(A \text{ AND } B) = \min\{U(A), U(B)\}$$

and

$$U(A \text{ OR } B) = \max\{U(A), U(B)\}.$$

It follows that if the uncertainties of A and B are equal, then also the uncertainty of 'A AND B' is equal to that of 'A OR B'. Suppose the uncertainty of 'Quincy is a boy' is equal to that of 'Quincy is a girl'; then this is also equal to the uncertainty that 'Quincy is a boy and a girl'!

For more discussion on the nature of uncertainty, see [Cooke, 1991], and [French, 1988].

2.3 Probability axioms

Saying that uncertainty is a positive normalized measure fixes the mathematical properties of uncertainty. These properties are usually specified by listing axioms which an uncertainty measure must satisfy. These axioms are presented in Chapter 3. For the present discussion we appeal to the intuitive notion of probability and introduce the fundamental notions from set theory:

- Σ, a non-empty set {outcomes, possible worlds};
- $\mathcal{P}\Sigma := \{A | A \subset \Sigma\}$ {power set of Σ};
- for $A \in \mathcal{P}\Sigma$, $A' = \{s \in \Sigma | s \notin A\}$ {complement of A in Σ};
- for $A, B \in \mathcal{P}\Sigma$, $A \backslash B = A \cap B'$;
- if $A \in \mathcal{P}\Sigma$, A finite, then $|A| = \#A$ = number of elements of A.

A set $A \subset \Sigma$ is called an 'event'. Events, being subsets, are subject to the operations of union or disjunction '$\cup$', intersection or conjunction '$\cap$' and complementation '''. We can also associate an event with a proposition asserting that the actual world is an element of the event. In this way of speaking events are properties, and union, intersection and complementation correspond with 'or' (inclusive), 'and' and 'not' respectively.

Example 2.1 *For $A, B \subset \Sigma$, let $s*$ denote the actual real world, then*

$s* \in A$ *is the same as 'A occurs'*,
$s* \in A'$ *is the same as 'A does not occur'*,
$s* \in A \cap B$ *is the same as 'A occurs and B occurs'* $\Leftrightarrow$ *'A and B occur'*,
$s* \in A \cup B$ *is the same as 'A occurs or B occurs, or both'*,
$s* \in A' \cup B$ *is the same as 'If A occurs then B occurs'*.

2.3.1 Interpretations

Three interpretations of the probability axioms have been defended with some vigor, namely the classical interpretation, the frequency interpretation, and the subjectivist interpretation. We discuss each briefly and give references for more details. A general introduction can be found in [Fine, 1973].

Classical interpretation The classical interpretation is the oldest interpretation, and betrays the origins of probability in games of chance. It is associated with Simon Laplace [Laplace, 1951]. Here

Σ = finite set of 'equally possible' outcomes of one experiment,
$P(A) = \#A/\#\Sigma$.

The 'experiment' is typically a roll of the dice or a spin of a roulette wheel. Probability is the ratio of 'favorable' cases to all cases – favorable presumably because one has bet on them. Thus the probability of finding an even number of dots on a roll with a fair die is $1/3$.

There is a long list of classical paradoxes associated with the classical interpretation. Most revolve around the notion of 'equally possible'. For example, consider throwing a pin of length 1 on a piece of paper on which

2.3 Probability axioms

coordinate axes are drawn. We might say that all angles with the x-axis are equally possible, or alternatively all x-coordinates of the pin are equally possible (translating the origin to the head of the pin). These lead to very different probabilities (see Exercise 1), yet angles and x-coordinates seem equally 'equally possible'.

Frequency interpretation This interpretation was championed by R. von Mises [von Mises, 1919, von Mises, 1981] and Reichenbach [Reichenbach, 1932]. Here

Σ = finite set of outcomes of an experiment that can be repeated indefinitely under 'identical conditions'.

Now, $P(A)$ is defined as $\lim_{n\to\infty}(1/n)\sum 1_{A_i}$; where 1_{A_i} is 1 if the outcome of experiment i is in A and is 0 otherwise.

The first concern of the frequentist is to demonstrate that $P(A)$ is well defined, that is, that the limit $\lim_{n\to\infty}(1/n)\sum 1_{A_i}$ exists. The following theorem of probability is usually enlisted for that purpose:

Theorem 2.1 (*Weak law of large numbers*). *Let* $\{A_i\}_{i=1,\ldots,\infty}$ *be independent with* $P(A_i) = p, i = 1,\ldots,\infty$; *then for all* $\delta > 0$

$$P\left\{\left|\frac{1}{n}\sum_{i=1}^{n}1_{A_i} - p\right| > \delta\right\} \to 0 \text{ as } n \to \infty.$$

Critics hasten to point out that the assumptions of the weak law of large numbers themselves appeal to probability, which must have been antecedently defined.

To avoid this evident circularity, frequentists must define the sequences of outcomes whose limiting frequencies define probabilities, and they must do so in a way which does not appeal to the notion of probability. Von Mises spoke of 'collectives'; that is sequences of outcomes whose limiting relative frequencies are invariant under selection of certain sub-sequences.[1] A satisfactory definition of collectives eluded Von Mises, but has been successfully developed in [Schnorr, 1970], [Martin-Löf, 1970] and [van Lambalgen, 1987]. Modern frequentists introduce 'random sequences' as a new primitive term; where randomness is defined in terms of stability under specified subsequence selection rules. Probability is then defined in terms of randomness, as limiting relative frequency in random sequences.

---

[1] For example $(1,0,1,0,\ldots)$ is not a collective although the limiting relative frequency of '1' in this sequence is $\frac{1}{2}$. The subsequence gotten by selecting every other outcome starting with the first has limiting frequency 1.

Subjective interpretation The subjective interpretation is widely applied in risk analysis, and it is therefore treated in more detail. The main features may be easily contrasted with the previous interpretations. Here

Σ = set of possible worlds, or set of possible states of the world (finite or infinite),

$P(A)$ = degree of belief, of one individual, in the occurrence of A, that is, degree of belief that the 'real' s is an element of A.

The foundations of the subjective interpretation were laid in [Borel, 1924], [Borel, 1965], [Ramsey, 1931], [De Finetti, 1964], [De Finetti, 1974] and [Savage, 1972]. De Groot [De Groot, 1970] formulated the foundations in a way amenable to statistical decision theory.

Because of its interest for risk modeling and risk assessment, we present the foundations of the subjectivist interpretation in more detail. Most authors introduce partial belief via 'bets' or 'lotteries'. This has led some people to conclude that subjective probability is restricted to betting. Savage however reduced the representation of partial belief to the representation of rational preference, and for this reason his approach is most suitable for applications in risk analysis.

2.4 Savage's theory of rational decision

Savage traces uncertainty, or partial belief, back to the notion of rational preference. He proves that rational preference can be uniquely represented in terms of a subjective probability and a utility measure.

Before sketching this theory, it is well to reflect briefly on the notion of preference. What is preference about? The naive answer is 'things'; we prefer some things to other things. With a little more reflection we might say that preference concerns events. Despite their initial plausibility, such notions ultimately frustrate clear operational definitions. What would it mean to say that I prefer winning the state lottery to inheriting a fortune? What observations are inferable from such statements?

If we wish to operationalize and measure partial belief through objective empirical procedures, then we must infer preference from choice behavior. What do we choose? Do we choose things or events? No, we choose to *do* things. In short we choose to perform actions. In Savage's theory preference is about actions. Partial belief is operationalized and measured by observing which actions a subject chooses to perform in given situations.

Now what is an act? It is a piece of intentional behavior undertaken for the sake of its consequences, where the consequences depend on the state of

2.4 Savage's theory of rational decision

the world. Since we don't fully know the state of the world, the consequences of our actions are uncertain.

What finally is a consequence? We might say that a consequence is a new state of the world. Savage chooses a different tack and defines a consequence as a 'state of the acting subject'. The act of investing in gold may lead to profit or loss, depending on the unknown state of the world; however, the consequences in Savage's sense are 'the awareness of having made a profit' and the 'awareness of having incurred loss'. An act is then simply a function mapping a state of the world into a state of the subject. Distinguishing states of the world from states of the subject in this way is essential for deriving a simple and elegant representation of preference.

Collecting the above, the primitive notions in Savage's theory of rational decision are:

Σ a non-empty set of possible worlds, or possible states of the world,

C a finite set of states of the acting subject,

F the set of acts, that is the set of all functions from Σ to **C**,

$\leq$: a preference relation over **F**,

where $f \leq g$ means that act g is at least as preferable as act f.

Savage proposes axioms to characterize 'rational preference'. The idea is that if someone exhibited choice behavior violating these axioms, we would call him/her irrational.

He proves that any rational preference can be uniquely represented as expected utility. More precisely, for any rational preference $\leq$, there exist a unique finitely additive probability P on $(\Sigma, \mathcal{P}\Sigma)$ and an affine unique utility function mapping **C** into the real numbers, such that for any acts f, g,

$$f \leq g \text{ (}g \text{ is at least as good as } f\text{)}$$
$$\text{if and only if}$$
$$\text{expected utility of } f \leq \text{expected utility of } g.$$

'Affine unique' means that if $U : \mathbf{C} \to \mathbb{R}$ satisfies the above statement, then so does $aU + b$, for any $a > 0$ and $b \in \mathbb{R}$.

We can also turn the above statement around: any probability on $(\Sigma, \mathcal{P}\Sigma)$ and any function $U : \mathbf{C} \to \mathbb{R}$ can be used to define a preference on the set of acts which is then rational in Savage's sense.

On the other hand, if our choice behavior is guided by a 'partial belief' which does not obey Kolmogorov's axioms, then this must lead to irrational preferences in Savage's sense.

2.4.1 Savage's axioms

The first axiom says that the rational decision maker can order his actions for the decision problem in terms of preference.

1. *Weak order:* For all $f, g \in \mathbf{F}$, $f \geq g$ or $g \geq f$ or both ($\geq$ is connected). For all $f, g, h \in \mathbf{F}$, if $f \geq g$, and $g \geq h$, then $f \geq h$ (transitivity).

Note that if $c \in \mathbf{C}$, then we may let $c \in \mathbf{F}$ denote the 'constant act' $c(s) \equiv c$. If $c, d \in \mathbf{C}$, then we may write '$c \geq d$' if this holds for the corresponding constant acts.

2. *Principle of definition:* There exist $g, b \in \mathbf{C}$ such that $g > b$ ($g =$ 'good', $b =$ 'bad'). For each $A \subset \Sigma$, define $r_A(s) = g$ if $s \in A$ and $r_A(s) = b$ otherwise. If $g', b' \in \mathbf{C}$ with $g' > b'$, let $r'_A(s) = g'$ if $s \in A$ and $r'_A(s) = b'$ otherwise, then

$$\text{for all } A, B \subset \Sigma, r_A \geq r_B \text{ if and only if } r'_A \geq r'_B.$$

The function r_A may be thought of as a lottery in which the prize g is won if A holds and b is obtained if A does not hold. The principle of definition says that preferences between different lotteries of this kind do not depend on the prize, as long as the 'winning prize' is more attractive than the 'losing prize'. See Exercise 2.

Definition 2.1 *For $A, B \subset \Sigma$, define $A \geq .B$ if $r_A \geq r_B$. We write $>., \simeq., <.$ with the obvious meaning.*

If $A \geq .B$, then we would rather have the reward 'g' conditional on A than on B. Intuitively, that means that we regard A as more likely than B. Note that A and B need not be exclusive.

The next axiom says essentially that preferences between actions can be determined by just considering those states on which they have different consequences (in statement of the axiom, the actions being compared have identical outcomes on A').

3. *Sure thing principle:* For all $A \subseteq \Sigma$, let $f, h, f^*, h^* \in \mathbf{F}$ be such that

$$f(s) = f^*(s) \text{ and } h(s) = h^*(s) \text{ for } s \in A,$$

but

$$f(s) = h(s) \text{ and } f^*(s) = h^*(s) \text{ for } s \notin A.$$

Then

$$f \geq h \text{ if and only if } f^* \geq h^*.$$

A straightforward consequence of the sure thing principle (which you will be asked to check in Exercise 3) is that '$\geq$.' is *additive*, that is, for all $A, B, C \subset \Sigma$, if $A \cap C = B \cap C = \emptyset$, then

$$A \geq .B \text{ if and only if } A \cup C \geq .B \cup C.$$

Definition 2.2 $A \subset \Sigma$ *is null if* $r_A \simeq r_\emptyset$.

The next axiom says that if, for every state of the world, the consequence of f is preferable to the consequence of g, then f is preferred to g.

4. *Dominance*: If for all $s \in \Sigma$, $f(s) \geq g(s)$, then $f \geq g$. If also $f(s) > g(s)$ for all $s \in B$, B non-null, then $f > g$.

A straightforward consequence (to be shown in Exercise 5) of the dominance principle is that for all $A \subset \Sigma$

$$\emptyset \leq .A \leq .\Sigma.$$

5. *Refinement*: For all $B, C \subset \Sigma$, if $B<.C$ then there exists a partition $\{A_i\}_{i=1,\ldots,n}$ of Σ such that

$$B \cup A_i <.C, \text{ for } i = 1, \ldots, n.$$

The refinement principle is a technical axiom which forbids granularity in our partial beliefs. Intuitively, if $B<.C$, then we can find a sufficiently small amount of 'partial belief' A_i, such that adding this to B still does not bridge the gap between B and C. Moreover since $\{A_i\}_{i=1,\ldots,n}$ is a partition, we can cover the entire set Σ with such small pieces.

Definition 2.3 *A relation* $<$ *on* $\mathscr{P}\Sigma \times \mathscr{P}\Sigma$ *is a* qualitative probability *if*

  (i) $<$ *is a weak order*,
  (ii) $<$ *is additive*,
  (iii) $\forall A \subset \Sigma, \emptyset < A < \Sigma$.

It is easy to check that the relation '$\geq$.' from Definition 2.1 is a qualitative probability. The principle of refinement is not really essential to the notion of qualitive probability, but it ensures that a qualitative probability can be uniquely represented by a quantitative probability. That is, there exists a unique probability P such that for all A, B

$$A \geq .B \text{ if and only if } P(A) \geq P(B).$$

With this probability, we can obtain a utility function and a unique representation of preference as expected utility.

The precise mathematical results are formulated below without proof. The original proofs in [Savage, 1972] have been simplified in [Cooke, 1991].

2.4.2 Quantitative probability

Theorem 2.2 *If $\geq$ on $\mathbf{F} \times \mathbf{F}$ satisfies axioms 1 through 5, then there exists a unique probability P on $(\Sigma, \mathcal{P}\Sigma)$, such that for all $A, B \subset \Sigma$,*

$$P(A) \geq P(B) \text{ if and only if } A \geq \cdot B.$$

2.4.3 Utility

Note that the foregoing derivation of P used the 'utility' of consequences only in the principle of definition. We need a utility function for the representation of preference. For this purpose we need one more axiom on the preference relation, and this axiom explicitly uses the P which was derived from the previous five axioms:

6. *Strengthened refinement principle:* For all $f, g \in \mathbf{F}$, if $f > g$, then there exists $\alpha > 0$, such that for all $A \subset \Sigma$ with $P(A) < \alpha$, and for all f', g' with

$$f(s) = f'(s) \text{ and } g(s) = g'(s) \text{ for } s \notin A,$$

we have $f' > g'$.

In other words, by changing the outcomes of f and g arbitrarily on a set of possible worlds with small enough probability, we would not change the preference between f and g.

Theorem 2.3 *If '$\leq$' satisfies axioms 1 through 6, then*
(i) *there exists a $U^* : \mathbf{C} \to \mathbb{R}$ such that for all $f, g \in \mathbf{F}$*

$$f \geq g \Leftrightarrow U^*(f) \geq U^*(g)$$

where $U^*(f) = \sum_{c \in \mathbf{C}} U^*(c) P(s \in \Sigma; f(s) = c)$, etc.,
(ii) *if $U : \mathbf{C} \to \mathbb{R}$ is another function for which*

$$f \geq g \Leftrightarrow U(f) \geq U(g),$$

then there exist $a, b \in \mathbb{R}$, $a > 0$, such that

$$U = aU^* + b.$$

2.4.4 Observation

Uncertainty concerns the outcomes of potential observations, and so uncertainty is reduced or removed by observation. A representation of uncertainty must also represent the role of observations, and explain the value of observation for decision making.

2.4 Savage's theory of rational decision

It is an easy matter to alter our partial belief to take account of observations. If our initial partial belief is given by a probability $P(\bullet)$, and we subsequently observe event A, our belief after observation is $P(\bullet|A)$.

More interesting is the question 'Why observe?' Isn't one subjective probability just as rational as every other? The value of observation does not devolve from the 'rationality' of the preference relation as defined by the axioms 1 through 6. Rather it depends on the acts at our disposal and on the utilities of consequences. The set **F** contains all functions from Σ to **C**. One of these acts takes the best consequence in all possible worlds. If this act is also available to us, there is no reason to observe anything.

In light of the preceding theorem we may simply assume that our acts are utility valued, and we write Ef instead of $U(f)$ for the expected utility of f.

As an example, suppose we are considering whether to invest in gold. There are two acts at our disposal, invest or not invest. Suppose we restrict attention to one uncertain event:

B: the price of gold rises next year.

Suppose before choosing whether to invest, we can perform a cost free observation. For example, we may ask an expert what (s)he thinks the gold price next year will be. The result of this observation is of course uncertain before it is performed. An observation must therefore be represented as a random variable $X : \Sigma \to \mathbb{R}$. Let $x_1, \ldots, x_n$ denote the possible values of X.

Before performing the observation, our degree of belief in B is $P(B)$. After observing $X = x_i$, our probability is $P(B|X_i = x_i)$ (the mathematical definition of conditional probability will be given in Chapter 3). If the observation of x_i is definitive, then $P(B|X_i = x_i) = 0$ or 1. We assume that none of the possible values x_i are definitive. Before actually performing the observation, we are uncertain what we will believe after the observation. We denote this by writing $P(B|X)$; if X takes values 0 and 1, this is a random variable taking values

$$P(B|X = 0) \text{ with probability } P(X = 0)$$

and

$$P(B|X = 1) \text{ with probability } P(X = 1).$$

Suppose that B really holds, then we should expect our belief in B to increase as a result of the observation. Similarly if B does not hold we expect our belief to decrease as a result of the observation. In other words,

$$E(P(B|X)|B) \geq P(B), \quad E(P(B|X)|B') \leq P(B).$$

These statements are proved in Section 13.3.

This result says that we expect to become more certain of B, if B is the case, by observing X. It shows how we expect our beliefs to change as a result of observation, but it does not answer the question 'Why observe?' To answer this question we must consider how the value of an observation is determined by the acts at our disposal.

Suppose $F \subset \mathbf{F}$ is a set of *available acts*, and we have to choose a single element of F.

Definition 2.4 *For $F \subset \mathbf{F}$ the value $v(F)$ of F is*

$$v(F) = \max_{f \in F} Ef.$$

Before choosing an element of F, we may consider the option of observing X. After observing $X = x_i$ our probability changes from $P(\bullet)$ to $P(\bullet|X = x_i)$, and the value of F would be computed with respect to $P(\bullet|X = x_i)$. Before observing we don't know which value of X will be observed, but we can consider the value of F as conditional on the value of X. Recall that the consequences are assumed to be utility valued.

Definition 2.5 *For $F \subset \mathbf{F}$ the value $v(F|X = x_i)$ of F given $X = x_i$ is*

$$v(F|X = x_i) = \max_{f \in F} E(f|x_i).$$

The value $v(F|X)$ of F given X is

$$v(F|X) = \sum_{i=1}^{n} v(F|X = x_i) P(X = x_i).$$

The following result is simple and important enough to prove here.

Theorem 2.4 *For any $X : \Sigma \to \mathbb{R}$, and any $F \subset \mathbf{F}$, $v(F|X) \geq v(F)$.*

Proof Fix a particular $f \in F$. Then

$$v(F|X) = \sum \max_{g \in F}[E(g|x_i)] P(X = x_i) \geq \sum E(f|x_i) P(X = x_i) = Ef.$$

Since this holds for all $f \in F$, we have $v(F|X) \geq \max_{f \in F} Ef = v(F)$. □

2.5 Measurement of subjective probabilities

It is sometimes said that subjective probabilities apply only to bets and lotteries. Savage shows that representing partial belief as subjective probability is inherent in the notion of rational preference. Moreover, his definitions provide for very general operational definitions of 'degree of belief'.

For example, suppose we offer a subject the choice between '$100 if rain tomorrow, otherwise $0' and '$100 if the Dow-Jones goes down tomorrow, otherwise $0' (these are examples of the acts r_A, r_B in the principle of definition). If the subject chooses the first, we may conclude that (s)he regards rain tomorrow as at least as likely as a falling Dow-Jones. By observing enough choices of this type, we can eventually pin down his/her subjective probability for any event to any desired accuracy.

Of course this method assumes that receiving $100 is better than receiving $0.

If we make more assumptions, we can design more efficient measurement procedures. Suppose for example that utility is linear in money, and that we observe indifference between the two options

(i) receive $300 if the Dow-Jones goes down tomorrow, receive $0 otherwise
(ii) receive tomorrow $100.

Knowing that preference is represented by expected utility, we infer that indifference in preference indicates equality of expected utility:

$$U(\$300)P(\text{Dow-Jones down}) + U(\$0)P(\text{Dow-Jones not down}) = U(\$100).$$

By linearity of utility,

$$U(\$300) = a300 + b \quad \text{and} \quad U(\$100) = a100 + b,$$

for some $a > 0, b \in \mathbb{R}$. We conclude

$$P\ (\text{Dow down})\ = 1/3.$$

Before leaving the representation of uncertainty, we do well to emphasize that no mathematical representation can ever do justice to all aspects of an informal concept like uncertainty. Savage's formal system is no exception. There are events about which we can be uncertain yet whose uncertainty cannot be adequately captured in Savage's system, and cannot be measured via measurement schemes like the above.

Consider for example the option

receive $100 if you do not die tomorrow, receive $0 otherwise.

For how much would you sell this option for a price agreed today, but paid tomorrow? It is not unreasonable to reason as follows: *If I die tomorrow, I don't care what happens, so I'm only concerned in the consequences if I do*

not die tomorrow. Therefore, I would not sell this lottery for less than $100.
Identifying indifference with equal expected utility, we would infer

$$U(\$100) \cdot P \text{ (not die tomorrow)} = U(\$100).$$

From this it follows that P(not die tomorrow) = 1. Proceeding in this way for the day after tomorrow, etc. we conclude that P(never die) = 1!

Is this 'subjectivist proof of immortality' really reasonable? Surely not. What goes wrong is the following. The 'constant' $100 on the right hand side above is not really constant. There is no way for me to receive the utility of $100 in the event that I die tomorrow. Savage assumes that we can formulate preference for all mathematically possible acts, including 'constant acts' which take the same utility consequence in every possible world. The above example shows that we cannot really offer such acts to anybody, and cannot really observe preference behavior for such acts.

Extreme events are those for which every transaction has the same subjective value. Death, total thermonuclear war, Buddhist enlightenment, the inconsistency of elementary arithmetic: these might be extreme events for various individuals. It follows that we cannot measure uncertainty in extreme events by observing preference behaviour.

First person events are those whose defining conditions involve decisions of the acting subject. Uncertainty regarding such events is *volitional* uncertainty. Cleaning my cellar next weekend is a first person event. Volitional uncertainties defy measurement via preference behavior. Consider the options:

(i) receive $1,000,000 if you clean your cellar next weekend, otherwise $0,
(ii) receive $1,000,000 if the Dow-Jones is lower at the end of the week, $0 otherwise.

Suppose you can choose one of these. You will have strong preference for the first, impling that P(clean cellar)$>$ P(Dow-Jones lower). However, changing the stakes, suppose you must choose between

(i) receive $1 if you clean your cellar next weekend, otherwise $0, and
(ii) receive $1 if the Dow-Jones is lower at the end of the week, $0 otherwise.

You might now prefer (ii), as this leaves the weekend free, implying P(clean cellar)$<$ P(Dow-Jones lower).

The point is that first person events violate Savage's *principle of definition*. Clearly, we can be uncertain about first person events, but we cannot measure this uncertainty by observing preference behavior. The act of measurement disturbs the thing we want to measure, recalling Heisenberg's principle in quantum mechanics.

For a fuller discussion of these and other problems with subjective probability, see [Cooke, 1986]. For psychometric literature on the measure and study of subjective probability, see [Kahneman *et al.*, 1982], [Hogarth, 1987] and [Wright and Ayton, 1994]. A good policy oriented discussion of uncertainty is given by Granger Morgan and Henrion [Granger Morgan and Henrion, 1990], and Bernstein [Bernstein, 1996] may be nominated as the official biographer of uncertainty.

2.6 Different types of uncertainty

Many sorts of uncertainty have been distinguished. Some of these types of uncertainty are quantifiable by subjective probabilities. Some are not.

Aleatory and epistemic uncertainties The two most frequently used 'types' of uncertainty are *aleatory* and *epistemic*. Aleatory uncertainties arise through natural variability in a system. Epistemic uncertainties arise, by contrast, through lack of knowledge of a system. Aleatory uncertainty could be quantified by measurements and statistical estimations, or by expert opinion. Epistemic uncertainty can in principle be quantified by experts, but cannot be measured.

The two sorts of uncertainty differ with respect to learning via the application of Bayes' theorem: epistemic uncertainties relate to those things about which we could learn if we were able; aleatory uncertainties are ones about which we either cannot or choose not to learn.

Many authors (ourselves included) take the view that the distinction between these uncertainties is of a more practical than a theoretical significance. Winkler [Winkler, 1996] points out that all uncertainty is quantified by probability, and that the classification of an uncertainty as aleatory or epistemic is somewhat arbitrary. The outcome of the toss of a fair coin might be viewed as an aleatory uncertainty, unless we take all the physical conditions (wind, initial placement of the coin, etc.) into account in which case we could apply the laws of physics to change our probability. In that case our uncertainty would clearly be epistemic.

The above example clearly shows that a categorization into aleatory and epistemic uncertainties is *for the purposes of a particular model*. The 'same' uncertainty in a different model with a different goal might by classified differently. Hora [Hora, 1996] gives a number of examples of this. The main aims of making the distinction between aleatory and epistemic are:

(i) to make modeling choices clear,
(ii) to provide the basis on which quantification will take place,
(iii) to demonstrate to the decision maker the effects of the epistemic uncertainties on the output, in particular the degree to which learning and reducing uncertainties would make a difference to the model outcome.

This last point is of particular importance when a system is under design. The epistemic uncertainty shows how much *could* still be controlled if needed.

We should note that the distinction between these two 'types' of uncertainty has been made regularly, and has been given several different names (for example, stochastic, type A, irreducible, and variability for aleatory, and subjective, type B, reducible, and state of knowledge for epistemic).

Parameter uncertainty This is uncertainty about the 'true' value of a parameter in a mathematical model. The relevance of this concept is shown by the wide ranges of values that one finds recommended in the literature for the parameters of physical models.

Often there is no strict physical or 'real-world' interpretation of the parameter in question. Hence the notion of a 'true' value is at best hazy. We have shown above that subjective probability only has a meaning when it represents uncertainty about the truth of some meaningful proposition.

Therefore parameter uncertainty, even of abstract parameters, can be given a meaning *if* we take that uncertainty to represent the uncertainty of the observer about the accuracy of model predictions on observable quantities. This view will be used in Chapter 16 where the process of probabilistic inversion will be used to capture the uncertainties in model output by translating these back into uncertainties over the model parameter values.

Model uncertainty Model uncertainty is sometimes described as 'uncertainty about the truth of the model'. However, since all models are false, this definition does not seem very useful.

Still, some false models are more useful than other false models. A model with a poor (or poorly understood) scientific basis can still give reasonable predictions. Indeed, for consequence analysis, the predictive quality of a model is the only thing that is important.

One way to give model uncertainty a meaning is to view it as a special case of parameter uncertainty, by introducing a new discrete parameter indicating which model is being used. The probabilistic inversion methodology discussed in Chapter 16 can then give results which include a probability for each model.

It should be stressed that the interpretation of the model probability is *not* as the probability that 'the model is correct'. Since the probabilities must sum to 1, this would mean we are assuming exactly one model to actually be correct. However, no model is exactly correct, and if we allow models to be 'approximately correct' then more than one model may satisfy this criterion.

Ambiguity As stressed above, ambiguity is not a form of uncertainty. There seems little point in trying to model ambiguity, since it can be removed by careful definition.

Volitional uncertainty Volitional uncertainty is the uncertainty that an individual has in whether or not he will do what he agreed to do. In risk analysis this is relevant, for example, in assessing the effectiveness of authorities in civil defense. As argued above, an individual cannot quantify his own volitional uncertainty using subjective probability. One individual could do it for another, of course.

2.7 Uncertainty about probabilities

The Rasmussen report [NRC, 1975] and other subsequent PRAs have frequently communicated not just probabilities of accidents, but also the *uncertainties* over these probabilities. Although this may seem to be a convenient *façon de parler*, it encourages sloppy thinking and misunderstanding. If, as in the Rasmussen report, probability is interpreted subjectively as degree of belief, then uncertainty over probability becomes uncertainty over an uncertainty. Critics rightfully ask: 'why not uncertainties over uncertainties over uncertainties?' and so on.

To escape this infinite regress of uncertainties we must recall the discussion at the beginning of this chapter. Uncertainty concerns possible observations. Therefore we can be uncertain about a probability only if this probability can be identified with results of possible observations. Some probabilities can be identified with (limiting) relative frequencies; and limiting relative frequencies can be observed with arbitrary accuracy by observing finite sequences. The question is, therefore, when can degrees of belief be identified with relative frequencies?

The answer to this question was given by De Finetti [De Finetti, 1974]. Given the conceptual pitfalls surrounding these issues, it is worth rehearsing De Finetti's theory carefully. Consider a sequence (finite or infinite) of random variables $X_1, \ldots, X_n, \ldots$ taking values in finite set $\mathscr{A}$. The variables are called *exchangeable* if the probability of any finite vector of outcomes is unchanged when the order of the outcomes is altered. In other words, the probability of

$$(X_1 = \alpha_1, X_2 = \alpha_2, \ldots, X_n = \alpha_n)$$

is unchanged by altering the order of the α_is. Only the number of occurrences of each outcome matters.

De Finetti's representation theorem states that if the sequence is infinite, then the distribution of $X_1, \ldots, X_n, \ldots$ can be uniquely represented as a mixture of independent distributions. If M denotes the set of probability distributions over $\mathscr{A}$, then for some unique distribution μ over M, we have for all n,

$$P(X_1 = \alpha_1, X_2 = \alpha_2, \ldots, X_n = \alpha_n) = \int_{p \in M} \prod_{i=1}^{n} p(\alpha_i) \, d\mu(p).$$

One says that $X_1, X_2, \ldots$ are *conditionally independent* given p. Further, when a large initial sequence is observed, the probability for an outcome p conditional on the observation is approximately equal to the observed relative frequency of α,

$$P(X_{n+1} = \alpha | X_1 = \alpha_1, \ldots, X_n = \alpha_n) \approx \frac{\#\text{occurrences of } \alpha \text{ in } \alpha_1, \ldots, \alpha_n}{n},$$

and we believe with probability 1 that the relative distribution converges to a distribution $p^* \in M$.

Hence, if one believes that the ordering of outcomes of $X_1, \ldots$ carries no information, then this is the same as believing that there is some unknown distribution, say p^*, over $\mathscr{A}$ under which the $X_1, \ldots$ are independent and identically distributed. Moreover, by conditionalizing our beliefs on large finite sequences, our beliefs track the observed relative frequencies and converge to p^*.

This result has an important consequence. If another Bayesian believes that the variables $X_1, X_2, \ldots$ follow a distribution $Q \neq P$, but also believes that the order of the outcomes carries no information, then if both Bayesians observe the same outcomes, their beliefs will converge. Hence observing outcomes of *exchangeable* variables induces agreement.

Suppose for example the outcomes of a coin are believed to be independent conditional on some unknown probability of heads $p \in (0,1)$. Suppose our prior belief about p is described by a uniform distribution over $(0,1)$. The probability of heads on one toss is

$$\int_0^1 p \, dp = \frac{1}{2}.$$

This is the same degree of belief in heads we would have if we knew for certain that $p = \frac{1}{2}$. However, these two beliefs are very different, as becomes evident when we consider two tosses. The probability of two heads on the first two tosses with a uniform prior belief is

$$\int_0^1 p^2 \, dp = \frac{1}{3}.$$

If we were certain that the probability of heads were $\frac{1}{2}$ then the probability of two successive heads would be $\frac{1}{4}$. With a uniform prior it is not difficult to show (see Exercise 6) that the probability of heads on tosses $1, \ldots, n$ is

$$\frac{\int_0^1 p^{n+1} \, dp}{\int_0^1 p^n \, dp} = \frac{n+1}{n+2}.$$

Such examples should not give the impression that 'you can show anything with probability theory'. Rather, they demonstrate the importance of a good understanding of uncertainty, and the need to make an informed choice of probabilistic model.

2.8 Exercises

2.1   Suppose that a pin of length 1 is thrown on a piece of paper on which coordinate axes are drawn. We assume that the angle made by the pin with the x-axis (measured in radians) is uniformly distributed (on $[0, 2\pi)$). Let (x_h, y_h) be the coordinates of the pin head, and (x_t, y_t) be those of the pin tip. Calculate the distribution of $x_t - x_h$ and show that this is not uniform on $[-1, 1]$.

2.2   Consider the two events

    A: Dow-Jones index goes up tomorrow,
    B: FTSE100 index goes up tomorrow.

Now consider the lotteries

$$L1 \begin{cases} \text{win 100 yen if } A, \\ \text{win 50 yen if not } A, \end{cases}$$

and

$$L2 \begin{cases} \text{win 100 yen if } B, \\ \text{win 50 yen if not } B. \end{cases}$$

If you had to choose between $L1$ and $L2$, which would you prefer? (Think carefully about your knowledge of the US and British economies before answering.)

Now suppose we change the 100 yen reward to x yen, and the 50 yen reward to y yen, for some x and y with $x > y$. Would your preference for $L1$ or $L2$ be changed if we changed x and y (whilst keeping $x > y$)? If not then you subscribe (in this case!) to Savage's Axiom 2 (principle of definition).

2.3 Check that the sure thing principle implies that $\geq.$ is additive, that is, for all $A, B, C \subset \Sigma$, if $A \cap C = B \cap C = \emptyset$, then

$$A \geq.B \text{ if and only if } A \cup C \geq.B \cup C.$$

(Hint: replace A by C' in the statement of the sure thing principle.)

2.4 The Allias counterexample consists of two choices between lotteries. Consider first the choice between

$A1$: \$ 1,000,000 with probability 1

and the lottery

$B1$: \$ 5,000,000 with probability 0.1,
$B2$: \$ 1,000,000 with probability 0.89,
$B3$: \$ 0 with probability 0.01.

The second lottery choice is between

$C1$: \$ 100,000 with probability 0.11
$C2$: \$ 0 with probability 0.89.

and

$D1$: \$ 5,000,000 with probability 0.1,
$D2$: \$ 0 with probability 0.9,

It has been observed that most people when confronted with these lotteries prefer lottery A over B, but prefer D over C. Explain why this violates the sure thing principle.

What are your preferences?

2.5 Use Savage's dominance axiom to show that for any $A \subset \Sigma$,

$$\emptyset \leq.A \leq.\Sigma.$$

2.6 Let p be the probability of heads by the toss of a coin. With a uniform prior for p, show that the probability of heads on tosses $1, \ldots, n$ is

$$\frac{\int_0^1 p^{n+1} \, dp}{\int_0^1 p^n \, dp} = \frac{n+1}{n+2}.$$

2.7 Suppose your prior belief in heads is $1/3$ with probability $1/2$ and $2/3$ with probability $1/2$. What is your prior probability of heads on the third toss given heads on the first two tosses?

3

Probabilistic methods

In this chapter we discuss some probabilistic methods required for quantifying component failure events. We begin by reviewing some elementary probability theory. This is a rapid review, and we assume that the reader has some previous knowledge of elementary probability. Good starting points for those with no previous knowledge are [Grimmett and Stirzaker, 1982] and [Ross, 1997].

3.1 Review of elementary probability theory

A probability is simply an assignment of a number between 0 and 1 to an event. A probability assignment has to follow certain mathematical rules. In order to define these rules it is convenient to require that probabilities be assigned to a whole collection of events called an *algebra* or equivalently a *field*. Recall that $\mathscr{P}\Omega$ is the collection of all subsets of the set Ω.

Definition 3.1 $\mathscr{F} \subset \mathscr{P}\Omega$ *is a* field *over* Ω *if*

(i) $\emptyset \in \mathscr{F}$,
(ii) $A \in \mathscr{F}$ *implies* $A' \in \mathscr{F}$,
(iii) $A \in \mathscr{F}, B \in \mathscr{F}$ *implies* $A \cup B \in \mathscr{F}$.

$\mathscr{F}$ *is a* σ-field *over* Ω *if* $\mathscr{F}$ *is a field over* Ω *and*

(iv) $A_1, A_2, \ldots \in \mathscr{F}$ *implies* $\bigcup A_i \in \mathscr{F}$.

If $\mathscr{A} \subset \mathscr{P}\Omega$, then $\mathscr{F}(\mathscr{A})$ denotes the smallest $(\sigma\text{-})$ field which contains all elements of $\mathscr{A}$. $\mathscr{F}(\mathscr{A})$ is called the $(\sigma\text{-})$ field generated by A:

$$\mathscr{F}(\mathscr{A}) = \cap \{\mathscr{G} | \mathscr{G} \text{ is a } (\sigma\text{-}) \text{ field and } \mathscr{G} \supseteq \mathscr{A}\}.$$

Definition 3.2 $P : \mathscr{F} \to [0, 1]$ *is a (finitely additive) positive normalized measure on* $(\Omega, \mathscr{F})$ *if*

(i) $\forall A \in \mathscr{F}, 0 \leq P(A) \leq 1$,
(ii) $P(\Omega) = 1$,
(iii) $\forall A, B \in \mathscr{F}, A \cap B = \emptyset$ implies $P(A \cup B) = P(A) + P(B)$.

Furthermore, P is countably additive or σ-additive if
(iv) whenever $A_1, A_2, \ldots \in \mathscr{F}, A_i \cap A_j = \emptyset$ for $i \neq j$, $P(\bigcup A_i) = \sum P(A_i)$.

(i), (ii), (iii) and (iv) are called Kolmogorov's axioms. When the elements of Ω are denoted 'possible worlds' or 'possible states of nature' it is customary to speak of $\mathscr{F}$ as a field of events and of P as a probability.

Important elementary properties of probability are

$$P(\emptyset) = 0, \tag{3.1}$$
$$P(A \cup B) = P(A) + P(B) - P(A \cap B). \tag{3.2}$$
$$P\left(\bigcup_{i=1}^{n} A_i\right) = \sum_{i=1}^{n} P(A_i) - \sum_{i<j} P(A_i \cap A_j)$$
$$+ \sum_{i<j<k} P(A_i \cap A_j \cap A_k) - \cdots + (-1)^{n+1} P(A_1 \cap \cdots \cap A_n). \tag{3.3}$$

The last relation is called the *inclusion–exclusion principle*.

If $P(B) > 0$, then the *conditional probability* of A with respect to B is

$$P(A|B) := P(A \cap B)/P(B).$$

This may be interpreted in terms of information. $P(A|B)$ is the probability of A given that we know B to hold. When information about B does not alter our uncertainty about the truth of A then

$$P(A|B) = P(A)$$

and we say that A and B are *independent*. Independence can equivalently be defined by the relation

$$P(A \cap B) = P(A)P(B).$$

Independence is a tricky concept to understand well. If A, B are independent then A' and B' have to be independent (see Exercise 2). But (same exercise) if A and B are independent, B and C are independent and A and C are independent, then it is not always the case that

$$P(A \cap B \cap C) = P(A)P(B)P(C).$$

Events $A_1, \ldots, A_n$ are said to be (mutually) independent if

$$P(A_{i_1} \cap A_{i_2} \cap \cdots \cap A_{i_k}) = P(A_{i_1}) \ldots P(A_{i_k})$$

for any choice of $i_1, \ldots, i_k$.

In this book we are not concerned with measure theoretic details, and we shall assume that familiar mathematical symbols, such as $\int$ and dx, are defined in the familiar way, and that assumptions sufficient to render these definitions meaningful are always in force. The interested reader is referred to a book on measure theory where these matters are treated rigorously (see for example [Halmos, 1974] or [Dudley, 1989]).

3.2 Random variables

Formally, a *random variable* is a real valued function on a probability space $X : \Omega \to \mathbb{R}$ with the property that for all $t \in \mathbb{R}$ we can assign a probability to the event $X \leq t$. This is formally equivalent to requiring that

$$\{\omega \in \Omega | X(\omega) \leq t\} \in \mathscr{F}.$$

Usually the underlying probability space Ω is suppressed, and only becomes important when we want to define how several variables behave simultaneously.

According to convention, random variables are denoted with capital letters: X, Y, Z, W etc. Lower case letters are used to denote possible values of random variables. Hence '$X = x$' says that the random variable X takes the value x. The *cumulative distribution function* of X is a function $F_X : \mathbb{R} \to [0, 1]$ defined as

$$F_X(t) = P(X \leq t) = P\{\omega \in \Omega | X(\omega) \leq t\}.$$

A variable X is called a *lifetime variable* if $X : \Omega \to [0, \infty)$. The distribution of such a random variable is then often given in terms of the *reliability* function,

$$R_X(t) = P(X > t) = 1 - F_X(t).$$

$R_X(t)$ is the probability that the component with lifetime X is still working at time t. (Sometimes R_X is denoted S_X and called the *survivor* function.) If F_X is differentiable, its derivative $f_X(t) = dF_X(t)/dt$ is called the *density* of X.

Example 3.1

(i) *Bernoulli variable with parameter p, $0 \leq p \leq 1$: Define $X = 1$ with probability p and $X = 0$ with probability $1 - p$.*

(ii) *Binomial variable with parameters n, p: This is simply the sum of n independent Bernoulli variables with parameter p. Alternatively, for $r =$*

$0, \ldots, n$, we can define

$$X = r \text{ with probability } \frac{n!}{(n-r)!r!} p^r (1-p)^{n-r}.$$

(iii) *Normal variable with parameters μ and $\sigma > 0$. A normal variable has a density function*

$$f_{\mu,\sigma}(t) = \frac{1}{\sqrt{2\pi}\sigma} \exp\left[-\frac{(t-\mu)^2}{2\sigma^2}\right], \quad -\infty < t < \infty,$$

and its distribution function is $F_{\mu,\sigma}(t) = \int_{-\infty}^{t} f_{\mu,\sigma}(s) \, ds$. The normal distribution with $\mu = 0$ and $\sigma = 1$ is called the standard normal distribution. Any normally distributed variable can be standardized: If X is normally distributed with parameters μ and σ then $\frac{X-\mu}{\sigma}$ has a standard normal distribution.

(iv) *Exponential variable with parameter $\lambda > 0$: An exponential variable has density function*

$$f_\lambda(t) = \lambda \exp(-\lambda t), \quad t \geq 0.$$

3.2.1 Moments

The *expectation* of X is defined as

$$E(X) = \int_{-\infty}^{\infty} t f_X(t) dt$$

if the right hand side is finite. For a non-negative variable X the expectation satisfies

$$E(X) = \int_0^\infty t f_X(t) \, dt = \int_0^\infty R_X(t) \, dt \qquad (3.4)$$

(see Exercise 5) and is often called the *mean time to failure, MTTF*. The *variance* of X and *standard deviation* σ_X are

$$\text{Var}(X) = E(X - E(X))^2 = E(X^2) - E(X)^2,$$
$$\sigma_X = (\text{Var}(X))^{\frac{1}{2}}.$$

Example 3.2

(i) *Bernoulli variable: The mean is p and variance is $p(1-p)$.*
(ii) *Binomial variable: The mean is np and the variance $np(1-p)$.*
(iii) *Normal variable: The mean is μ and the variance σ^2.*
(iv) *Exponential variable: The mean is λ and the variance is $1/\lambda^2$.*

More generally, the expectation $E(X^r)$ is called the *rth moment*, and the expectation $E[(X - E(X))^r]$ is called the *centralized rth moment*.

3.2.2 Several random variables

When we want to compute with several random variables at once we have to consider their *joint distribution*. The *joint distribution function* of $X_1, \ldots, X_n$ is the function

$$F_{X_1\ldots X_n}(t_1, \ldots, t_n) = P(X_1 \leq t_1, \ldots, X_n \leq t_n).$$

If this function is differentiable with respect to each variable, then we can speak about the *joint density*,

$$f_{X_1\ldots X_n}(t_1, \ldots, t_n) = \frac{d}{dt_1} \cdots \frac{d}{dt_n} F_{X_1\ldots X_n}(t_1, \ldots, t_n).$$

Example 3.3

(i) *Standard bivariate normal distribution*: This is a joint distribution of two standard normal variables X and Y, and has density function

$$f(x, y) = \frac{1}{2\pi\sqrt{1-\rho^2}} \exp\left[-\frac{(x^2 - 2\rho xy + y^2)}{2(1-\rho^2)}\right], \quad -\infty < x, y < \infty.$$

Here, ρ is a parameter between -1 and 1 which equals the product moment correlation of X and Y (see subsection 3.2.3).

(ii) *Multivariate normal distribution* $N(\mu, V)$: This is a joint distribution of n normally distributed variables $\bar{X}_1, \ldots, X_n$. Using vector notation $\underline{x} = (x_1, \ldots, x_n)$ we can write its joint density function

$$f(\underline{x}) = \frac{1}{\sqrt{(2\pi)^n |V|}} \exp[-\frac{1}{2}(\underline{x} - \underline{\mu})V^{-1}(\underline{x} - \underline{\mu})^T], \quad \underline{x} \in \mathbb{R}^n,$$

where $\underline{\mu}$ is a vector, and V is a positive definite symmetric matrix with determinant $|V|$.

Two random variables are mutually *independent* when for every pair of events A_1 and A_2,

$$P(X_1 \in A_1, X_2 \in A_2) = P(X_1 \in A_1)P(X_2 \in A_2).$$

This is equivalent to the condition that

$$F_{X_1, X_2}(t_1, t_2) = F_{X_1}(t_1)F_{X_2}(t_2),$$

and, when the densities exist, to the condition that

$$f_{X_1,X_2}(t_1, t_2) = f_{X_1}(t_1) f_{X_2}(t_2).$$

Random variables $X_1, \ldots, X_n$ are independent when the joint distribution (or equivalently the joint density) factorizes in the analogous way.

If the value of one random variable is observed, $X_n = x_n$ say, then we can conditionalize the distributions of the others on that observed value:

$$f_{X_1 \ldots X_{n-1} | X_n = x_n}(t_1, \ldots, t_{n-1}) \propto f_{X_1 \ldots X_n}(t_1, \ldots, t_{n-1}, x_n),$$

where the $\propto$ sign means that the right hand side has to be normalized to ensure that the integral of the density function equals 1.

Example 3.4 *Take the standard bivariate normal distribution from the last example, and condition X on Y = 2. The conditional density for X given Y = 2 is proportional to $f(x, 2)$, which itself is proportional to*

$$\exp\left[-\frac{(x^2 - 4\rho x + 4)}{2(1 - \rho^2)}\right] = c \cdot \exp\left[-\frac{(x - 2\rho)^2}{2(1 - \rho^2)}\right]$$

for some constant c. Comparing this density with the general formula for the normal density, we see that conditional on Y = 2, X is normally distributed with mean 2ρ and variance $(1 - \rho^2)$.

3.2.3 Correlations

If X and Y are random variables for which the expectations and variances exist, then the *covariance* and *product moment correlation coefficient* ρ are defined respectively as

$$\text{Cov}(X, Y) = E(XY) - E(X)E(Y), \qquad (3.5)$$
$$\rho(X, Y) = \text{Cov}(X, Y)/(\sigma_X \sigma_Y). \qquad (3.6)$$

The product moment correlation coefficient takes values between -1 and 1, and measures the degree of linear relationship between X and Y: $\rho(X, Y) = 1$ if and only if $Y = aX + b$ for some $a > 0$. Similarly $\rho(X, Y) = -1$ if and only if $Y = aX + b$ for some $a < 0$. If X and Y are independent then $\rho(X, Y) = 0$ (although the converse is not true).

Example 3.5

(i) *Standard bivariate normal distribution:* The parameter ρ is the product moment correlation of X and Y.

(ii) *Multivariate normal distribution $N(\underline{\mu}, V)$:* Recall that $\underline{\mu}$ is a vector, and V is a positive definite symmetric matrix. The interpretation of these parameters is that $E(X_i) = \mu_i$ for all i, and $V = (v_{ij}$ is the covariance matrix, i.e. $v_{ij} = \text{Cov}(X_i, X_j)$.

When two variables X and Y are functionally related, but in a *non-linear* way (for example $Y = X^{11}$), then the product moment correlation may be small. For this reason the rank correlation is often used as a measure of the degree of *monotone* relationship (that is, rank correlation measures the extent to which large values of X occur with large values of Y and small values of X occur with small values of Y).

In order to define the rank correlation we have to first define the quantile function. If X is a random variable and $g : \mathbb{R} \to \mathbb{R}$ is some function, then $g(X)$ is also a random variable. We are free to choose $g = F_X$. Then the random variable $F_X(X)$ is called the *quantile function* of X. The quantile function for any random variable with a continuous invertible cumulative distribution function is uniformly distributed on the [0,1] interval:

$$P(F_X(X) \leq t) = P(X \leq F_X^{-1}(t)) = F_X[F_X^{-1}(t)] = t. \qquad (3.7)$$

The (Spearman) *rank correlation* of X and Y is defined as the product moment correlation of the quantile functions of X and Y:

$$\tau(X, Y) = \rho(F_X(X), F_Y(Y)). \qquad (3.8)$$

It is straightforward (or at least it will seem that way *after* you have done Exercise 3) to show that if h and g are strictly increasing continuous functions, then

$$\tau(X, Y) = \tau(g(X), h(Y)).$$

In particular, if $Y = X^{11}$ and $h(Y) = Y^{1/11}$ then

$$\tau(X, Y) = \tau(X, h(Y)) = \tau(X, X) = 1.$$

3.2.4 Failure rates

Let X be a life variable, then the *failure rate, hazard rate,* or *force of mortality* of X, $h_X(t)$, is defined as

$$h_X(t) = \lim_{\Delta \to 0} \frac{P(X \leq t + \Delta | X > t)}{\Delta},$$

so that

$$h_X(t)\Delta \approx P(X \leq t + \Delta | X > t)$$

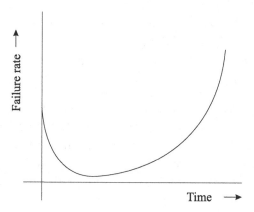

Fig. 3.1. A schematic representation of a bathtub curve

for small $\Delta > 0$. The failure rate tells us the probability of a failure just after time t given that the component is still functioning at time t. If X has a density $f(t)$ and distribution function $F(t)$, we may write

$$h_X(t) = \frac{f(t)}{(1-F(t))} = \frac{f(t)}{R(t)} = -\frac{dR}{dt}\frac{1}{R(t)}. \tag{3.9}$$

The reliability function, and thus also the distribution function, can be written in terms of the failure rate (you are asked to prove this in Exercise 6),

$$R_X(t) = \exp\left\{-\int_0^t h_X(s)\,ds\right\}. \tag{3.10}$$

This shows that the failure rate and the distribution function are equivalent ways of specifying the probabilistic behaviour of a life variable. It is frequently assumed that the failure rate of a life variable follows the general shape of the bathtub curve, see Figure 3.1. The decreasing failure rate at the beginning shows the effect of badly made components failing early. After a period with a more or less constant failure rate, the rate starts to increase again due to aging. [Kumamoto and Henley, 1996] give some population statistics showing that the failure rate for humans follows an approximate bathtub curve!

We say that a life variable X is IFR, *increasing failure rate*, if h_X is a monotone increasing function. If h_X is a monotone decreasing function then X is DFR, *decreasing failure rate*.

We now review some of the lifetime distributions commonly used in reliability theory. One of the most important distributions, the Weibull distribution, will be discussed in a separate chapter.

3.3 The exponential life distribution

A non-negative random variable X is said to have an *exponential life distribution with parameter* λ if the cumulative distribution function $F_X(t)$ is given by

$$F_X(t) = P\{X \leq t\} = 1 - e^{-\lambda t}, \quad t \geq 0.$$

It is easy to see that X is exponentially distributed with parameter λ if and only if its failure rate is constant and $h_X(t) \equiv \lambda$ for all t. For components whose failure rate follows a bathtub curve, the exponential model might be appropriate if the early failures have been removed (e.g. by 'burning in') and the components are replaced (or restored) before aging sets in.

An important property of the exponential distribution is the following: let X be an exponential variable with parameter λ, and let $x, y > 0$; then

$$P(X > x + y | X > y) = P(X > x). \tag{3.11}$$

Equation 3.11 expresses the so-called 'memoryless' property of the exponential distribution: the conditional distribution of $X - y$ given that $X > y$ is the same as the unconditioned distribution of X. Hence, for exponential components, inspecting and verifying that they have not expired is equivalent to replacement with a new component. An exponential life variable may be conceived intuitively as one which expires only as a result of random insults from outside ... it is not subject to 'aging'. More generally, when a life random variable Z has the property of being almost memoryless then its distribution is close to being exponential. Define

$$h(x, y) = P(Z > x + y | Z > y) - P(Z > x),$$

to measure the degree to which the distribution of Z departs from being memoryless. Clearly Z has the exponential distribution if and only if $h(x, y) = 0$ for all x, y. The following theorem is representative of results on the stability of the exponential distribution:

Theorem 3.1 *[Azalarov and Volodin, 1986] If there exists ϵ, $0 \leq \epsilon < 1$, such that*

$$\sup_{x,y} |h(x, y)| \leq \epsilon,$$

then $1/\lambda = E(Z) < \infty$ and

$$\sup_{x \geq 0} |F_Z(x) - (1 - e^{-x\lambda})| \leq 2\epsilon.$$

Another important property of the exponential distribution is that a *series* system, built with components having independent exponentially distributed lifetimes, also has an exponentially distributed lifetime. A series system fails when one of its components fails. Hence the lifetime of a series system is the *minimum* of the lifetimes of its components.

Theorem 3.2 *Let $X_1, \ldots, X_n$ be independent exponential variables with failure rates $\lambda_1, \ldots, \lambda_n$. Define $X = \min\{X_1, \ldots, X_n\}$, then X is exponentially distributed with failure rate $\lambda = \lambda_1 + \cdots + \lambda_n$.*

Proof Consider the reliability function of X,

$$\begin{aligned} P(X > t) &= P(X_1 > t, \ldots, X_n > t) \\ &= P(X_1 > t) \ldots P(X_n > t) \\ &= e^{-\lambda_1 t} \ldots e^{-\lambda_n t} \\ &= e^{-(\lambda_1 + \cdots + \lambda_n)t}. \end{aligned}$$

This is the reliability function of the exponential distribution with failure rate λ, as claimed. □

Another property of the exponential distribution that seems surprising at first is that a *mixture* of exponentials is DFR. This means that whenever a set of components have individual exponential life distributions, but with different parameters, then the population life statistics will show DFR behaviour. This is straightforward to understand: Those components with a higher failure rate will tend to fail first, while those remaining have not aged because of the memoryless property. The net result is a decreasing overall failure rate.

Another property of the exponential distribution that forms the basis for many statistical tests is the following. If $X_1, \ldots, X_n$ are independent samples from the same exponential distribution then the quantities

$$Z_j = S_j/S_n \quad (j = 1, \ldots, n-1)$$

where $S_j = X_1 + \cdots + X_j$ have the same multivariate distribution as the order statistics of a sample of size $n - 1$ from a population of variables uniformly distributed in $[0, 1]$ [Cox and Lewis, 1966, Chapter 3]. You are asked to check the simplest case of this result in Exercise 12.

3.3.1 Constant test intervals

Suppose that a component with an exponential life distribution, with parameter λ, is inspected at regular intervals of length I. If discovered failed

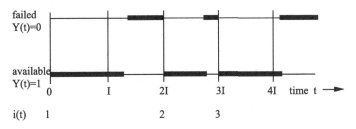

Fig. 3.2. Availability of a component under constant test intervals

during a test, the component is immediately replaced with an independent copy as good as new. Let the test times be

$$0 = t_0 < t_1 < t_2 < \cdots$$

with $t_{i+1} - t_i = I$ for $i = 1, 2, \ldots$, and define the index $i(t)$ as the number of the last component installed before time t (see Figure 3.2). Then we may consider the process

$$Y(t) = \begin{cases} 1 & \text{if component } i(t) \text{ works at time } t, \\ 0 & \text{otherwise.} \end{cases}$$

Define the *availability of an exponential component with parameter* λ *and test interval* I as

$$A(t) = P(Y(t) = 1)$$

and the *equilibrium availability* as

$$\lim_{T \to \infty} (1/T) \int_0^T A(t) dt,$$

when this limit exists.

Theorem 3.3 *The equilibrium availability of an exponential component inspected at regular intervals I is*

$$\frac{1}{\lambda I}\left(1 - e^{-I\lambda}\right) \approx 1 - \frac{\lambda I}{2} \quad \text{for } \lambda I < 0.1.$$

Proof First note that $A(t) = e^{-\lambda t}$ for $0 \le t \le I$, and that by the memoryless property of the exponential distribution, $A(t) = A(t+I)$ for all $t \ge 0$. Hence

$$\lim_{T \to \infty} (1/T) \int_0^T A(t)dt = (1/I) \int_0^I e^{-\lambda t} dt = \frac{1}{\lambda I}(1 - e^{-\lambda I}).$$

The approximation to $1 - \frac{\lambda I}{2}$ comes by taking the first terms of the Taylor

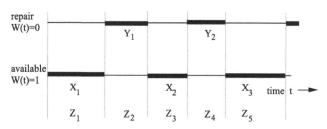

Fig. 3.3. Exponential failure and repair

expansion. Hence, for $\lambda I < 0.1$, the probability that a given regularly inspected exponential component with inspection interval I is unavailable is approximately

$$\lambda I/2.$$

□

3.3.2 Exponential failure and repair

Consider an exponential component with parameter λ. When the component fails it is taken out of service and repaired; the length of time required to repair the component is independent of the component's life distribution, and is exponentially distributed with parameter μ. To model this process we consider a set of exponential life variables $X_1, X_2, \ldots$ each with parameter λ, and a set of repair variables $Y_1, Y_2, \ldots$, each exponential with parameter μ, where all variables are independent. Define

$$Z_j = \begin{cases} Y_{j/2} & \text{if } j \text{ even,} \\ X_{(j+1)/2} & \text{if } j \text{ odd} \end{cases}$$

and define the process

$$W(t) = \begin{cases} 1 & \text{if the least } j \text{ such that } \sum_{i=1}^{j} Z_i \geq t \text{ is odd,} \\ 0 & \text{otherwise.} \end{cases}$$

Intuitively, $W(t) = 1$ means that the component is in service at time t, and $W(t) = 0$ means that the component is in repair at time t. See Figure 3.3.

Define the *equilibrium availability of an exponential component with parameter λ, under exponential repair with repair rate μ,* as

$$\lim_{T \to \infty} (1/T) \int_0^T P(W(t) = 1) dt. \tag{3.12}$$

The limit in 3.12 exists under more general conditions than those assumed here. We shall not prove existence here; however, if we assume the process W is in its equilibrium state, it is easy to determine what that state is.

Theorem 3.4 *If W is in its equilibrium state at time $t = 0$, that is if*

$$P(W(0) = 1) = \lim_{T \to \infty} (1/T) \int_0^T P(W(t) = 1) dt$$

then

$$P(W(0) = 1) = \frac{\mu}{\lambda + \mu}.$$

Proof Assume that the limit in 3.12 exists and equals P_G. Then

P(component available at dt)
$= P$(component available at $t = 0$ and not failed in dt)
$+ P$(component unavailable at $t = 0$ and repaired in dt)
$= P_G(1 - \lambda dt) + (1 - P_G)\mu dt = P_G - P_G(\lambda + \mu)dt + \mu dt.$

Hence

$$\frac{P(\text{component available at } dt) - P_G}{dt} = -P_G(\lambda + \mu) + \mu.$$

Since the component is in equilibrium at $t = 0$, the probability of being available at t is constant as a function of t, and the left hand side must equal 0. It follows that

$$P_G = \frac{\mu}{\lambda + \mu}.$$

□

3.4 The Poisson distribution

A variable Z taking values in the set $\{0, 1, 2, \ldots\}$ of non-negative integers has a *Poisson distribution with parameter λ* if

$$P(Z = k) = e^{-\lambda} \frac{\lambda^k}{k!}$$

for all k. It is easy to check that $E(Z)$ is equal to λ.

The importance of the Poisson distribution in reliability is that it describes the number of failures which have occurred up to time t in a socket. A new component is plugged into a socket and functions until failure. When it fails then it is immediately replaced by a new component which also functions until failure, etc. When the sequence of components have independent and identical exponential lifetimes then the Poisson distribution describes the number of failures:

Theorem 3.5 *Let $X_1, X_2, \ldots$ be independent and exponentially distributed with parameter λ. Define*

$$N(t) = \text{smallest } k \text{ such that } X_1 + X_2 + \cdots + X_{k+1} > t;$$

then $N(t)$ has a Poisson distribution with parameter λt.

Proof See Exercise 13. □

More generally, one can show that the random function $N(t)$ (a function of time t) follows a *Poisson process*. The definition of a Poisson process is given in Section 3.8.

3.5 The gamma distribution

The gamma distribution is sometimes used as a lifetime distribution.

The *gamma density with parameters* α, ν is defined as

$$f_{\alpha,\nu}(t) = \frac{\alpha^\nu t^{\nu-1} e^{-\alpha t}}{\Gamma(\nu)}$$

where $t > 0$, $\alpha > 0$ and $\nu > 0$. Here the *gamma function* is

$$\Gamma(\nu) = \int_0^\infty t^{\nu-1} e^{-t} dt. \tag{3.13}$$

Integration by parts shows that $\Gamma(\nu) = (\nu-1)\Gamma(\nu-1)$. Hence for $n = 1, 2, \ldots$, $\Gamma(n) = (n-1)!$ The mean of the gamma density $f_{\alpha,\nu}$ is

$$\int f_{\alpha,\nu}(t) t \, dt = \int \frac{\alpha^\nu t^\nu e^{-\alpha t}}{\Gamma(\nu)} dt = \frac{\Gamma(\nu+1)}{\alpha \Gamma(\nu)} = \frac{\nu}{\alpha}.$$

One can show similarly that the variance is ν/α^2.

The exponential distribution is obtained as a special case of the gamma distribution by taking $\nu = 1$. For $\nu \geq 1$ the gamma distribution is IFR, and for $\nu \leq 1$ the gamma distribution is DFR.

The gamma distribution plays an important role for exponential lifetimes since the sum of a fixed number of independent exponentially distributed variables is gamma distributed.

For independent non-negative random variables X, Y with densities f and g, the *convolution* $f * g$ is the density of $X + Y$ and is given by

$$(f * g)(x) = \int_0^x f(x-y) g(y) dy.$$

Theorem 3.6 *For $\alpha, \nu, \mu > 0$ we have*

$$(f_{\alpha,\nu} * f_{\alpha,\mu})(x) = f_{\alpha,\nu+\mu}(x).$$

Proof

$$\begin{aligned}(f_{\alpha,\nu} * f_{\alpha,\mu})(x) &= \int_0^x f_{\alpha,\nu}(x-y) f_{\alpha,\mu}(y) dy \\ &= \frac{\alpha^{\nu+\mu} e^{-\alpha x}}{\Gamma(\nu)\Gamma(\mu)} \int_0^x (x-y)^{\nu-1} y^{\mu-1} dy. \end{aligned} \qquad (3.14)$$

Substitute $y = xt, dy = xdt$; then the left hand side is proportional to $f_{\alpha,\nu+\mu}(x)$. However, since both are densities the proportionality constant must be 1. □

By straightforward induction we derive

Corollary 3.7 *If $X_1, \ldots, X_n$ are independent and identically distributed random variables whose distribution is gamma (with parameters α, ν) then their sum $X_1 + \ldots + X_n$ is gamma distributed (with parameters $\alpha, n\nu$).*

A more important use of the gamma distribution in reliability analysis is as a tool to model uncertainty in an unknown constant failure rate. This will be dealt with in more detail in Chapter 4.

3.6 The beta distribution

The beta distribution is of importance in modeling uncertainty about an unknown discrete failure probability, that is, the uncertainty in a probability p in a binomial or Bernoulli model. It is a continuous distribution on the interval $[0,1]$ (although by making a simple linear transform it may easily be defined on any closed interval).

Definition 3.3 *The beta distribution with parameters α and β has density function*

$$f_{\alpha,\beta} = \frac{1}{B(\alpha,\beta)} x^{\alpha-1}(1-x)^{\beta-1} 1_{[0,1]}(x),$$

where $\alpha, \beta > 0$, and $B(\alpha, \beta)$ is the beta integral

$$B(\alpha,\beta) = \int_0^1 x^{\alpha-1}(1-x)^{\beta-1} dx.$$

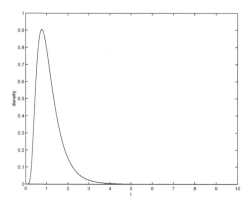

Fig. 3.4. A lognormal density

It follows from the proof of Theorem 3.6 that

$$B(\alpha, \beta) = \frac{\Gamma(\alpha)\Gamma(\beta)}{\Gamma(\alpha + \beta)}$$

from which we can then easily derive simple formulae for the mean and variance. The mean of the beta distribution is

$$\frac{B(\alpha + 1, \beta)}{B(\alpha, \beta)} = \frac{\alpha}{\alpha + \beta},$$

and the variance similarly shown to be

$$\frac{\alpha\beta}{(\alpha + \beta + 1)(\alpha + \beta)^2}.$$

When $\alpha = \beta = 1$ we have the uniform distribution as a special case.

3.7 The lognormal distribution

Another distribution sometimes used in modeling stochastic lifetimes is the lognormal distribution. We say that X is lognormally distributed $LN(\mu, \sigma)$ if $Y = \log X$ is normally distributed $N(\mu, \sigma)$ where log denotes the natural logarithm. The density function of X is

$$f(t; \mu, \sigma) = \frac{1}{\sqrt{2\pi}\sigma t} \exp\left\{-\frac{1}{2}\left(\frac{\log t - \mu}{\sigma}\right)^2\right\},$$

and is shown in Figure 3.4 with $\mu = 0$, $\sigma = 0.5$. The failure rate of the lognormal distribution shown in Figure 3.4 is illustrated in Figure 3.5.

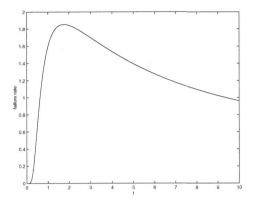

Fig. 3.5. A lognormal failure rate

The mean value of the lognormal distribution is

$$E(X) = \exp\left(\mu + \frac{\sigma^2}{2}\right),$$

and the standard deviation is

$$\sqrt{(\exp(2\mu + 2\sigma^2) - \exp(2\mu + \sigma^2))}.$$

3.8 Stochastic processes

When we want to consider a sequence of random events occurring through time, then we model this as a stochastic process. The prototype example is of failure data from a single socket. A new component is plugged into a socket and functions until failure. When it fails then it is immediately replaced by a new component which also functions until failure, etc. The sequence of failure times observed are modeled as a stochastic counting process.

Definition 3.4 *A* counting process *is a random function* $N : \mathbb{R}^+ \to \mathbb{R}$ *such that (a)* $N(0) = 0$ *and* $N(t) \in \{0, 1, 2, 3, \ldots\}$ *for all t, and (b) if* $s < t$ *then* $N(s) \leq N(t)$.

For the failure data example above, $N(t)$ simply counts the number of failures up to time t. When the failure times have independent and identically exponential distributions with parameter λ then (as remarked earlier in this chapter) the process $N(t)$ is a Poisson process.

Definition 3.5 *A counting process* $N(t)$ *is a* Poisson process *if*

(i) *for all $t, h > 0$,*

$$P(N(t+h) = n + m | N(t) = n) = \begin{cases} \lambda h + o(h) & \text{if } m = 1, \\ o(h) & \text{if } m > 1, \\ 1 - \lambda h + o(h) & \text{if } m = 0, \end{cases}$$

(ii) *if $s < t$ then $N(t) - N(s)$ is independent of all events prior to s.*

To distinguish the Poisson process from a more general class of models that are going to be described soon we call it a *homogeneous Poisson process with intensity λ*. In reliability applications the intensity is often also called the *rate of occurrence of failure, ROCOF*. This must not be confused with the failure rate of a single random variable. The ROCOF is particularly important in the modeling of repairable systems (see [Ascher and Feingold, 1984] for an entertaining discussion of the differences in modeling repairable systems and non-repairable systems).

Of course, there is no reason why a process should be a homogeneous Poisson process. The general definition of the ROCOF is

Definition 3.6 *For a counting process with $N(t) = $ the number of failures in $(0, t]$, the mean number of failures to time t is*

$$M(t) = E(N(t)).$$

The rate of occurrence of failure, ROCOF, or intensity is

$$m(t) = \frac{d}{dt} M(t).$$

For the homogeneous Poisson process (HPP) the ROCOF is constant. More generally, when $m(t)$ is not necessarily constant and the numbers of failures in disjoint intervals are independent, then we have a *non-homogeneous Poisson process (NHPP)*. The number of failures in a time interval $(t_1, t_2]$ has a Poisson distribution with mean $M(t_2) - M(t_1)$, that is

$$P(k \text{ failures in } (t_1, t_2]) = \frac{(M(t_2) - M(t_1))^k}{k!} \exp(-(M(t_2) - M(t_1))).$$

The assumption that a failure pattern is generated by an HPP implies that the inter-failure times X_i are independent and identically distributed exponential variables. By contrast, under the assumption of an NHPP the X_i are neither independent nor identically distributed. Other models of failure patterns exist in which the assumption that the X_i are independent is weakened (see for example the branching Poisson process model in [Ascher and Feingold, 1984]) but will not be treated here. Such models might be appropriate when secondary failures arising from a primary failure are registered and included in the data.

An intuitive understanding of the difference between the HPP and the NHPP models can be gained by considering how they may be simulated.

Simulation of the HPP Given the parameter of the HPP, λ, generate a sequence $X_1, X_2, \ldots$ of independent exponentially distributed variables with parameter λ. The set of points $0, X_1, X_1 + X_2, X_1 + X_2 + X_3, \ldots$ form the points of the realization of the process.

Simulation of the NHPP A realization of an NHPP can be obtained from a realization of an HPP simply by adjusting the time scale: Let $\psi : \mathbb{R} \to \mathbb{R}$ be the functional inverse of M,

$$\psi(x) = \inf\{t | M(t) \geq x\}.$$

Then if $0, X_1, X_1 + X_2, X_1 + X_2 + X_3, \ldots$ is the realization of an HPP with $\lambda = 1$, the set $\psi(0), \psi(X_1), \psi(X_1 + X_2), \psi(X_1 + X_2 + X_3), \ldots$ is a realization of the NHPP with intensity $m(t)$.

Two choices of $v(t)$ are commonly made in the reliability literature:

$$v_1(t) = \exp(\beta_0 + \beta_1 t) \quad \text{LOG-LINEAR}, \tag{3.15}$$
$$v_2(t) = \gamma \delta t^{\delta - 1} \quad \text{POWER LAW}. \tag{3.16}$$

We remark that amongst the class of NHPPs, the HPP is the only stationary process, that is, it is the only one whose counting process has stationary increments.

Definition 3.7 *A counting process $N(t)$ has* stationary increments *if for any times $t_1 < s_1, \ldots, t_n < s_n$, the distribution of*

$$(N(s_1 + h) - N(t_1 + h), N(s_2 + h) - N(t_2 + h), \ldots, N(s_n + h) - N(t_n + h))$$

does not depend on $h > 0$.

Intuitively, this means that if we only look at the pattern of failures from time h onwards, then this should have the same distribution as when we looked from time 0 onwards. When a counting process has stationary increments then clearly

$$M(s_1 + h) - M(t_1 + h) = E(N(s_1 + h) - N(t_1 + h))$$

does not depend on h. In particular,

$$M(a + b) = M(a) + [M(a + b) - M(a)] = M(a) + M(b),$$

and since M is a non-decreasing function we conclude (using Exercise 15) that $M(a) = \lambda a$ for some $\lambda > 0$. We conclude that if the data is from a general NHPP and is stationary, then it must be from an HPP.

Hence if one has data coming from an HPP and (linearly) plots the cumulative number of failures against time then one should see approximately a straight line. However, an approximate straight line is in itself only an indication of *stationarity*, that is other stationary models (such as renewal processes) cannot be excluded from consideration (see also Exercise 16).

3.9 Approximating distributions

Many results in probability and statistics are about ways of approximating one (unknown, or difficult to calculate) distribution by others. Approximation is usually measured in terms of the nearness of the distribution functions to one another. There is a corresponding notion of convergence for random variables:

Definition 3.8 *A sequence of random variables X_1, X_2, ..., with distribution functions F_1, F_2, ..., converges in distribution to X (with distribution function F_X) if*

$$F_n(t) \to F_X(t)$$

for all t at which F_X is continuous.

The practical significance of this definition is that, for large n, $P(X_n \leq t)$ can be replaced by $P(X \leq t)$ in calculations.

The most famous example of convergence in distribution is the central limit theorem (CLT). This says that for *any* random variable X with finite variance σ^2, the mean of n independent samples of X is approximately normally distributed with mean $E(X)$ and variance σ^2/n.

Other examples important for risk analysis applications are the two limits of the binomial distribution. Suppose that X_n has a binomial (n, p) distribution, that is, X_n is the number of failures in n independent trials each of which has probability p of failure. As $n \to \infty$ with p fixed then the distribution of X_n is approximately normal with mean np and variance $np(1-p)$. On the other hand, if $n \to \infty$ and $p \to 0$ so that $np \to \lambda$, then the distribution of X_n is approximately Poisson with mean λ.

3.10 Exercises

3.1 Verify the following:

(a) $P(\emptyset) = 0$.
(b) $P(A \cup B) = P(A) + P(B) - P(A \cap B)$.
(c) $P(A \cap B \cap C) \geq P(A) + P(B) + P(C) - 2$.
(d) $P\left(\bigcup_{i=1}^{n} A_i\right) = \sum_{i=1}^{n} P(A_i) - \sum_{i<j} P(A_i \cap A_j) + \sum_{i<j<k} P(A_i \cap A_j \cap A_k)$
$- \cdots + (-1)^{n+1} P(A_1 \cap \cdots \cap A_n)$.

3.2 If A, B are independent, are A' and B' necessarily independent? If A and B are independent, B and C are independent and A and C are independent, is necessarily

$$P(A \cap B \cap C) = P(A)P(B)P(C)?$$

(Hint: think of three events involving two coin tosses.)

3.3 Verify the following.

(a) $\text{Var}(X) = E(X^2) - E(X)^2$.
(b) If $X' = aX + b$, $Y' = cY + d$, with $a, c > 0$, then $\rho(X, Y) = \rho(X', Y')$.
(c) If h and g are strictly increasing continuous functions, then

$$\tau(X, Y) = \tau(g(X), h(Y)).$$

(Hint: show first that, for any random variable X and strictly increasing function g,

$$F_X(t) = F_{g(X)}(g(t))$$

holds for all t.)

3.4 The Cauchy–Schwarz inequality is: For any random variables X and Y we have $E(XY)^2 \leq E(X^2)E(Y^2)$, with equality if and only if for some c and d (not both zero) $P(cX = dY) = 1$. Use this inequality to show that $\rho(X, Y) = 1$ implies that $Y = aX + b$ for some $a > 0$ and b.

3.5 Prove the claim in Equation 3.4 using integration by parts.
3.6 Prove that Equations 3.9 and 3.10 hold.
3.7 Show, for $F(x) = 1 - \exp(-\lambda x)$, that $F(x) \approx \lambda x$ when $\lambda x < 0.1$.
3.8 Let X be exponentially distributed. Show that $E(X^n) = n!/\lambda^n$. (Hint: use Equation 3.13.)
3.9 Let X be an exponential variable with parameter λ, and let $s > t$; then show

$$P(X \leq s | X > t) = P(X \leq s - t).$$

3.10 Let X and X' be independent exponential variables with parameters λ and λ' respectively. Determine the distribution functions of $\min\{X,X'\}$ and $\max\{X,X'\}$.

3.11 Let $X_1,\ldots,X_n$ be independent exponential variables with parameter λ. Show that

$$E(\min\{X_1,\ldots,X_n\}) = (1/n)E(X_1)$$

and

$$E(\max\{X_1,\ldots,X_n\}) = \left(\sum_{i=1}^{n}\frac{1}{i}\right)E(X_1).$$

(Hint: for the second problem, try to apply the memoryless property repeatedly to get each term in the sum.)

3.12 Let X_1 and X_2 be independent exponential random variables with the same parameter λ. Define $S = X_1 + X_2$. Determine the probability density function for S, and show that, conditional upon $S = s$, X_1 is uniformly distributed on $[0,s]$.

3.13 Prove Theorem 3.5: Let $X_1, X_2,\ldots$ be independent and exponentially distributed with parameter λ. Define

$$N(t) = \text{smallest } k \text{ such that } X_1 + X_2 + \cdots + X_{k+1} > t;$$

then $N(t)$ is a Poisson process with parameter λ. (Hint: use Corollary 3.7.)

3.14 Plant records show that on a large number N of tests, a component has been found failed n times. The tests occur at regular intervals I. Assume that the component has an exponential life distribution with parameter λ, $I\lambda < 0.1$. Estimate to first order the probability that the component will fail on demand (interpret the probability of failure on demand as the equilibrium *un*availability of the component).

3.15 If $M(t)$ is a non-decreasing function, and $M(a+b) = M(a) + M(b)$ holds for all a, b, show that $M(a) = \lambda a$ for some $\lambda > 0$. (Hint: show this first for rational numbers a and b, then extend to any a and b.)

3.16 Define a counting process as follows. Let $f : [0,\infty) \to \mathbb{R}$ be the function that assigns to a point t the largest integer smaller than t (so f is just a 'staircase' function). Now let U be a uniform variable in $[0,1]$, and define $N(t) = f(t+U)$. Show that N is a counting process with stationary increments, and that $M(t) = t$.

4

Statistical inference

The general problem of statistical inference is one in which, given observations of some random phenomenon, we try to make an inference about the probability distribution describing it. Much of statistics is devoted to the problem of inference. Usually we will suppose that the distribution is one of a family of distributions $f(t|\theta)$ parameterized by θ, and we try to make an assessment of the likely values taken by θ. An example is the exponential distribution $f(t|\lambda) = \lambda \exp(-\lambda t)$, but also the joint distribution of n independent samples from the same exponential, $f(t_1,\ldots,t_n|\lambda) = \lambda^n \exp(-\lambda(t_1 + \cdots + t_n))$, falls into the same category and is relevant when making inference on the basis of n independent samples.

Unfortunately, statisticians are not in agreement about the ways in which statistical inference should be carried out. There is a plethora of estimation methods which give rise to different estimates. Statisticians are not even in agreement about the *principles* that should be used to judge the quality of estimation techniques. The various creeds of statistician, of which the most important categories are Bayesian and frequentist, differ largely in the choice of principles to which they subscribe. (An entertaining guide to the differences is given in the paper of Bradley Efron 'Why isn't everyone a Bayesian?' and the heated discussion that follows, [Efron, 1986].) To some extent the question is whether one thinks that statistical inference should be inductive or deductive. Both approaches are possible, and the choice is largely a matter of personal taste.

4.1 Foundations

We begin with a discussion of the principles underlying statistical inference, to show what types of inferential rules can be supported from an axiomatic basis. We shall conclude that the likelihood principle is the best basis for statistical inference. This says (very loosely) that statistical inference should be based on the likelihood of a parameter given the data. The likelihood principle supports both Bayesian and (frequentist) maximum likelihood methods. These methods will be described after this short section on foundations, and will be used throughout the book, although sometimes other statistical techniques will be mentioned where they are widely used.

The following notion of *sufficiency* is highly important. Intuitively, given a distribution class with parameter θ, a sufficient statistic summarizes everything that the data can say about θ.

Definition 4.1 *If X is distributed with density function $f(x|\theta)$, a function S of X (called a statistic) is sufficient if the distribution of X conditional on $S(X)$ does not depend on θ.*

An example of a sufficient statistic for the failure rate λ of an exponential distribution is given by the sample mean,

$$S(X_1,\ldots,X_n) = \frac{1}{n}(X_1 + \ldots + X_n).$$

This is because the joint density function is

$$f(X_1 = t_1,\ldots,X_n = t_n|\lambda) = \prod_{i=1}^{n} f(t_i|\lambda) = \lambda^n \exp\left(-\lambda \sum_{1}^{n} t_i\right),$$

so that if we further condition $X_1,\ldots,X_n$ on $S = r$, that is $\sum t_i = nr$, then the conditional density no longer depends on $t_1,\ldots, t_n$, and so is uniform on the feasible $(t_1,\ldots,t_n)$ (that is those $(t_1,\ldots,t_n)$ satisfying $\sum t_i = nr$ and $t_i \geq 0$). In particular, it does not depend on λ.

The first, highly reasonable, principle in the foundations of statistical inference is the *sufficiency principle* which says that whenever two observations lead to the same value of a sufficient statistic then they must lead to the same inference on θ.

Another principle that seems very reasonable is the *conditionality principle*. This says that if two experiments E_1 and E_2 are available to provide information on the parameter θ, and if the experiment to be carried out is chosen by the toss of a fair coin, then the inference on θ should only depend on the experiment that was selected, and not on the one that was not selected.

Birnbaum [Birnbaum, 1962] showed that the sufficiency principle and the conditionality principle together imply the *likelihood principle*.

Definition 4.2 *The likelihood function $L(\theta|x)$ for θ given data x is equal to the probability density of x given θ, $f(x|\theta)$.*

The *likelihood principle* says that (i) the information gained about θ from an observation X is contained entirely in the likelihood function $L(\theta|X)$, and (ii) if X_1 and X_2 are two observations depending on the same parameter θ and there is a constant c such that

$$L(\theta|X_1) = cL(\theta|X_2)$$

for all θ, then X_1 and X_2 give the same information about θ and must lead to the same inferences.

The likelihood principle supports both the (frequentist) maximum likelihood procedure, and Bayesian inference.

Further support for Bayesian inferential methods is gained by considering the relation with decision theory (see [Robert, 1994] for a lengthy discussion of the axiomatic basis of Bayesian inference). Bayesian inference allows, via the use of a loss function, the analyst to take account of the fact that different inferences might have different cost consequences. We shall discuss decision theory in Chapter 13. The other main pragmatic reason for Bayesian inference is that it allows the use of prior knowledge, which is very important when there is little data (as is often the case). Prior knowledge might come from old data which is not quite distributed in the same way (for example from before a design revision of a system), or from some form of expert judgement.

A good reason for using maximum likelihood methods is that these are often much easier to use than Bayesian methods. In many cases a simple formula can be used to calculate the maximum likelihood estimate (although in other cases a complex optimization procedure might be necessary). It is not necessary to determine a prior distribution, as in the Bayesian technique. Furthermore, when there is a reasonable amount of data, the posterior distribution given by the Bayesian method will tend to 'forget' the prior, and converge to a point mass distribution at the true value of the parameter. As the maximum likelihood estimator also converges to the true value of the parameter, the two methods will tend to give similar answers. It would then seem reasonable to use the method that is easiest. (This is what Good, a committed Bayesian, advocates and calls Type 2 rationality [Good, 1983].)

4.2 Bayesian inference

In the previous chapter we described some of the most common parametric distributions used in reliability modeling. Since there are many uncertainties in modeling component lifetimes, one important question that needs to be discussed is: 'Which uncertainties are we modeling?'.

When using a single distribution (for example the exponential with $\lambda = 2.76$) to model the lifetime of a flashlight we are assuming that all flashlights for which the distribution will be used have exactly the same λ. The flashlights are assumed to be manufactured in an identical way and used in the same way. The randomness enters through unknown effects such as the intensity of external insults.

Recall however that $1/\lambda$ is the mean of the exponential distribution. Hence when taking the mean life of a very large sample of flashlights we would expect to get approximately $1/\lambda$. Thus $1/\lambda$ (and hence also λ) is actually an observable quantity about whose values we can be uncertain (recall Sentence (VII) from Chapter 2). In the Bayesian approach we model that uncertainty with a probability distribution for λ. This distribution, the *a priori* or *prior distribution*, contains all of the knowledge and expertise built up by the model maker. Using the theorem of Bayes one can combine the *prior* distribution with the results of an experiment and a likelihood function to obtain the *posterior* distribution. Bayes' Theorem is the mathematically correct way to make deductions. The only problem is that one has to be certain about what it is one is uncertain about: we are not allowed to look at the data and decide (for example) that the exponential model is after all not good.

This Bayesian viewpoint is very popular in the risk analysis community. Data from many different sources is combined to produce *generic* data, mostly in the form of a generic (average) failure rate. The parameter is considered to be stochastic in order to represent all the various sources of uncertainty that we have in specifying the failure rate. More important, possibly, is that the idea of updating a prior distribution to obtain a posterior distribution gives an important role to the engineer. He or she can use their expertise to decide on the form of the prior distribution. Experiments can then be used to update the prior. This combination of giving weight to experts but still allowing for the scientific evidence makes Bayesian reliability very popular amongst the experts.

4.2.1 Bayes' Theorem

Bayes' Theorem provides the technique for calculating a posterior distribution from the prior and the likelihood function. We first present it in the form used for discrete distributions.

Definition 4.3 *A* partition *is a collection of events* $A_1, \ldots, A_n \subset \Omega$ *such that* $A_i \cap A_j = \emptyset$ *whenever* $i \neq j$, *and* $A_1 \cup \cdots \cup A_n \subset \Omega$.

4.2 Bayesian inference

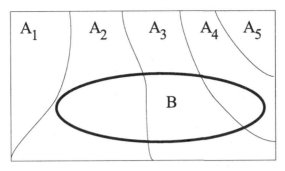

Fig. 4.1.

The idea of the theorem is the following. Suppose a point is chosen at random, and we want to know in which of the A_i it lies. Even if we cannot observe this, we may be able to observe whether or not it lies in another event B, and by taking account of the degree to which B overlaps each of the A_i we may make a new inference about the probability of x lying in A_i. See Figure 4.1.

Theorem 4.1 *Suppose that B is an event and $A_1, \ldots, A_n$ a partition. Then*

$$P(A_i|B) = \frac{P(B|A_i)P(A_i)}{\sum_{j=1}^{n} P(B|A_j)P(A_j)}.$$

Proof Using the definition of conditional probability we can write

$$P(A_i|B) = \frac{P(B|A_i)P(A_i)}{P(B)}.$$

It remains to expand the $P(B)$ term. We have

$$P(B) = \sum_{j=1}^{n} P(B \cap A_j)$$
$$= \sum_{j=1}^{n} P(B|A_j)P(A_j),$$

where the first equality holds because the A_j form a partition, and the second follows from the definition of conditional probability. □

We remark that the expression

$$P(B) = \sum_{j=1}^{n} P(B|A_j)P(A_j) \qquad (4.1)$$

is called the law (or theorem) of total probability.

The initial probability $P(A_i)$ is called the *prior probability*, and the updated probability $P(A_i|B)$ is the *posterior probability*. The term $P(B|A_i)$ is called the *likelihood*. Bayes' theorem is often written in the form $P(A_i|B) \propto P(B|A_i)P(A_i)$, and then $1/\sum_{j=1}^{n} P(B|A_j)P(A_j)$ is the constant of proportionality required to ensure that the total probability equals 1.

Example 4.1 *Components produced on a production line are either good, g, or bad, b, with probabilities $P(g) = 0.99$, $P(b) = 0.01$. Visual inspection can be used to identify good and bad components, and is rather effective in the sense that good components are identified as good with probability 0.99, and bad components are identified as bad with probability 0.99.*

Suppose that a component is identified as bad under the visual inspection. What is the probability that it is actually bad?

The prior probability that it is bad is 0.01. We need to determine the probability $P(b|\text{inspec } b)$. By Bayes' theorem we have:

$$P(b|\text{inspec } b) = \frac{P(\text{inspec } b|b)P(b)}{P(\text{inspec } b|b)P(b) + P(\text{inspec } b|g)P(g)}$$

$$= \frac{0.99 \times 0.01}{0.99 \times 0.01 + 0.01 \times 0.99} = \frac{1}{2}.$$

The version of Bayes' theorem for continuous variables, or vectors of continuous variables, is the following.

Theorem 4.2 *Let $\underline{X}$ and $\underline{\Theta}$ be continuous random vectors with joint probability density function $f(\underline{x}, \underline{\theta})$. Let $f(\underline{x}|\underline{\theta})$ and $f(\underline{\theta}|\underline{x})$ be the corresponding conditional densities, and $f(\underline{\theta}) = \int f(\underline{x}, \underline{\theta}) \, d\underline{x}$ the marginal density of $\underline{\Theta}$. Then*

$$f(\underline{\theta}|\underline{x}) = \frac{f(\underline{x}|\underline{\theta})f(\underline{\theta})}{\int f(\underline{x}|\underline{\theta})f(\underline{\theta}) \, d\underline{\theta}}$$

Proof By definition of the conditional probability density,

$$f(\underline{x}|\underline{\theta})f(\underline{\theta}) = f(\underline{x}, \underline{\theta}) = f(\underline{\theta}|\underline{x})f(\underline{x}).$$

Hence if $f(\underline{\theta}) > 0$ and $f(\underline{x}) > 0$,

$$f(\underline{\theta}|\underline{x}) = \frac{f(\underline{x}|\underline{\theta})f(\underline{\theta})}{f(\underline{x})} = \frac{f(\underline{x}|\underline{\theta})f(\underline{\theta})}{\int f(\underline{x}|\underline{\theta})f(\underline{\theta}) \, d\underline{\theta}}.$$

□

The application of this theorem is as follows. The *prior* distribution on the parameter Θ is $f_0(\theta)$. If we have a sample of n lifetimes $x_1, \ldots, x_n$ with joint density $f(x_1,\ldots,x_n|\theta)$, then the *posterior* density of Θ given the observations is

$$f_1(\theta|x_1,\ldots,x_n) = \frac{f(x_1,\ldots,x_n|\theta)f_0(\theta)}{\int f(x_1,\ldots,x_n|\theta)f_0(\theta)\,d\theta}. \qquad (4.2)$$

The function $f(x_1,\ldots,x_n|\theta)$ considered as a function of θ is also called the *likelihood* of θ given the observations $x_1,\ldots,x_n$, and is written

$$L(\theta|x_1,\ldots,x_n).$$

One can write the above expression as

$$f_1(\theta|x_1,\ldots,x_n) \propto L(\theta|x_1,\ldots,x_n)f_0(\theta). \qquad (4.3)$$

The constant of proportionality is determined by the requirement that f_1 is a probability density function (i.e. the integral is 1).

4.2.2 An example with the exponential distribution

In an experiment 10 components are tested until failure. The lifetime of each component is noted, and this gives data $t_1,\ldots,t_{10}$.

The data $t_1,\ldots,t_{10}$ is considered as observations of random variables $T_1,\ldots,T_{10}$ which are assume to be exponentially distributed and independent given the parameter λ. The joint density of the observed values is thus

$$f(t_1,\ldots,t_{10}|\lambda) = \prod_{i=1}^{10} f(t_i|\lambda) = \lambda^{10}\exp\left(-\lambda\sum_1^{10} t_i\right).$$

This gives as the likelihood function

$$L(\lambda|t_1,\ldots,t_{10}) = \lambda^{10}\exp\left(-\lambda\sum_1^{10} t_i\right).$$

Suppose now that the sample mean is 1.7. In the classical approach a point estimator for the unknown parameter λ is found. For example the maximum likelihood estimator (to be discussed in subsection 4.3.1) of λ is that value $\hat{\lambda}$ for which the likelihood function is maximized, $\hat{\lambda} = 10(\sum t_i)^{-1} = 0.59$. However, we wish to express our uncertainty about λ by giving a distribution rather than a point estimate.

Beforehand an expert had given an uncertainty estimate for the unknown failure rate. Measuring time in units of months the expert has assessed the 5% quantile for the failure rate as 0.23, and the 95% quantile as 1.29. It is decided to approximate the expert information using a gamma distribution with parameters $\alpha = 6$ and $\nu = 4$.

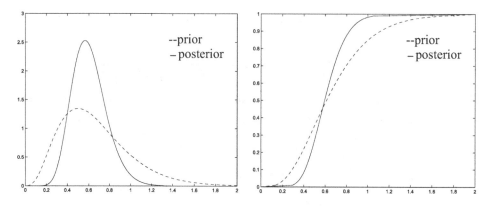

Fig. 4.2. Prior and posterior density and distribution functions

According to Bayes' theorem, the posterior density of λ given $t_1, \ldots, t_{10}$ is

$$\begin{aligned} f_1(\lambda|t_1,\ldots,t_{10}) &\propto L(\lambda|t_1,\ldots,t_{10})f_0(\lambda) \\ &\propto \lambda^{10}\exp(-17\lambda)\lambda^3\exp(-6\lambda) \\ &= \lambda^{13}\exp(-23\lambda). \end{aligned}$$

The posterior distribution is thus again a gamma distribution, but now with parameters $\alpha = 23$ and $v = 14$. The 5% and 95% quantiles of this distribution are 0.37 and 0.9 respectively. See Figure 4.2. This now shows how our uncertainty in the parameter λ has changed.

Suppose that we want to make a prediction about the lifetime of a new component. We assume that the new component also has an exponentially distributed lifetime T with parameter λ, and that T is independent of other component lifetimes given λ. The *predictive* distribution for T is given by averaging out the uncertainty in λ. We have

$$\begin{aligned} f(t|t_1,\ldots,t_{10}) &= \int_0^\infty f(t|\lambda,t_1,\ldots,t_{10})f(\lambda|t_1,\ldots,t_{10})\,d\lambda \\ &= \int_0^\infty f(t|\lambda)f(\lambda|t_1,\ldots,t_{10})\,d\lambda \\ &= \int_0^\infty \lambda e^{-\lambda t}\frac{23^{14}}{\Gamma(14)}\lambda^{13}e^{-23\lambda}\,d\lambda \\ &= \frac{23^{14}}{\Gamma(14)}\int_0^\infty \lambda^{14}e^{-(23+t)\lambda}\,d\lambda \\ &= \frac{23^{14}}{\Gamma(14)(23+t)^{15}}\int_0^\infty \tilde{\lambda}^{14}e^{-\tilde{\lambda}}\,d\tilde{\lambda} \\ &= \frac{\Gamma(14)\times 23^{14}}{\Gamma(13)(23+t)^{15}} = \frac{14\times 23^{14}}{(23+t)^{15}}, \end{aligned}$$

4.2 Bayesian inference

where $\tilde{\lambda} = (23 + t)\lambda$. This is a so-called Pareto distribution. More generally, upon integrating out a gamma (α, v) distribution for λ we obtain a Pareto (α, v) distribution for T, with density function

$$v\alpha^v/(t+\alpha)^{v+1}.$$

The mean and variance are

$$\frac{v\alpha}{v-1} \quad \text{and} \quad \frac{v\alpha^2}{(v-1)^2(v-2)}$$

respectively. These give the predictive mean and variance for the unknown lifetime.

4.2.3 Conjugate distributions

In the above example we saw that a prior gamma distribution was transformed into a posterior gamma distribution. This occurs more generally. Suppose that λ has a gamma prior distribution with parameters α and v. If we make n independent observations of failure times $t_1, \ldots, t_n$ we can use this information to update the prior. We have

$$\begin{aligned} f_1(\lambda|t_1,\ldots,t_n) &\propto \lambda^n \exp\left(-\lambda \sum t_i\right) \lambda^{v-1} \exp(-\alpha\lambda) \\ &= \lambda^{n+v-1} \exp\left(-\left(\alpha + \sum t_i\right)\lambda\right). \end{aligned}$$

Thus the new posterior density is also gamma but now with shape parameter $n+v$ and scale parameter $\alpha + \sum t_i$. Whenever we start with a gamma prior and combine this with new data about exponentially distributed lifetimes, we obtain a gamma posterior, with different parameters.

When the likelihood function comes from such a class of distributions that the distributional classes of the prior and the posterior are the same (as happened in the above example) then the two classes of distributions are said to be *conjugate*.

Another example of a conjugate pair of distributions is that of the beta and binomial distributions. See [O'Hagan, 1994] for the general form of conjugate distributions for exponential families (including the exponential, gamma, and normal families).

The use of conjugate distributions makes it relatively simple to carry out the process of Bayesian updating. One only has to check how a few parameters change. However, given a likelihood function, you might not want to model the prior with the conjugate distribution because it simply does not reflect your prior opinion.

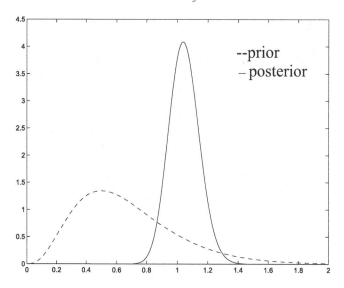

Fig. 4.3. Prior and posterior density and distribution functions (100 observations)

We remark also that in the above example it is easy to see the *consistency* of the procedure. This means that as more information becomes available, the estimate given by the statistical procedure will converge to the 'true' value. If the data $t_1, t_2, \ldots$ really is independent realizations from an exponential distribution with failure rate λ^*, then since $1/\lambda^*$ is the population mean we have $n/\sum_1^n t_i \to \lambda^*$ as $n \to \infty$ with probability 1. The sequence of posterior distributions converge also to a point mass at λ^* since

$$E(\lambda|t_1,\ldots,t_n) = \frac{v+n}{\alpha+\sum_1^n t_i}, \quad \text{and}$$
$$\text{Var}(\lambda|t_1,\ldots,t_n) = \frac{v+n}{(\alpha+\sum_1^n t_i)^2},$$

so we see that $E(\lambda|t_1,\ldots,t_n) \to \lambda^*$ and $\text{Var}(\lambda|t_1,\ldots,t_n) \to 0$.

Figure 4.3 shows the posterior density when $n = 100$ and $\sum_1^{100} t_i = 186$.

4.2.4 First find your prior

The principal difference between Bayesian and maximum likelihood methods (discussed later) is that the Bayesian method requires the specification of a prior distribution. The problem of specifying that prior is not one that the Bayesian paradigm can help us with, as the existence of the prior is a supposition.

The Bayesian paradigm uses the notion of subjective probability defined in the theory of rational decision making. In this theory every rational

4.2 Bayesian inference

individual is a free agent and is not constrained to choose the same prior as another rational individual. One of the principal objections from non-Bayesians to the Bayesian paradigm is that the choice of prior is subjective. In their opinion this means that it is impossible to use the theory to make objectively sustainable inferences. In practice many Bayesians try to find a prior over which there is consensus. Apart from the methods described below, expert judgement studies (see Chapter 10) and generic data from reliability databases (see Chapter 9) are popular ways of determining a prior over which a 'rational consensus' can exist.

In this subsection we shall look at various different ways that have been suggested for determining a prior distribution.

4.2.4.1 Improper priors

To illustrate that idea of an improper prior we return to the example of subsection 4.2.2. In that example we used quantiles from an expert to estimate a prior. Without this input we might have decided that we had 'no idea' about the prior. Some statisticians would say that complete lack of information in this case could be modeled by a uniform prior density on λ,

$$f_0(\lambda) = 1$$

for all $\lambda > 0$. This is not the density of a probability, as the integral is infinite. Such a prior is called an *improper prior*. It may be regarded in this case as the limit of uniform distributions

$$f_0(\lambda) = \begin{cases} 1/M & \text{for } 0 \le \lambda \le M, \\ 0 & \text{otherwise,} \end{cases} \quad (4.4)$$

as $M \to \infty$. Assuming the prior density in Equation 4.4 and using the exponential form of the lifetimes, we have the following form for the posterior density:

$$f_1(\lambda|t_1,\ldots,t_{10}) = \frac{M^{-1}\lambda^{10}\exp(-\lambda\sum t_i)}{\int_0^M M^{-1}\lambda^{10}\exp(-\lambda\sum t_i)\,d\lambda}.$$

Now, since M is supposed to be large we can use the identity

$$\int_0^\infty c^{n+1} u^n \exp(-cu)\,du = \Gamma(n+1) := n!, \quad c \ge 0,$$

to see that in the limit $M \to \infty$,

$$f_1(\lambda|t_1,\ldots,t_{10}) = \frac{(\sum t_i)^{11}\lambda^{10}\exp(-\lambda\sum t_i)}{\Gamma(11)}.$$

This is again a gamma density.

Improper priors have been widely used, but do not always give a proper posterior after updating. There are consistency problems with the use of improper priors (see [Efron, 1978] for a nice practical example).

4.2.4.2 Non-informative priors

The improper prior defined above was supposed to represent the total lack of information on λ. More generally, for any parameter θ we can just take a uniform distribution over θ. Sometimes this will be improper, but it may also be proper (for example, if we are uncertain about the probability p of getting heads by the toss of a coin, then a non-informative distribution for p would be uniform on $[0, 1]$).

Whether the non-informative distribution is proper or not, there is still a fundamental problem with the notion of non-informativeness. This is that 'non-informativeness' is not a coordinate-free concept. To illustrate this consider again the exponential distribution. We have used the failure rate λ as a parameter. Many texts on statistics use however the mean value μ as parameter, $\mu = 1/\lambda$. Now clearly a uniform distribution for λ does not stay uniform when we transform to μ.

One approach proposed in [Box and Tiao, 1992] is to adopt a transformation of the prior parameter under which the likelihood function is 'data translated'.

Another practical approach is to always try to use parameters that have a physical interpretation. For only then is it possible to make an informed choice about the appropriate form of non-informativeness that should be applied.

We believe also that in most practical applications it is possible to do much better than non-informative priors. We claim that, when the parameters have a physical interpretation, there are no cases in which the analyst has completely no idea. In the exponential example given above, it might be clear that a failure rate of above 100 failures per unit time is not physically reasonable. So why use a non-informative prior that gives positive weight to this possibility?

In fact, the choice of non-informative priors is more usually dictated by the perceived need to have an 'objective' choice of prior, rather than because it is really necessary to model total lack of information. But, thinking about the exponential example above, why should a prior distribution that everyone will agree is wrong be judged 'objective'?

4.2.4.3 Maximum entropy priors

Maximum entropy priors were forcefully advocated by Jaynes [Jaynes, 1983]. Entropy is an important measure of unpredictability in a system, and is a

4.2 Bayesian inference

fundamental notion in physics. In statistics it is closely related to Kullback–Liebler divergence (see [O'Hagan, 1994], [Kullback, 1959]), also called the information.

The application of the maximum entropy principle can best be explained with an example. Suppose a random quantity N can just take values $1, \ldots, n$. We want to determine a discrete probability vector $\underline{p} = (p_1, \ldots, p_n)$ (where naturally $p_i \geq 0$ for all i, and $\sum_i p_i = 1$) to describe the distribution of N. However, we also want the expected value of the function $g_1(N)$ to equal k_1, $E(g_2(N)) = k_2, \ldots, E(g_m(N)) = k_m$. The probability vector satisfying these constraints is the one solving the following convex optimisation problem:

$$
\begin{aligned}
\max \quad & -\sum_i p_i \log p_i \\
\text{such that} \quad & \sum_i p_i g_1(i) = k_1, \\
& \quad \vdots \\
& \sum_i p_i g_m(i) = k_m, \\
& \sum_i p_i = 1, \\
& p_1, \ldots, p_n \geq 0.
\end{aligned}
$$

The solution is easy to determine by using Lagrange multipliers. It always has the form

$$ p_i = \frac{\exp(-\sum_j \mu_j g_j(i))}{\sum_{k=1}^n \exp(-\sum_j \mu_j g_j(k))}, $$

for certain real numbers μ_j (the Lagrange multipliers) whose values can be obtained by substituting the expression for p_i back into the constraints.

It is also possible to apply the theory for continuous variables (as opposed to the discrete N discussed above). In Chapter 17 we shall discuss an algorithm for determining maximum entropy (or equivalently, minimal information) distributions.

Bernardo [Bernardo, 1979] has developed the entropy approach further, and has shown that it is possible to develop a theory which covers the maximum entropy approach illustrated above and the older approach of Jeffreys (see [O'Hagan, 1994] for a discussion).

4.2.4.4 Empirical Bayes priors

Empirical Bayesians are prepared to use some real data to construct a prior. This method is controversial within the Bayesian community as, according to the Bayesian paradigm, the only coherent way to use observations is via Bayes' theorem. However, as remarked above, the Bayesian paradigm does not say how we should determine the prior (the true Bayesian has fixed his prior at conception and has been updating ever since; those of us who neglected to sort out priors at conception are faced with the non-Bayesian problem of picking one later). It might seem reasonable for a moderately rational individual to look at some data in order to build up his 'initial idea' of the distribution.

The empirical Bayes approach is used when the prior distribution for the unknown parameter θ is supposed to be from a parameterized family of distributions. For example, a constant failure rate λ might have a gamma distribution, which has parameters α and ν. These are called the *hyperparameters*. Empirical Bayes uses data, and a classical esimation procedure, to obtain a single estimate for the hyperparameters. For more details, the reader is referred to [O'Hagan, 1994].

4.2.5 Point estimators from the parameter distribution

The above approach is based on the philosophy that our uncertainty about a particular parameter can be reflected through the choice of a probability distribution on the parameter space.

If one really needs to work with a single value of the parameter then this may be obtained from the distribution on the parameter space. In decision theory one often uses the idea of a loss function. A function $\ell(\tilde{\theta}, \theta)$ is defined as the loss arising from choosing $\tilde{\theta}$ when the true value is θ. We choose $\tilde{\theta}$ so that the expected loss is minimized,

$$\min E\ell(\tilde{\theta}, \theta).$$

The precise solution depends on the form chosen for ℓ. In particular, if one takes $\ell(\tilde{\theta}, \theta) = K(\tilde{\theta} - \theta)^2$ for $K > 0$ a constant, then the solution for $\tilde{\theta}$ is the mean of θ with respect to the posterior distribution. If one takes $\ell(\tilde{\theta}, \theta) = |\tilde{\theta} - \theta|$, then the solution is the posterior median. See Exercise 3.

4.2.6 Asymptotic behaviour of the posterior

Suppose that we have modeled our uncertainty in a parameter θ using a prior, and that this uncertainty has been updated by a sequence of observations. The *consistency* property says that if the observations were really generated by a fixed θ_0, then the posterior probability should concentrate around θ_0. In the limit of infinite observations the posterior mass is concentrated at θ_0.

The Bayesian approach is consistent, except when the prior gives zero probability to the true value θ_0. If an improper prior is used then consistency holds if, after a finite number of applications of Bayes' theorem, the posterior becomes proper.

Furthermore, it is possible to show that (under the appropriate conditions: see [Cox and Hinkley, 1974] or [O'Hagan, 1994]) the posterior is *asymptotically normal*. Suppose for simplicity that θ is a real number. Asymptotic normality means that after n observations, the posterior is concentrated around the true value θ_0 approximately in the form of a normal distribution with mean θ_0 and variance v/n, where $v = \int \frac{\partial^2 \log f(x|\theta)}{\partial \theta^2} f(x|\theta_0)\, dx$ is the *Fisher information*.

4.3 Classical statistical inference

In the frequentist approach, it is not permissible to talk about the probability of a certain parameter value as the parameter is not considered a random variable. The classical approach is therefore to make a single estimate in the value of a parameter. If necessary a confidence interval will be built around the estimate telling us something (non-probabilistic) about the degree of certainty in the estimate. We shall briefly discuss some of the most important methods in frequentist statistics. For more information we refer to [Mood *et al.*, 1987].

4.3.1 Estimation of parameters

The maximum likelihood method of estimation is the most important classical estimator satisfying the likelihood principle.

Definition 4.4 *Given a likelihood function $L(\theta|x)$, a maximum likelihood estimator, MLE, for θ is a value $\hat{\theta}$ such that*

$$L(\hat{\theta}|x) = \max_\theta L(\theta|x).$$

Intuitively, the MLE is that value of the parameter for which the data observed has the highest probability. Note that this is not the same as saying that the MLE has the highest probability. Although one often speaks of 'the' maximum likelihood estimator, it does not need to be unique as the likelihood function may have more than one maximum. The MLE has the advantage of a simple definition, and can often be written in a closed form.

The asymptotic normality that we discussed above for the posterior distribution in the Bayesian approach also holds for the MLE. Confidence intervals can be calculated for the MLE based on the normal approximation.

For ease of calculation it is standard practice to maximize the log of the likelihood. When the likelihood is differentiable and unimodal (that is, it has

4 Statistical inference

one maximum) then the MLE may be found by setting the derivative of the log-likelihood equal to zero.

4.3.1.1 Binomial distribution

If X is binomial with parameters (n, p), then the MLE for p based on r events in n trials is r/n.

To see this, note that the likelihood function is

$$L(p|r, n) = \frac{n!}{r!(n-r)!} p^r (1-p)^{n-r}.$$

Taking logarithms and setting the derivative d/dp equal to zero gives

$$\frac{r}{p} - \frac{n-r}{1-p} = 0.$$

Solving for p gives $p = r/n$ as claimed.

4.3.1.2 Exponential distribution

If T is distributed exponentially with failure rate λ, then the MLE estimate of λ based on n observations $t_1, \ldots, t_n$, is

$$\frac{n}{\sum t_i}.$$

This is known as the *total time on test statistic*. To see this note that the likelihood is

$$L(\lambda|t_1, \ldots, t_n) = \lambda^n \exp\left(-\lambda \left(\sum_{i=1}^n t_i\right)\right).$$

Taking logarithms and setting the derivative $d/d\lambda$ equal to zero gives

$$\frac{n}{\lambda} - \sum_{i=1}^n t_i = 0.$$

4.3.1.3 The log-linear NHPP

Consider the log-linear NHPP with intensity function

$$v_1(t) = \exp(\beta_0 + \beta_1 t)$$

from Equation 3.15. Suppose that the system is observed for the time interval $(0, t_0)$, with events occurring at $t_1, t_2, \ldots, t_n$. Then the ML estimate for β_1, $\hat{\beta}_1$ is obtained by numerically solving

$$\sum_{i=1}^n t_i + n\hat{\beta}_1^{-1} - nt_0\{1 - \exp(-\hat{\beta}_1 t_0)\}^{-1} = 0.$$

The ML estimate of $\hat{\beta}_0$ is now

$$\hat{\beta}_0 = \log\left\{\frac{n\hat{\beta}_1}{\exp(\hat{\beta}_1 t_0) - 1}\right\}.$$

4.3.1.4 The power law NHPP

Now consider the power law NHPP with intensity function (originally defined in Equation 3.16)

$$v_2(t) = \gamma \delta t^{\delta-1}.$$

Suppose that the system is observed for the time interval $(0, t_0)$, with events occurring at $t_1, t_2, \ldots, t_n$. Then the ML estimates for the parameters δ and γ are

$$\hat{\delta} = \frac{n}{n \log t_0 - \sum_{i=1}^{n} \log t_i},$$
$$\hat{\gamma} = \frac{n}{t_0^{\hat{\delta}}}.$$

4.3.2 Non-parametric estimation

Given a set of observations $t_1, t_2, \ldots, t_n$ of a lifetime variable, we can construct the *empirical distribution function*,

$$\hat{F}_n(t) = \frac{1}{n}\|\{i : t_i \le t\}$$

This is a random function (because it depends on the random observations), and is a non-parametric estimate of the true distribution function F (non-parametric means that we do not assume that it belongs to any particular family of distributions).

The famous Glivenko–Cantelli theorem (see for example [Pollard, 1984]) says that, with probability,

$$\sup_t |F(t) - \hat{F}_n(t)| \to 0$$

as $n \to \infty$. This is a form of consistency for the empirical distribution function.

The empirical distribution function is not the only way to approximate a distribution function – another way based on order statistics will be discussed in Chapter 5.

4.3.3 Confidence intervals

Within the frequentist statistical philosophy it is not permissible to give parameters a probability distribution. Frequentists do however feel the need to acknowledge that some point estimates are more acceptable than others are. The degree of acceptability in a parameter estimate can be conveyed using a confidence interval.

The basic idea of a confidence interval is to look at how likely it is to obtain an estimator value 'more extreme' than that actually computed for various parameter values of the underlying population. If, for a given parameter value, the observed value becomes extremely unlikely, then that parameter value may be considered itself unlikely.

4.3.3.1 Binomial distribution

Recall that the maximum likelihood estimator (MLE) of p, given r events in n trials, is r/n. For r events in n observations, the central 90% statistical confidence bounds are given by two values p_ℓ and p_u such that

- p_ℓ = largest p such that $P(r$ or more failures in n trials $\mid p) \leq 0.05$,
- p_u = least p such that $P(r$ or fewer failures in n trials $\mid p) \leq 0.05$.

We CANNOT say that there is a 90% probability that the true value falls within $[p_\ell, p_u]$; i.e. p_ℓ and p_u are not quantiles for the uncertainty distribution of p. In contrast, by taking the 5% and 95% quantiles, q_ℓ and q_u, of his posterior distribution, the Bayesian *can* claim that there is (for him) a 90% probability that the true parameter value falls within $[q_\ell, q_u]$.

4.3.3.2 Exponential distribution

Let $X_1, \ldots, X_n$ be independent exponential variables with parameter λ. To find statistical confidence bands, recall (from Corollary 3.7) that $\sum X_i$ follows a gamma distribution with shape n and scale λ. Substituting $Z = 2\lambda \sum_{i=1}^{n} X_i$ we find that the density of Z is

$$f(z) = \frac{2^{-n} z^{n-1} e^{-z/2}}{\Gamma(n)},$$

which is the *chi-square-density with 2n degrees of freedom*. The chi-square-distribution is available both in many statistics textbooks, and as a built-in function in most spreadsheet and statistical software. The 90% statistical confidence band is given by two values λ_ℓ and λ_u:

- λ_ℓ = lowest λ such that $P(\sum X_i \leq \sum x_i \mid \lambda) \leq 5\%$;
- λ_u = largest λ such that $P\{\sum X_i \geq \sum x_i \mid \lambda\} \leq 5\%$.

4.3.4 Hypothesis testing

Hypothesis testing is, like estimation, a major part of classical statistics. We shall only consider tests of simple hypotheses, that is, hypotheses in which the distribution of the random variable is fully specified. For example 'the distribution of T is exponential with parameter 1' is simple, but 'the distribution of T is exponential with parameter larger than 1' is not.

The standard approach to hypothesis testing is the use of *test statistic*. This is just some real valued function of the data whose distribution, under the hypothesis, is known. If the value of the test statistic based on the data is extreme (in a probabilistic sense that will be made precise shortly), then the hypothesis is rejected.

Hypothesis testing involves the idea of *significance level*. Assume for now that the test statistic is typically not large under the hypothesis. We reject the hypothesis at the $m\%$ significance level if the value of the test statistic calculated from the data is in the upper $m\%$ tail of the distribution. Typically one chooses m to be 10, 5, 2.5 or 1, depending on the strength of the test required.

One problem is that the *exact* distribution of the test statistic is usually not known (although it may be simulated by computer). More often the *asymptotic* distribution of the test statistic is known, and is approximately valid for the case that the number of observations is large.

We give a few examples.

4.3.4.1 The chi-squared test for the multinomial distribution

Suppose that a random quantity can take values in one of n distinct categories. We want to test the hypothesis that the probability of being in category k is equal to p_k. We obtain a random sample of N independent observations, and form the empirical probabilities

$$s_k = \frac{1}{N} \#(\text{observations in category } k).$$

Intuitively, a large deviation between the s_ks and p_ks should not occur if the hypothesis is true. A test statistic is given by the relative information,

$$I(s;p) = \sum_{1}^{n} s_i \log\left(\frac{s_i}{p_i}\right), \tag{4.5}$$

which is always non-negative and takes minimal value 0 if and only if $s = p$ (Exercise 7). It is well known that for large N the distribution of the relative information is approximately chi-squared distributed with $n - 1$ degrees of freedom. So if the calculated value of $I(s;p)$ is in the upper tail of the χ^2_{n-1} distribution then the hypothesis is rejected.

4.3.4.2 The Kolmogorov–Smirnov test

As mentioned in subsection 4.3.2, the empirical distribution function $\hat{F}_n$ converges towards the true distribution function F. In fact this happens so that the largest vertical distance between the graphs of $\hat{F}_n$ and F is of order $1/\sqrt{n}$. One can show that

$$D_n = \sqrt{n} \sup_t |F(t) - \hat{F}_n(t)|$$

converges in distribution to a random variable with the Kolmogorov–Smirnov distribution. Hence, if D_n is above a critical value of the KS statistic, then the hypothesis that the true distribution is F can be rejected. Critical values of the KS statistic can be found in statistics textbooks or built into spreadsheet software.

The KS test is a distribution-free test, because it depends on the statistical properties of $F(T)$ and $\hat{F}_n(T)$. For this reason the KS test is a goodness-of-fit test that is applicable to all continuous distributions.

Other goodness-of-fit tests based on measuring the distance between the empirical distribution function and the hypothesized distribution function are also available, see [Stephens, 1986].

4.3.4.3 Laplace's test for the log-linear NHPP

It is of special interest to test the hypothesis that $\beta_1 = 0$, corresponding to the special case of the HPP. A commonly used test, known as Laplace's test, is based on the fact that the statistic

$$U = \frac{\sum_{i=1}^{n} t_i - \frac{1}{2}nt_0}{t_0(n/12)^{\frac{1}{2}}}$$

has approximately a standard normal distribution under the hypothesis that $\beta_1 = 0$.

4.3.4.4 The chi-squared test for the power law NHPP

Again, one may wish to test the hypothesis that the data comes from an HPP against the power law NHPP, that is a test with null hypothesis $\delta = 1$. A suitable test statistic is

$$V = 2 \sum_{i=1}^{n} \log\left(\frac{t_0}{t_i}\right)$$

which under the hypothesis $\delta = 1$ has a chi-squared distribution with $2n$ degrees of freedom (if $t_0 = t_n$, that is, the observations are ended with a failure time, then V has the chi-squared distribution with $2n - 2$ degrees of freedom).

4.4 Exercises

4.1 A coin is thought to be biased, with probability p of heads. The prior distribution for p is from the class of beta distributions, with density proportional to $p^{10}(1-p)^{14}$ on the unit interval $[0,1]$. To obtain more certainty about p, the coin is tossed five times. Three tosses give heads and two give tails. Write down the likelihood function associated with this experiment, and calculate the posterior distribution for p.

4.2 Suppose that X has a normal $N(\mu, 1)$ distribution, where μ is unknown. Take as prior for μ a normal $N(0, 2)$ distribution. Calculate the posterior distribution for μ after a single observation of $X = 3$, and show that the posterior is normal. Calculate general formulae when X has distribution $N(\mu, \sigma)$ and μ has distribution $N(\tilde{\mu}, \tilde{\sigma})$.

4.3 Show that if one takes a loss function $\ell(\tilde{\theta}, \theta) = K(\tilde{\theta} - \theta)^2$ for $K > 0$ a constant, then the Bayes estimate for $\tilde{\theta}$ is the mean of θ with respect to the posterior distribution. If one takes $\ell(\tilde{\theta}, \theta) = |\tilde{\theta} - \theta|$, then the Bayes estimate is the posterior median.

4.4 (Based on an example in [Howard, 1988]) Suppose that there are two identical machines that can fail on demand. The probability of failure is unknown, but is the same for both machines. We choose to model the uncertainty in the failure probability p by a density $f(p)$. Machine 1 is tested and fails on demand. Use Bayes' theorem to show that the probability of failure of machine 2 now satisfies

$$P(M2 \text{ fails}|M1 \text{ failed}) = \frac{E(p^2)}{E(p)}.$$

Using the formula for variance conclude that the above expression equals

$$E(p) + \frac{\text{var}(p)}{E(p)}.$$

Take a numerical example, with $f(p)$ the density equal to 249 between 0 and $\frac{1}{250}$ and $\frac{1}{249}$ between $\frac{1}{250}$ and 1. Show that the prior probability of failure of $M2$ (the mean of f) is $\frac{1}{250}$, but that the posterior probability after the failure of $M1$ is more than $\frac{1}{3}$.

4.5 Suppose that X is uniformly distributed on an unknown interval $[a, b]$. What is the MLE of b given realizations $x_1, \ldots, x_n$?

4.6 A component fails on demand two times in five tests. Determine the MLE and a 90% confidence interval for the probability p of failure.

4.7 Show that the relative information function defined in Equation 4.5 is always non-negative. (Hint: use concavity of the logarithm to show that $\log(x) \leq x - 1$.)

5

Weibull Analysis

The Weibull distribution finds wide application in reliability theory, and is useful in analyzing failure and maintenance data. Its popularity arises from the fact that it offers flexibility in modeling failure rates, is easy to calculate, and most importantly, adequately describes many physical life processes. Examples include electronic components, ball bearings, semi-conductors, motors, various biological organisms, fatigued materials, corrosion and leakage of batteries. Classical and Bayesian techniques of Weibull estimation are described in [Abernathy *et al.*, 1983] and [Kapur and Lamberson, 1977].

Components with constant failure rates (i.e. exponential life distributions) need not be maintained, only inspected. If they are found unfailed on inspection, they are 'as good as new'. Components with increasing failure rates usually require *preventive maintenance*. Life data on such components is often heavily censored, as components are removed from service for many reasons other than failure. Weibull methods are therefore discussed in relation to censoring. In this chapter we assume that the censoring process is independent or random; that is, the censoring process is independent of the failure process. This may arise, for example, when components undergo planned revision, or are failed by overload caused by some upstream failures. In other cases a service sojourn may terminate for a reason which is not itself a failure, but is related to failure. When components are removed during preventive maintenance to repair degraded performance, we certainly may not assume that the censoring is independent.

In this chapter various techniques are illustrated with one very small data set taken from a Swedish nuclear power plant. The data concerns a pressure relief valve which is required to open if demanded. Although there are 20 identical component sockets in the 314 pressure relief system, we focus only on the first (V001). Table 5.1 shows the data. The socket 314 V001 of plant X starts service on 01/01/75. The first service sojourn terminates on 17/09/80

Table 5.1. *Swedish nuclear power plant data failure fields*

PLANT	SYS	SOCKET	SUBCOMP	FAILURE DETECT'N	DETECT'N DATE	FAILURE EFFECT	TYPE OF FAILURE	DAYS BETWEEN FUNCL FAILURE	SERVICE SOJOURN
X	314	V001	*****	*****	01/01/75	START	*****		
X	314	V001	VALVE	PM	17/09/80	FUNCTIONAL	INT'L LEAK	2086	2086
X	314	V001	VALVE	PM	01/08/82	NONFUNCTIONAL	EXT'L LEAK		683
X	314	V001	VALVE	PM	11/07/83	NONFUNCTIONAL	CORROSION		344
X	314	V001	SENSOR	OPERATOR	07/10/85	SPURIOUS SIGNAL	CONTROL SYS		819
X	314	V001	SENSOR	TEST	18/08/86	SIGNAL FAILURE	MISCALIBRT'N		315
X	314	V001	VALVE	INSPECTION	14/09/92	FUNCTIONAL	INT'L LEAK	4380	2219
X	314	V001	SENSOR	TEST	03/07/93	FUNCTIONAL	NICK	292	292
X	314	V001	VALVE	OPERATOR	06/09/93	SPURIOUS SIGNAL	MISCALIBRT'N		65
X	314	V001	VALVE	OPERATOR	13/05/95	FUNCTIONAL	EXT'L LEAK	679	614

when an internal leak caused a functional failure of V001. There were 2086 days of service. The next service sojourn terminates on 01/08/82; an external leak caused a non-functional failure. In other words, the component could have functioned if demanded, but was taken off line anyway. There were 683 days in this service sojourn. The next functional failure occurs on 14/09/92, when again internal leak rendered the valve non-functional. The time of outage (not shown) is negligible, and components are always returned to service as good as new.

If we had assumed a constant failure rate, then, as we shall see in Chapter 9, we could have filtered the data retaining only the functional failures. We would have found that the second (functional) sojourn ended on 14/09/92, lasting 4380 days. In total we would have found 4 functional failures in 7437 days, averaging 1859 days between functional failure.

This way of treating the data is quite wrong if we do not assume a constant failure rate. We must then consider that the sojourn ending on 14/09/92 in fact started with a component as good as new on 18/08/86. The sojourn ending on 14/09/92 lasted only 2219 days. In analyzing this data we shall assume that failure effects which are not called 'functional' would not disable the component, and therefore may be grouped as 'non-functional'. Hence we have 4 sojourns terminating in a functional failure, and 5 terminating in non-functional outage.

Weibulls are popular in reliability modeling because they arise as one of the classes of extreme value distributions, and because they can model various forms of failure rate functions. The Bayesian and maximum likelihood methods for parameter estimation discussed in the previous chapter are rather cumbersome for the Weibull distribution. We therefore start with the more popular graphical methods. This is followed by a discussion of the Kaplan–Meier estimator, maximum likelihood and Bayesian methods. The final substantive section addresses the relation to extreme value theory.

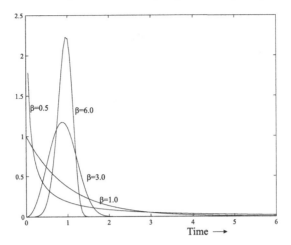

Fig. 5.1. Densities for the Weibull distribution

5.1 Definitions

A random life variable T is said to have a Weibull distribution with shape parameter β and scale parameter η, $\beta > 0, \eta > 0$, if its cumulative distribution function is given by

$$F(t) = 1 - e^{-(t/\eta)^\beta}. \tag{5.1}$$

Sometimes t is replaced by $t - t_0$, if the component enters service at t_0. This is called the *three parameter Weibull distribution*.

The failure rate for a Weibull (η, β) is

$$h(t) = f(t)/(1 - F(t)) = f(t)/R(t) = \beta t^{\beta-1} \eta^{-\beta}. \tag{5.2}$$

The failure rate $h(t)$ is either decreasing, constant, or increasing if β is less than, equal to or greater than 1 respectively. The transformation $Y = T^\beta$ yields an exponential variable Y with parameter $\eta^{-\beta}$.

The rth moment of the Weibull (η, β) is given by

$$E(T^r) = \eta^r \Gamma((r + \beta)/\beta), \tag{5.3}$$

see Exercise 1.

Figure 5.1 shows densities for various values of β.

5.2 Graphical methods for parameter fitting

In this section we demonstrate some plotting methods. These methods can be applied for other distributions besides Weibull.

The easiest and most popular methods for analyzing data which one hopes to model with a Weibull distribution involve plotting the failure data on 'Weibull paper' and graphically extracting the parameters η and β. We have

$$\begin{aligned} 1 - F(t) &= e^{-(t/\eta)^\beta}, \\ \log(1 - F(t))^{-1} &= (t/\eta)^\beta, \\ \log[\log(1 - F(t))^{-1}] &= \beta \log(t) - \beta \log(\eta). \end{aligned} \qquad (5.4)$$

Suppose we are fortunate enough to have failure data of the following form. A large population of n components is brought into service at $t = 0$, and the observation is continued until all components fail. Let $\hat{F}$ denote the empirical cumulative distribution function, that is

$$\hat{F}(t) = \% \text{ of population which fails before time } t. \qquad (5.5)$$

Note that $\hat{F}(t_n) = 1$, by definition. We now make the transformation

$$\begin{aligned} x_i &= \log(t_i), \\ y_i &= \log[\log(1 - \hat{F}(t_i))^{-1}], \ i = 1, \ldots, n. \end{aligned} \qquad (5.6)$$

From Equation 5.4 we see that the values (x_i, y_i) should lie approximately on a straight line, if indeed the population represents independent samples from a Weibull distribution:

$$y = Bx + A, \text{ with } B = \beta, A = -\beta \log(\eta). \qquad (5.7)$$

When data points are graphed, a 'best fitting' line is determined by least square fitting, or more commonly, by oculation. Notice that this method gives no way of determining whether the fit is 'good enough'. From Equation 5.4 we see that the parameter β is the slope of the line. The other parameter η can be determined from the intercept. Since, however, for any value of β the equation

$$P(T \leq \eta) = 1 - e^{-1} \approx 0.632$$

holds, the easiest way to estimate η is to determine the value of t with $P(T \leq t) = 0.632$. The number η is sometimes called the *characteristic life*. Comparison with Equation 5.4 shows that this corresponds to putting $y = 0$, solving Equation 5.7 for x, and then finding $\eta = \exp(x)$.

5.2.1 Rank order methods

The above method gives bad results on small data sets, as $\hat{F}$ approximates F rather poorly. Consider the data given in Table 5.2, based on the times

5.2 Graphical methods for parameter fitting

Table 5.2. *A small mortality table for failure data*

Time t till failure (hr)	Proportion $\hat{F}$ of population failed before ($\leq$) time t	$\log(t)$	$\log[\log(1-\hat{F})^{-1}]$
292	0.25	5.677	−1.2459
614	0.50	6.420	−0.3665
2086	0.75	7.643	0.3266
2219	1.00	7.705	∞

between functional failure in Table 5.1. It is very unreasonable to suppose that F (2219 days) $= 1$. In such cases 'median rank methods' give better results. We sketch the idea behind these methods and illustrate how to use the appropriate tables.

Table 5.2 shows the failure ranked from lowest to highest. These times are denoted $t_{(1)}, \ldots, t_{(4)}$, and (assuming the samples are drawn independently from the same distribution) $t_{(i)}$ is called the i-th order statistic from four samples. Consider the question 'which quantile of the distribution F is realized by the lowest value of the order statistic with four elements?' The question asks for the value of the random variable $F(t_{(1)})$.

Now recall (from Equation 3.7) that, for an arbitrary random variable X with continuous distribution function F, the random variable $F(X)$ is uniformly distributed on the unit interval. Of course, if we sample n failure times independently, then the distribution function applied to the lowest of these, $F(t_{(1)})$, does not yield a uniform variable. It is however fairly straightforward to compute this distribution, the density function for the ith value $p = F(t_{(i)})$ of n samples being the beta density $f(p)$ (see Exercise 4),

$$f(p) = \frac{n!}{(i-1)!(n-i)!} p^{i-1}(1-p)^{n-i}, \quad 0 \leq p \leq 1$$

(note however that this is called the *rank distribution* in the context of data plotting). In analyzing small sets of life data, it is customary to choose a 'representative value' from this distribution. One could choose the mean $i/(n+1)$, but it is more customary to choose the median. This is because the beta distributions for high and low values of i are skewed, so that the means $i/(n+1)$ tend to be 'too low' for high i and 'too large' for small i. The effect on the plot is to rotate the points clockwise leading to a low estimate of the slope. A 'median rank table' is given in Table 5.3. (An approximation to the median rank is given by the formula $(i-0.3)/(n+0.4)$.) Using this table, the

Table 5.3. *Median ranks*

Rank order	Sample size									
	1	2	3	4	5	6	7	8	9	10
1	50.0	29.2	20.6	15.9	12.9	10.9	9.4	8.3	7.4	6.6
2		70.7	50.0	38.5	31.3	26.4	22.8	20.1	17.9	16.2
3			79.3	61.4	50.0	42.1	36.4	32.0	28.6	25.8
4				84.0	68.6	57.8	50.0	44.0	39.3	35.5
5					87.0	73.5	63.5	55.9	50.0	45.1
6						89.0	77.1	67.9	60.6	54.8
7							90.5	79.8	71.3	64.4
8								91.7	82.0	74.1
9									92.5	83.7
10										93.3

Table 5.4. *A small mortality table with median rank estimates*

Time t till failure (hr)	Median rank estimate F_M of $F(t)$	log(t)	log[log$(1-F_M)^{-1}$]
292	0.159	5.6768	−1.7535
614	0.385	6.4200	−0.7213
2086	0.614	7.6430	−0.0493
2219	0.840	7.7048	0.6057

failure data in Table 5.2 can be associated with estimates of the quantiles realized by the failure times as shown in Table 5.4.

A Weibull plot of the data from Table 5.4 is shown in Figure 5.2.

The rank distributions can be used to build confidence limits for the distribution. We refer to [Kapur and Lamberson, 1977] for more details.

5.2.2 Suspended or censored items

One speaks of *censored data* when the observations give some information about the value of a random variable, but when the value itself cannot be observed. One speaks of 'suspensions' in life data when a component is observed to live up to a given time, but is then withdrawn from observation. Suspension is thus a form of censoring, called 'right censoring'. 'Left censoring' refers to a situation in which we see a component expire, but do not know when the component started service. With 'interval censoring', one observes only an interval in which the expiration of a component takes place. For example, a component may be found failed during an inspection, hence we know only that it expired between two inspection times.

The subject of censoring will be discussed in more detail in Chapter 9. For the present we simply present and motivate a popular way of analyzing random right censored Weibull data.

5.2 Graphical methods for parameter fitting

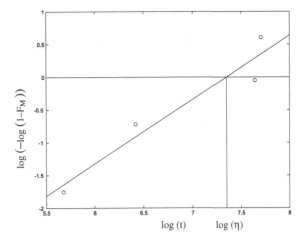

Fig. 5.2. A Weibull plot of the data in Table 5.4

Table 5.5. *Failure times and suspensions*

Rank	Time (days)
1	65 (suspension)
2	292
3	315 (suspension)
4	344 (suspension)
5	614
6	683 (suspension)
7	819 (suspension)
8	2086
9	2219

Suppose the data in Table 5.4 really included three additional items which were suspended without expiring. The suspensions may result from failure due to an unintended failure cause – perhaps the components were failed because they were overloaded by an upstream failed component. Table 5.5 presents the data ranked along with the suspension times.

It is clear that the suspensions contain additional information about the underlying life distribution, and an adequate analysis of the data should extract and use this information.

Consider sojourn ranked 3. We know that it functioned for 315 days, before being retired for some reason other than failure. Had it been allowed to fail, then it would have failed later than 315 days, and its failure time would have appeared later in the table.

Table 5.6. *Revised median rank estimates, with suspensions*

Rank	Time (days)	Adjusted rank	Revised $F(t)$, F_R
1	65 (suspension)	–	–
2	292	1.1111	0.08628
3	315 (suspension)	–	–
4	344 (suspension)	–	–
5	614	2.5926	0.24389
6	683 (suspension)	–	–
7	(suspension)	–	–
8	2086	5.06173	0.50657
9	2219	6.70782	0.68168

Let us assume that the process causing the suspension is independent of the life process. Before day 315, all permutations of failure times for sojourns are equally likely. Because of independence, the knowledge that a given component was suspended at day 315 does not alter that fact. It follows that it is equally likely that its failure time would be in the intervals $(t_{(4)}, t_{(5)}]$, $(t_{(5)}, t_{(6)}]$, $(t_{(6)}, t_{(7)}]$, $(t_{(7)}, t_{(8)}]$, $(t_{(8)}, t_{(9)}]$, $(t_{(9)}, \infty]$. With some smaller probability it would fall in the interval $(t_{(3)}, t_{(4)}]$. Each of these possibilities would have led to different rankings and would have caused us to produce a different estimate for $F(t_{(4)}), \ldots, F(t_{(9)})$. A revised estimate of the latter six values can be gotten by computing the expected rank of the succeeding failure times under the above six possibilities. From this the associated quantile is obtained with the median rank table. Fractional ranks can be handled by interpolating tabulated values, or by the approximation given below.

The method presented below, due to Johnson [Johnson, 1964], results from implementing this idea. For each suspension, revised estimates for the quantiles of the *succeeding* failure times *prior to the next suspension* are calculated.

Let N be the total number of ranked times ($N = 9$ in Table 5.5) and let k and k' be ranks for suspended components, with no suspensions between k and k'. For rank j, $k < j < k'$, (j is necessarily the rank of a failure time), compute

rank addition for j
$$= \frac{(N+1) - \text{revised rank for last failed component before } k}{1 + N - k}. \quad (5.8)$$

This leads to Table 5.6.

5.2 Graphical methods for parameter fitting

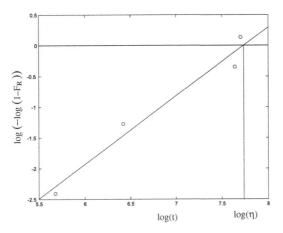

Fig. 5.3. Revised Weibull plot for Table 5.6

Note that only the failure times get their ranks revised. We illustrate the calculation for ranks 2 and 5:

rank addition for $2 = (9 + 1 - 0)/(1 + 9 - 1) = 1.1111$;
adjusted rank for $2 = 0 + 1.1111$;
rank addition for $5 = (9 + 1 - 1.1111)/(1 + 9 - 4) = 1.4815$;
adjusted rank for $5 = 1.1111 + 1.4815 = 2.5926$.

In interpolating the median rank table, the sample size $N(= 9)$ must be used. A common approximation also gives acceptable results:

$$F(t_{(i)}) = \frac{i' - 0.3}{N + 0.4}; \quad i' = \text{adjusted rank for } i. \tag{5.9}$$

This approximation was used in Table 5.6. Plotting this revised data on Weibull paper produces Figure 5.3.

We emphasize that this method assumes the censoring process and the life processes are independent.

5.2.3 The Kaplan–Meier estimator

Although the above estimation method is well established in the reliability literature, a slightly different method is used more frequently in the biostatistics literature. This method, called both the Kaplan–Meier estimator and the product–limit estimator, can be easily calculated by using a recursive formula. Let r_i be the *reverse rank* of the ith failure time, and $\hat{R}_i$ the estimate of the reliability at failure time t_i. The recursive formula used in the

Table 5.7. *Kaplan–Meier estimates*

Rank	Reverse rank	Time (days)	$\hat{R}(t)$	$\hat{R}(t+)$
1	9	65 (suspension)	–	–
2	8	292	1	0.875
3	7	315 (suspension)	–	–
4	6	344 (suspension)	–	–
5	5	614	0.875	0.7
6	4	683 (suspension)	–	–
7	3	819 (suspension)	–	–
8	2	2086	0.7	0.35
9	1	2219	0.35	0

Kaplan–Meier estimator is

$$\hat{R}_i = \frac{r_i - 1}{r_i} \hat{R}_{i-1}$$

(this is an empirical version of $P(T > t_i) = P(T > t_i | T > t_{i-1}) P(T > t_{i-1})$).

Iterating the above we obtain the Kaplan–Meier estimator for the reliability function at time t:

$$\hat{R}(t) = \prod_{i | t_i < t} \left(1 - \frac{1}{t_i}\right)$$

where the product is taken over the i such that the *failure time* $t_i < t$ (hence there are no terms arising from the censoring times). We illustrate this on the data set used above in Table 5.7. Since the estimator jumps at the failure times, we have also given $\hat{R}(t+)$, the value of the estimator just after time t.

By contrast, the method used in the last section is based on the almost identical recursion

$$R_i = \frac{r_i}{r_i + 1} R_{i-1}$$

(see [Johnson, 1964], [Nelson, 1982]).

Note that the Kaplan–Meier estimator can be considered a non-parametric maximum likelihood estimator of a distribution function under *independent* right censoring. When no censoring is present in the data set, the estimator gives the usual empirical distribution function.

5.3 Maximum likelihood methods for parameter estimation

In some cases the method of maximum likelihood can be used to estimate parameters of life distributions in a risk/reliability context. To recall, the like-

5.3 Maximum likelihood methods for parameter estimation

lihood $L(\theta|D)$ of parameter value θ given data D is the probability (density) $P(D|\theta)$ of the data, when distributional parameter value θ is assumed.

The maximum likelihood estimate (MLE) $\hat{\theta}$ of θ is defined as

$$\hat{\theta} = \arg\max_{\theta} L(\theta|D). \tag{5.10}$$

Suppose the lifetimes $t_1, \ldots, t_n$ are assumed to be independent realizations of a Weibull life distribution. Then (Exercise 5) by setting the derivatives of the log-likelihood (with respect to β and η) equal to zero, we obtain two equations,

$$\hat{\eta} = \left(\frac{\sum t_i^{\beta}}{n}\right)^{1/\beta},$$

$$0 = 1 + \frac{\beta}{n} + \sum_{i=1}^{n} \log t_i - \beta^c \frac{\sum_{i=1}^{n} \log t_i^{\beta-1}}{\sum_{i=1}^{n} \log t_i^{\beta}}.$$

The second equation must be solved first for β (this must be done numerically) before substituting the value of β obtained into the first equation to get the estimate for η.

MLE methods can be used with censored data when the censoring and life processes are independent. Let X and Y be independent life variables, and assume that $\xi = 1$ if $X < Y$ and $\xi = 0$ if $Y \geq X$. Suppose we observe k independent realizations of $(X \wedge Y, \xi)$. In other words, we see the smaller of X and Y, and we see which it is (either an X or a Y). Hence the data can be represented as realizations $D = x_1, \ldots, x_n, y_1, \ldots, y_m$ ($n + m = k$).

Let F and f be the cumulative distribution function and density, respectively, for X, and G and g the same for Y. For an observation x_i the likelihood is

$$f(x_i)(1 - G(x_i)),$$

since we observe X failing at x_i, and we simultaneously observe the independent event that Y is larger than x_i. In the same way, for an observation y_j the likelihood is

$$(1 - F(y_j))g(y_j).$$

The overall likelihood is now

$$\prod_i f(x_i)(1 - G(x_i)) \prod_j (1 - F(y_j))g(y_j).$$

Assuming now that F is a Weibull distribution, then when we take the log of the likelihood function and differentiate with respect to η and β, the terms with G and g disappear. Using numerical methods one can again estimate η and β. For obvious reasons, MLE methods can only be used when the assumption of independence between life and censoring processes is reasonable. In some cases, this may be reasonable. Unfortunately, in reliability applications, the assumption of independence is often questionable. A good example is a component which is right censored by preventive maintenance. When the maintenance personnel believe the component is starting to wear out, it is taken off-line and revised. Such suspensions are obviously tightly coupled to the life process.

We return to the problem of analyzing right censored data under possibly dependent distributions in Chapter 9.

5.4 Bayesian estimation

The Weibull distribution does not lend itself easily to Bayesian updating. The main problem is that there is no distribution class on the parameters that is preserved under Bayesian updating. (Formally, there is no conjugate distribution – the Weibull distribution does not belong to an 'exponential family' – see [Robert, 1994].) This means that only simulation methods can be used to determine the updated distributions. Robert [Robert, 1994] suggests the use of the Metropolis simulation algorithm in conjunction with a particular choice of prior.

5.5 Extreme value theory

The reliability of large series or parallel systems can be estimated by the use of certain limit distributions known as extreme value distributions. There are many different areas of application of these distributions. One of the earliest applications was in predicting the size and frequency of floods. Other areas of application are in determining the strength of materials or of maximum wind gusts. A chain is a series system which fails when one of its links breaks. A metal beam fails when one of the many tiny flaws fails critically.

Suppose that we have a certain series system S which consists of n components with lifetimes $T_1, \ldots, T_n$. These lifetimes are assumed to be independent and identically distributed with reliability function $R(t)$. Let T_{sys} be the lifetime of the system. Since the system is in series we have

$$T_{sys} = \min_{1 \leq i \leq n} T_i.$$

The reliability of the system R_{sys} is therefore

$$\begin{aligned} R_{sys}(t) &= P(T_{sys} > t) \\ &= P(T_1 > t, T_2 > t, \ldots, T_n > t) \\ &= \prod_i P(T_i > t) \\ &= R^n(t). \end{aligned}$$

5.5 Extreme value theory

By taking the limit $n \to \infty$ we can gain insight into the behaviour of R_{sys}. Clearly if $R(t) < 1$ we have $R_{sys}(t) = R^n(t) \to 0$ if $n \to \infty$. Often more can be said about R_{sys}. This is done by an affine change of time scale, that is by choosing two sequences $a_n > 0$ and b_n such that the sequence of stochastic variables

$$\frac{T_{sys} - b_n}{a_n}$$

converges in distribution.

We give an example to show how the parameters a_n and b_n and the limit distribution can be found.

Example 5.1 *A series system consists of n components which all have reliability*

$$R(t) = \frac{1}{(1+t^2)^3} \quad (t > 0).$$

The lifetime distributions of the components are mutually independent. Let T_{sys} be the lifetime of the system. Then

$$P\left(\frac{T_{sys} - b_n}{a_n} > t\right) = P(T_{sys} > a_n t + b_n)$$
$$= R^n(a_n t + b_n)$$
$$= \frac{1}{(1 + (a_n t)^2 + b_n)^{3n}}.$$

We have to choose a_n and b_n so that this probability converges to a non-trivial limit. Taking $a_n = \sqrt{\frac{1}{3n}}$ and $b_n = 0$ we have

$$P\left(\frac{T_{sys}}{1/3n} > t\right) = \frac{1}{(1+\frac{t^2}{3n})^{3n}}$$
$$\to e^{-t^2} \quad \text{as } n \to \infty,$$

where we have used the well known limit $(1 + \frac{x}{m})^m \to e^x$ as $m \to \infty$. For large n we have

$$P(T_{sys} > t) \approx e^{-3nt^2}.$$

This shows that the reliability of this system is approximately Weibull distributed for large n.

It can be shown (see for example [Galambos, 1978]) that only three *types* of distributions can occur as limit distributions for extreme values, when we consider affine time rescalings as above. These three types are called the *extreme value distributions*, and are shown in Table 5.8. A *type* of distribution

Table 5.8. *Extreme value distributions*

Name	Distribution function	
Type I (Gumbel)	$F(t) = \exp(-e^{-t})$	$(-\infty < t < \infty)$
Type II (Weibull)	$F(t) = \exp(-t^{-\alpha})$	$(0 < t < \infty, \alpha > 0)$
Type III	$F(t) = \exp(-(-t^{\alpha}))$	$(-\infty < t < 0, \alpha > 0)$

is just a class of distributions which are the same up to an affine rescaling. Hence it is possible to have $F(at + b)$ as a limit distribution, where F is any of the distributions shown in Table 5.8.

The Type I distribution is frequently called the Gumbel distribution, and (confusingly) is often called *the extreme value distribution* in reliability texts. It is related to the Weibull distribution in the same way as the normal and log-normal distributions are related: if $\log(T)$ is a Gumbel variable then T is a Weibull variable. Hence by exponentiating Gumbel data we get Weibull data, and can use the above graphical and MLE methods for data estimation.

5.6 Exercises

5.1 Show that the rth moment of the Weibull (η, β) is given by

$$E(T^r) = \eta^r \Gamma((r + \beta)/\beta).$$

5.2 Choose $p \in (0, 1)$, and consider independent uniform $[0, 1]$ variables $U_1, \ldots, U_n$. What is the probability of having i of the $U_1, \ldots, U_n$ less than p, and $n - i$ greater than p? Use this to derive the beta density.

5.3 If Y is exponential with parameter λ, then

$$X = Y^{1/\beta} \qquad (5.11)$$

is Weibull with parameters β and $\eta = \lambda^{-\beta}$.

5.4 Suppose two independent observations are made of T_1, T_2, from a distribution F. Write $T_{(1)} = \min\{T_1, T_2\}$, $T_{(2)} = \max\{T_1, T_2\}$. These are the *order statistics*. Show that the pair $(U_{(1)}, U_{(2)}) = (F(T_{(1)}), F(T_{(2)}))$ is uniformly distributed on the triangle with corners $(0, 0)$, $(0, 1)$, and $(1, 1)$. Calculate the median and mean values of $U_{(1)}$ and $U_{(2)}$.

5.5 Derive the equations for the maximum likelihood estimate of η and β in 5.11.

Part III

System analysis and quantification

6

Fault and event trees

6.1 Fault and event trees

Fault and event trees are modeling tools used as part of a quantitative analysis of a system. Other semi-quantitative or qualitative tools such as failure modes and effects analysis (FMEA) are often performed in preparation for a more exact analysis. Such tools are outside the (quantitative) scope of this book, and the interested reader is referred to [Kumamoto and Henley, 1996], [Andrews and Moss, 1993]. These books also provide further information and more examples on fault tree modeling as does the *Fault Tree Handbook* [Vesely *et al.*, 1981].

Fault tree and event tree analyses are two of the basic tools in system analysis. Both methodologies give rise to a pictorial representation of a statement in Boolean logic. We shall concentrate on fault tree analysis, but briefly explain the difference in the situations modeled by event trees and fault trees.

Event trees use 'forward logic'. They begin with an *initiating event* (an abnormal incident) and 'propagate' this event through the system under study by considering all possible ways in which it can effect the behaviour of the (sub)system. The nodes of an event tree represent the possible functioning or malfunctioning of a (sub)system. If a sufficient set of such systems functions normally then the plant will return to normal operating conditions. A path through an event tree resulting in an accident is called an *accident sequence*. A schematic event tree is shown in Figure 6.1. The event tree consists of an initiating event and two safety systems which are designed to mitigate the initiating event. If either of the safety systems functions properly then the plant returns to a normal operating state. Otherwise an accident occurs. An event tree (taken from [Frank, 1999]) produced as part of a study of the Cassini mission (a space vehicle carrying a plutonium load) is shown in

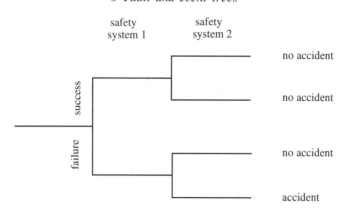

Fig. 6.1. An event tree

Figure 6.2. An event tree is similar in structure to the so-called decision trees used in decision analysis (see Chapter 13).

A fault tree works with 'backward logic'. Given a particular failure of a system, the *top event*, one seeks the component failures which contribute to the system failure. Using the Boolean operations *and*, *or* and *not* one writes down which (combinations of) component faults give rise to the top event. A simple example of a fault tree is given in Figure 6.3.

6.2 The aim of a fault-tree analysis

In a fault-tree analysis one attempts to develop a deterministic description of the occurrence of an event, the *top event*, in terms of the occurrence or non-occurrence of other (intermediate) events. Intermediate events are also described further until, at the finest level of detail, the *basic events* are reached. When the top event is the failure of a system then the basic events are usually failures of components. A fault tree can be considered as an expression in Boolean logic. In itself a fault tree does not give enough information to allow an estimate to be made of system reliability. Fault trees are however used together with reliability data for the basic events to make estimates of system reliability.

Since a fault tree is an expression in Boolean logic, when using them as a modeling tool we must assume that the top event and all basic events are *binary*, that is, true or false. This makes it difficult to model situations in which more than two states are important, for example, a pump may give low pressure, normal pressure, or high pressure (by contrast in multi-state reliability theory [Natvig, 1985], one may model a number of states greater

Space Craft Reentry	Steep Path Angle? (>160) St	FOS GPHS Orientation? F	Aeroshell Failure at Altitude? Af	GIS Release? Gr	SOS/ns or Near End On GIS Orientation? So	GIS Failure at Altitude? Gf	Clad Melt? M	Terra Firma Impact? T	End State	Nominal Sequence Conditional Probability
9.5E-07	0.89	0.85	1.00	0.90	0.67	0.97		0.23	A	5.2E-01
						0.03	Assume clad melts 0.23	0.23	B	3.8E-03
						0.00		0.23	A	0.0E+00
					0.33	1.00	Assume clad melts 0.23	0.23	B	6.0E-02
						conservative assumption owing to lack of data		0.23	A	8.9E-02
				0.10				0.23	D	0.0E+00
			0.00						Xfer to 1	
	Shallow 0.11	Non-FOS 0.15	0.62	0.90	0.67	0.00		0.23	A	0.0E+00
						1.00	0.50	0.23	B	4.1E-03
							0.50	0.23	C	4.1E-03
					0.33	GIS intact until impact	0.50	0.23	B	2.0E-03
							0.50	0.23	C	2.0E-03
				0.10		0.00		0.23	A	0.0E+00
			0.38	If not FOS, then GPHS survives to impact				0.23	D	8.3E-03
		0.15						0.23	D	3.8E-03

Legend
OI = Ocean Impact
A = At altitude release
B = GIS impact on Terra Firma/ clad melt
C = GIS impact on Terra Firma/ clad OK
D = GPHS impact on Terra Firma
--- = event not relevant
△ = point of transfer

Total Air Release = 5.8E-07
Total Land Release = 8.4E-08
Total Ocean Impact = 2.9E-07

Fig. 6.2. The Cassini event tree

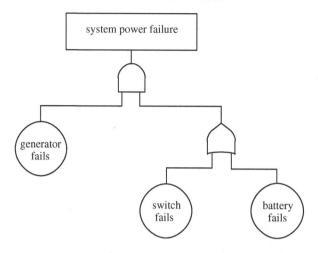

Fig. 6.3. Security system

than two). Furthermore, we often would like our representation of a system to be *coherent*, that is, the system as a whole cannot improve when one or more subsystems fail (this is necessary for the definitions of minimal cut and path sets – fundamental notions for the quantification of fault trees – to make sense). Later we shall look at the problem of fault tree quantification when the tree is not coherent. Here is an example of a non-coherent system:

Example 6.1 *A pumping system consists of a tank, a pump, and a pipe from the tank to the pump. The system works when water is removed from the tank at a rate between 1 and $2 \mathrm{l s}^{-1}$. The pump can fail by not pumping hard enough or by pumping too hard. The pipe can fail by leaking. Now, if the pump is pumping too hard enough and the pipe is leaking then the system is not in the failed state, but if the pump then starts to work normally the pressure will drop and the system fails. Hence the system is not coherent.*

The standard way to model this system would be to make the assumption that it is unlikely that the system will be saved by two components failing together and therefore *define* the system to have failed when the pump is pumping too hard and the pipe is leaking.

It may not be possible to realistically model a system in a coherent way. An example of such a system will be given in Chapter 7 in a discussion on algorithms for non-coherent systems.

An elementary example of a fault tree analysis is the following.

Example 6.2 *A security system in a hospital is powered by a generator. If*

Fig. 6.4. AND and OR gates

the generator fails then a switching system switches over to a battery. The top event, T, is the event that the security system fails through lack of power.

We first introduce some notation to describe the most important events. Let G be the event that the generator fails, let S be the event that the switch fails and let B be the event that the battery fails. The development of the fault tree is as follows. The event T that the security system has no power occurs if the generator fails and if the back-up system fails. The back-up system can fail in two ways – either because the switch has failed or because the battery does not work. Denoting an AND gate and an OR gate as in Figure 6.4, we can represent the top event by the fault tree shown in Figure 6.3.

6.3 The definition of a system and of a top event

Since fault trees are representations of a statement in logic it is necessary to make a very precise definition of the system and top event under discussion. Indeed, definition of the system is probably the most difficult part of fault tree construction.

The main problem is to put boundaries on the system under discussion. The most important boundaries are *external, internal* and *temporal*. In the case of the security system discussed above we have defined the system as the collection

{generator, switch, battery}.

This definition is obviously somewhat arbitrary. It might be of importance to include wires linking the various components, or to consider subcomponents of the generator. Furthermore the time at which components fail is important, especially if they are subject to regular inspection and repair. These three examples are examples of external, internal and temporal boundaries. Usually the choice of boundaries is determined by the purpose of the study.

6.3.1 External boundaries

The external boundary of a system depends on what factors could be of influence in the top event, and in particular on a subjective judgement of how important the various factors are. In the security system example, should we include the oil tank which supplies oil to the generator? If so should we include the oil transporter which is used to supply the generator? Should one consider the possibility that the generator and the switch are both struck by lightning? For this particular example it is clear that these problems are not so important and can therefore be ignored (unless the generator has been built on top of a hill in a region with lots of thunderstorms).

6.3.2 Internal boundaries

In how much detail should we study our system? In the security system example, should we consider the generator as a single unit, or is it valuable to split the generator up into its subcomponents so that one can make a more detailed assessment of the causes of system failure? The choice of *resolution* depends on two factors. Firstly, if one wishes to make design changes to the system as a result of the analysis, then the basic components should be those which can move or replace in the new design. Secondly, if one wishes to make an estimate of the reliability of the system, then one should try to choose the level of resolution in such a way that one has good data about the reliability of the components. This second factor can also be thought of as a choice of probabilistic boundaries for the problem.

6.3.3 Temporal boundaries

Many systems change through the course of time, for example through maintenance or because the sub-systems behave differently under different circumstances. In the security system example it is clear that the charge on the battery is dependent on whether the generator is working or not. It is reasonable to suppose that the failure rate of the battery depends on whether the battery is being used. Databases of failure rates often give two failure rates corresponding to the states 'standby' and 'active' (the latter state may also be split into different states depending on the type of use of a component). For these reasons it is often important to specify not only a component but also its state when constructing a fault tree.

6.4 What classes of faults can occur?

It is often helpful to make a qualitative classification of faults. This is important in deciding where the boundaries of the problem should be set. This is particularly important because it gives some idea of where the probabilistic cut-off can be carried out.

6.4.1 Active and passive components

The first classification is into *active* and *passive* components. The difference in failure rates between components which actively do something (for example a pump) and components which passively contribute to the operation of the system (for example a pipe, or a tank) can be large. In the WASH-1400 study [NRC, 1975] it was observed that active components often have a failure rate of at least two or three orders of magnitude higher than those of passive components.

6.4.2 Primary, secondary and command faults

A *primary fault* occurs when a component fails in a task it was designed for: for example if a bicycle tire leaks under normal loading conditions. A *secondary fault* occurs if a component fails in situation for which it was not designed: for example if a bicycle tire bursts because you accidentally heat it with a blow torch. A *command failure* occurs when the component operates in the correct way but at the wrong time or in the wrong place, or when it does not get the command to work. An example is a pump on standby which does not get the signal to start.

6.4.3 Failure modes, effects and mechanisms

This sort of classification is useful in constructing a fault tree because it allows the analyst to determine the way in which components and subcomponents interact with each other.

As an example consider a pump. The *effect* of failure of the pump is lack of pressure. The *modes* of failure could be failure to start or failure to run. One *mechanism* by which the first failure mode could occur is leaking oil. For a further discussion of these concepts see Chapter 9.

From this example it is clear that in the construction of a fault tree the connections to events 'higher' in the tree are made via the failure effects. The failure modes are the fault tree events that concern the component under analysis, and the failure mechanisms are the connections to events 'lower' in the tree.

Whether a particular event is a failure mechanism, mode or effect depends on which component/(sub)system is being considered. For example, if the pump considered above has to supply cool water to another system, then the failure effect 'lack of pressure' is, *at the level of the system*, a failure mechanism. On the other hand, the failure mechanism 'leaking oil' (for the pump) is a failure effect for the component consisting of the pump casing (for if the casing fails, then oil will leak).

6.5 Symbols for fault trees

We have already introduced symbols for AND and OR gates. In principle it is possible to construct every fault tree from a combination of AND and OR gates. In practice it is helpful to introduce other gates (as a sort of shorthand) and also to distinguish between different sorts of events within the tree. Figure 6.5 shows commonly used gates and states. An example of a fault tree using these symbols is given in Figure 6.7.

6.6 Fault tree construction

A fault tree begins with a top event which is a particular system failure mode. The analyst develops the tree by finding necessary and sufficient causes for the top event. These causes are supposed to be the *immediate* causes – not the basic causes. The causes should be faults, and should be clearly stated. The two most important things to be stated here are *what* has happened and *when* it happened. An example of such a statement is

> motor fails to start when power is applied.

As we move down through the fault tree, failure events become more detailed. The following heuristic guidelines for developing a fault tree are given in [Kumamoto and Henley, 1996].

(i) Replace an abstract event by a less abstract event. Example: 'motor operates too long' versus 'current to motor too long'.
(ii) Classify an event into more elementary events. Example: 'explosion of tank' versus 'explosion by overfilling' or 'explosion by runaway reaction'.
(iii) Identify distinct causes for an event. Example: 'runaway reaction' versus 'excessive feed' and 'loss of cooling'.
(iv) Couple trigger event with 'no protective action'. Example: 'overheating' versus 'loss of cooling' coupled with 'no system shutdown'.
(v) Find co-operative causes for an event. Example: 'fire' versus 'leak of flammable fluid' and 'relay sparks'.
(vi) Pinpoint a component failure event. Example: 'no current to motor' versus 'no current in wire'. Another example is 'no cooling water' versus 'main valve is closed' coupled with 'bypass valve is not opened'.

As the tree is developed, more events can be developed in terms of *state-of-component events*. These are events defined in terms of the failure of a particular component, which are caused only by failures internal to

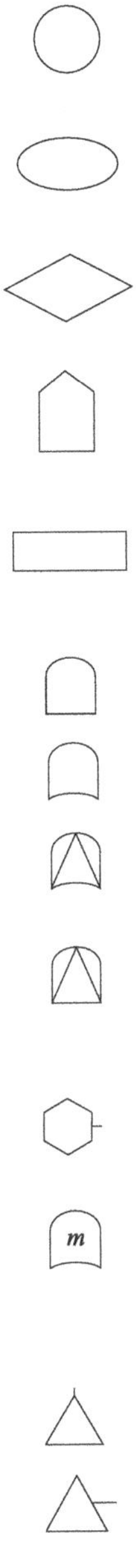

6.6 Fault tree construction

Basic event: failure at the lowest level (no further development necessary).

Conditioning event: specific restrictions or conditions applicable to a given gate.

Undeveloped event: an event that is not developed further because there is no information, or because it is not necessary.

External event: an event that usually occurs

Intermediate event

AND gate

OR gate

Exclusive OR gate: output if and only if exactly one of the inputs active.

Priority AND gate: output if and only if the inputs occur in a given order. The order is indicated with a conditioning event.

Inhibit: output occurs if and only if the single input occurs in the presence of a conditioning event.

m Out of n gate: Output if and only if at least m of the n inputs are active.

Transfer in: the tree is developed further elsewhere. The place where the development takes place is indicated with a Transfer out symbol.

Transfer out

Fig. 6.5. Common gates and states

that component. State-of-component events are important for later analysis. An event which is not a state-of-component event is a *state-of-system event*.

The *Fault Tree Handbook* [Vesely et al., 1981] gives three more rules for fault tree generation:

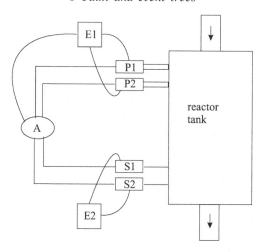

Fig. 6.6. Schematic diagram for the reactor protection system

- *No miracles.* If the normal functioning of a component propagates a fault sequence then assume that the component functions normally.
- *Complete the gate.* Fully describe all inputs to a particular gate before developing any of the inputs further.
- *No gate-to-gate connections.* The input to a gate should always be a properly defined fault event and not another gate.

Of these three rules, the first helps to make fault trees coherent (see the discussion below), and the last two are 'good practice' that enable the tree to be easily checked.

6.7 Examples

6.7.1 Reactor vessel

The first example (based on an example from [CPR, 1985]) is of a protection system for a continuous reaction process. The system has two sensors $S1$ and $S2$ that can independently detect an abnormal temperature. They pass an electrical signal to an electrical OR-gate A, which passes a signal on to both the protection units $P1$ and $P2$ if either sensor gives a signal. The protection units and the OR-gate get electrical power from electric power unit $E1$, while the sensors get power from electric power unit $E2$, see Figure 6.6. The fault tree in Figure 6.7 describes the top event 'protection system not available' in terms of the unavailability of the basic events.

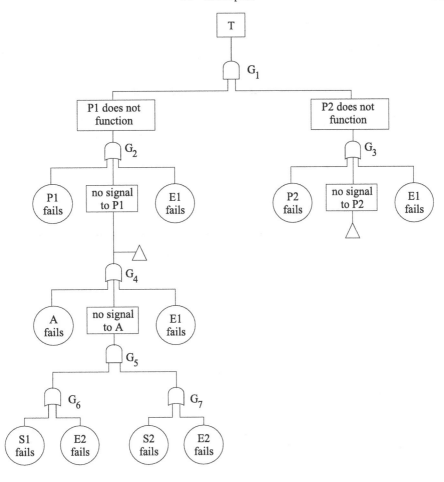

Fig. 6.7. Fault tree for the reactor protection system

6.7.2 New Waterway barrier

The Storm Surge Barrier in the New Waterway near Rotterdam is designed to protect the local population against high water.[1] Because of the choice of design (two floating doors mounted on a ball hinge), the barrier is not able to withstand high pressure from the river. If the river water level is more than one meter higher than the sea level then the barrier will fail.

A water measurement system has been installed to measure the water levels on both sides of the barrier, so that the barrier can be opened before the water-level difference is too high. The barrier is equipped with four water-level measurement posts (WMP), each of which contains two water-

---

[1] We are indebted to Mart Janssen of the Ministry of Water Management, the Netherlands, for this example.

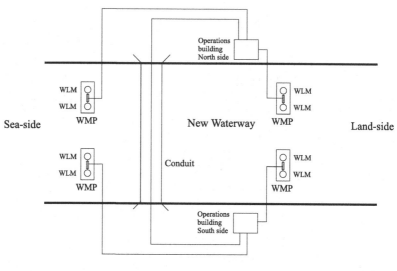

Fig. 6.8. Schematic diagram for the New Waterway water-level measurement system

level meters (WLM), see Figure 6.8. There is a water-level measurement post by each bank, on both the sea and the land side. On each bank of the New Waterway is an Operations building which combine, the water-level measurements. A cable through a conduit under the river is used to pass information between the Operations buildings. Only one of the Operations buildings is required to determine the water-level difference.

The fault tree in Figure 6.9 describes the top event 'no water-level difference measurement' in terms of the unavailability of the cable sections. The basic events are:

NS, cable on north bank/sea side failed,
SS, cable on south bank/sea side failed,
NL, cable on north bank/land side failed,
SL, cable on south bank/land side failed,
C, cable in conduit failed.

6.8 Minimal path and cut sets for coherent systems
6.8.1 Cut sets

A *cut set* is a collection of basic events such that if these events occur together then the top event will certainly occur. A *minimal* cut set is a collection of

6.8 Minimal path and cut sets for coherent systems

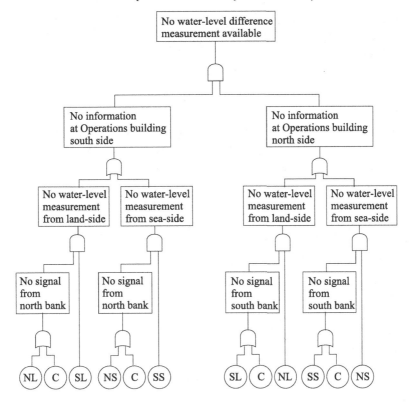

Fig. 6.9. Fault tree for the New Waterway water-level measurement system

basic events forming a cut set such that if any of the basic events is removed then the remaining set is no longer a cut set. As an example consider the security system fault tree from Figure 6.3. Here we have a top event T (system power failure) and three basic events

G, the generator fails,
S, the switch fails,
B, the battery fails.

Clearly if everything fails then the system loses its power supply. An example of a cut set is therefore given by

$$\{G, S, B\}.$$

Is this cut set minimal? The answer is no since the system still fails if the switch works, i.e. $\{G, B\}$ is a cut set. This set is a minimal cut set since the only non-empty proper subsets are $\{G\}$ and $\{B\}$, and these are clearly not cut sets. With similar reasoning one sees that $\{G, S\}$ is a minimal cut set.

112   6 Fault and event trees

These two minimal cut sets are in fact the only minimal cut sets because the system cannot fail if the generator does not fail. Hence G is always in a (minimal) cut set.

6.8.2 Path sets

A *path set* is a collection of basic events such that if *none* of these events occur then the top event will certainly *not* occur. A *minimal* path set is a collection of basic events forming a path set such that if any of the basic events is removed then the remaining set is no longer a path set. It should be clear from the definition of cut and path sets that there is *duality* between them. This relationship will be formalized later. In the example of Figure 6.3 the set $\{G\}$ is a path because if G does not occur (i.e. if the generator does not fail) then the top event cannot occur. The set $\{S\}$ is not a path because the top event could still occur even if the switch does not fail. The minimal paths here are

$$\{G\} \text{ and } \{S, B\}.$$

6.9 Set theoretic description of cut and path sets

We can gain insight into the relationship between cut and path sets by using set theory or Boolean algebra.

6.9.1 Boolean algebra

A basic event in a fault tree can be represented by a Boolean variable, i.e. a variable taking one of the values *true* and *false*, in an obvious way. In Boolean algebra there are two binary operators, *and* and *or*, together with one unary operator *not*. These operators are denoted in the following ways:

$$X \text{ and } Y, \quad X \cdot Y;$$
$$X \text{ or } Y, \quad X + Y;$$
$$\text{not } X, \quad X'.$$

The confusing choice of the symbol $+$ to represent *or* is explained by the equivalent set-theoretic notation:

$$X \text{ and } Y, \quad X \cap Y;$$
$$X \text{ or } Y, \quad X \cup Y;$$
$$\text{not } X, \quad X^c.$$

6.9 Set theoretic description of cut and path sets

Table 6.1. *Laws of Boolean algebra*

Commutative laws	$X \cdot Y = Y \cdot X$
	$X + Y = Y + X$
Associative laws	$X \cdot (Y \cdot Z) = (X \cdot Y) \cdot Z$
	$X + (Y + Z) = (X + Y) + Z$
Distributive laws	$X \cdot (Y + Z) = X \cdot Y + X \cdot Z$
Idempotent laws	$X \cdot X = X$
	$X + X = X$
Absorption law	$X + X \cdot Y = X$
Complementation	$X + X' = \Omega$
	$(X')' = X$
De Morgan's laws	$(X \cdot Y)' = X' + Y'$
	$(X + Y)' = X' \cdot Y'$
Empty set/universal set	$\emptyset' = \Omega$

Some of the laws of Boolean algebra are given in Table 6.1.

A fault tree is simply a pictorial representation of a Boolean expression. For example, consider the reactor protection example given above whose fault tree was shown in Figure 6.7. If we use A to denote the event that component A fails etc., and G_i to denote output at gate i, then the top event is represented by

$$\begin{aligned}
T &= G_1 \\
&= G_2 \cdot G_3 \\
&= (P1 + E1 + G_4) \cdot (P2 + E1 + G_4) \\
&= (P1 + E1 + A + G_5) \cdot (P2 + E1 + A + G_5) \\
&= (P1 + E1 + A + G_6 \cdot G_7) \cdot (P2 + E1 + A + G_6 \cdot G_7) \\
&= (P1 + E1 + A + (S1 + E2) \cdot (S2 + E2)) \\
&\quad \cdot (P2 + E1 + A + (S1 + E2) \cdot (S2 + E2)).
\end{aligned}$$

Applying the rules of Boolean logic we can multiply out the brackets. Then by further applying the reduction rules (for example $E2 \cdot S1 + E2 = E2$) we obtain the equivalent expression

$$T = P1 \cdot P2 + E1 + E2 + A + S1 \cdot S2. \tag{6.1}$$

The fault tree representing this Boolean formula is shown in Figure 6.10. This form, in which a single OR gate is connected to a number of AND gates at the same level, is called a *cut set representation*.

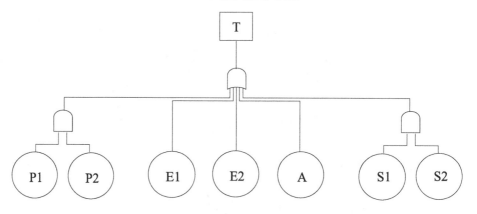

Fig. 6.10. Cut set fault tree representation for the reactor protection example

6.9.2 Cut set representation

We denote a basic event in a fault tree for a coherent system by X and the top event by T. It is then clear from the definition of a cut set that an event C is a cut set if and only if

$$T = C + A \quad \text{for some event } A,$$

and

$$C = X_1 X_2 \ldots X_n$$

where the X_i are the basic events in the cut set. If a particular fault tree has minimal cut sets $M_1, \ldots, M_k$ then we can write

$$\begin{aligned} T &= M_1 + \cdots + M_k, \quad \text{where} \\ M_i &= X_{1,i} \cdot X_{2,i} \ldots X_{n_i,i} \quad \text{for each } i. \end{aligned} \quad (6.2)$$

Note that an expression for the top event in the general form of Expression 6.2 does not have to give the minimal cut sets. For example suppose that the basic events are X_1, X_2 and X_3, and that minimal cut sets are $\{X_1, X_2\}$ and $\{X_3\}$. Then the minimal cut set representation is

$$T = (X_1 \cdot X_2) + X_3,$$

but we can also write

$$T = (X_1 \cdot X_2) + (X_1 \cdot X_3) + X_3.$$

The number of cut sets typically grows very quickly with the number of components. A general bound is given by a theorem of Sperner (who proved

6.9 Set theoretic description of cut and path sets

the result in the context of enumerating 'antichains' of a finite set). The notation $\lfloor x \rfloor$ is used to denote the largest integer smaller than x.

Theorem 6.1 *[Sperner, 1928]* *For a system with n components, the number of minimal cut sets is at most*

$$\binom{n}{\lfloor \frac{n}{2} \rfloor}.$$

6.9.3 Path set representation

As above we write T for the top event and X for a basic event in the tree. If the minimal path events are $P_1, \ldots, P_m$ then we can write

$$T' = P_1 + \ldots + P_m$$

where for each i,

$$P_i = X'_{1,i} \cdot X'_{2,i} \ldots X'_{r_i,i}$$

(these are not necessarily the same Xs appearing in the description of the cut sets!). In this case a path set is

$$\{X_{1,i}, X_{2,i}, \ldots, X_{r_i,i}\}.$$

Note that the minimal cut set representation of a general system can be thought of as a description of the system in terms of a number of subsystems (the minimal sets) arranged in series. Similarly the *working* of the system (i.e. the complement of the top event) can be represented as a series system in which the subsystems are the minimal path sets. For the reactor protection example, where the path sets are $\{P1, E1, E2, A, S1\}$, $\{P1, E1, E2, A, S2\}$, $\{P2, E1, E2, A, S1\}$ and $\{P2, E1, E2, A, S2\}$, this would be as shown in Figure 6.11.

6.9.4 Minimal cut set/path set duality

The *dual tree* of a given fault-tree is the tree for occurrence of T' obtained by replacing all AND gates by OR gates and vice versa and replacing all events by their negation. The dual tree for the reactor protection is shown in Figure 6.12. We have noted before that a fault tree is a pictorial representation of a Boolean expression. In Exercise 4 you are asked to check that De Morgan's law (see Table 6.1) implies that the process of interchanging gates etc. gives a true representation of T'. It is easy to see that the dual tree of a dual tree is again the original tree. From the above expressions for T in terms of minimal path or cut sets we see immediately

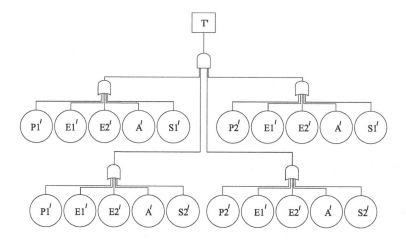

Fig. 6.11. Path set fault tree representation for the reactor protection example

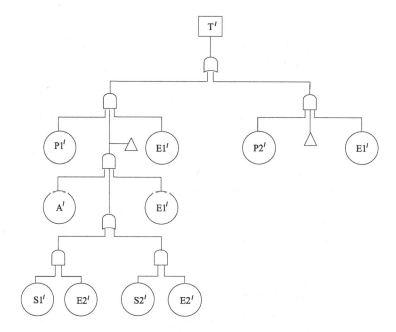

Fig. 6.12. Dual tree for the reactor protection example

Theorem 6.2 *A (minimal) path set for a coherent tree is a (minimal) cut set for the dual tree and vice versa.*

A consequence of this theorem is that any algorithm for finding minimal cut sets can also be used to find minimal path sets.

6.9.5 Parallel and series systems

The simplest types of systems are parallel and series systems.

Definition 6.1 *A system is a* parallel *system if it functions whenever at least one of its components functions. A system is a* series *system if it functions only when all of its components function.*

Suppose that there are n basic components and that the event that the ith component fails is X_i. Denoting the top event that the system fails by T, we see that the following expressions hold:

$$\text{parallel system} \quad T = X_1 \cdot X_2 \ldots X_n;$$
$$\text{series system} \quad T = X_1 + X_2 + \cdots + X_n.$$

We conclude that the following holds.

Theorem 6.3 *A parallel system has one minimal cut $\{X_1, X_2, \ldots, X_n\}$ and n minimal path sets $\{X_1\}, \{X_2\}, \ldots, \{X_n\}$. A series system has n minimal cut sets $\{X_1\}, \{X_2\}, \ldots, \{X_n\}$ and one minimal path set $\{X_1, X_2, \ldots, X_n\}$. The fault tree for a parallel system has as dual tree the fault tree of a series system, and vice versa.*

6.10 Estimating the probability of the top event

The minimal cut and path sets tell us something qualitative about the way in which the top event can occur. Usually, though, it is necessary to quantify the probability of the top event. For this the fault tree representation can be used.

Suppose that T is the top event, and that $C_1, \ldots, C_n$ are the minimal cut sets. We know that

$$T = C_1 \cup C_2 \cup \ldots \cup C_n,$$

so that by applying the *inclusion–exclusion* law (3.3), we have the exact expression

$$P(T) = \sum_{i=1}^{n} P(C_i) - \sum_{i<j} P(C_i \cap C_j)$$
$$+ \sum_{i<j<k} P(C_i \cap C_j \cap C_k) - \cdots + (-1)^{n+1} P(C_1 \cap C_2 \cap C_n). \quad (6.3)$$

Furthermore, it is easy to show that when the expression on the right hand side of 6.3 is broken off after a positive term we get an upper bound for

$P(T)$, and when it is broken off after a negative term we get a lower bound. In particular this gives

$$\sum_{i=1}^{n} P(C_i) - \sum_{i<j} P(C_i \cap C_j) \le P(T) \le \sum_{i=1}^{n} P(C_i).$$

The approximation

$$P(T) \approx \sum_{i=1}^{n} P(C_i)$$

is called the *rare event approximation*. It is based on the idea that the simultaneous occurrence of two minimal cut sets should occur with a probability which is an order of magnitude smaller than the probability that one of the two cut sets occurs. This assumption is obviously problematic when there is a large degree of overlap between the elements of the cut sets, but it is always a *conservative* assumption, that is, an assumption which will not make our calculations more optimistic than the true situation.

In order to determine the probability of the top event we now have to determine the probability of each minimal cut set. Usually it is assumed that the basic events in a particular minimal cut set occur independently. This means that the probability of the cut set is simply the product of the probabilities of each of the basic events.

For the very simple fault tree shown in Figure 6.13 we compare the exact probability of the top event with the rare event approximation. The events A, B and C occur independently with probability 0.1. The cut sets are $\{A, B\}$ and $\{A, C\}$, so the rare event approximation gives

$$P(T) \approx P(A \cap B) + P(A \cap C) = 0.02.$$

The exact probability is

$$\begin{aligned} P(T) &= P((A \cap B) \cup (A \cap C)) \\ &= P(A \cap B) + P(A \cap C) - P(A \cap B \cap C) = 0.019. \end{aligned}$$

In general, when the probabilities of the basic events are very small and the basic events occur independently then the rare event approximation is good. In Chapter 7 we discuss new methods that allow for rapid exact calculations.

6.10.1 Common cause

If the events in a minimal cut set are not independent then the probability of the cut set event occurring (i.e. of the simultaneous occurrence of the basic

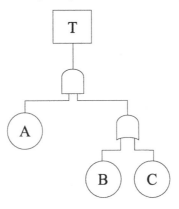

Fig. 6.13. Very simple fault tree

events in the minimal cut set) is not equal to the product of the probabilities. Engineers call such a minimal cut set a *common mode cut set*, because there is some common mode which is causing the events not to be independent.

Example 6.3 *Consider the reactor protection example. It is decided to improve the reliability of the system by laying 10 wires in parallel between the sensors and the OR-gate so that even if 9 of the wires break then the OR-gate will still operate. Unfortunately all 10 wires pass through the same conduit which has been built in a hallway containing old paper and a dangerous old electric heater. If the heater causes a fire then the paper will burn and all 10 wires will be destroyed at the same time.*[1]

The most important reasons for common-mode failures are:

- **Environment:** temperature, grit, stress.
- **System:** loss of energy source, vibration, same subcomponents.
- **Personnel:** same installation crew, same maintenance crew, same operator, faulty test procedures.
- **Aging:** components of the same materials.

Common-mode failures are often very important contributors to system unreliability since they can take out several redundant safety systems at once. More will be said about the modeling of common-mode failures in Chapter 8.

---

[1] If this seems a silly example, bear in mind that the Browns Ferry nuclear power plant fire of 1975 was caused when an operator was using a naked candle to check for leaks. Polyurethane foam in a cabling duct caught fire damaging much of the control cabling for the plant, most of which ran through the duct [Kirwan, 1994].

6.11 Exercises

6.1 A system contains six components A, B, C, D, E, and F and has minimal cut sets

$$\{A,B,C\}, \{A,D,F\}, \{B,C,E\}, \{D,E,F\}.$$

(a) Draw a fault tree in which each basic event occurs once.
(b) Give the minimal paths of the system.

6.2 Give an example of a system with 4 components and $\binom{4}{2} = 6$ minimal cut sets (this is the maximum number possible, by Sperner's Theorem).

6.3 Determine path and cut sets for the New Waterway fault tree in Section 6.7.2 by inspection.

6.4 Check that De Morgan's law (see Table 6.1) implies that the process of interchanging gates etc. to build the dual tree gives a true representation of T'.

6.5 Extend the New Waterway fault tree by including new basic events to represent the failures of the water-level meters. Assume that each water-level meter either fails or sends a correct measurement, and that only one measurement from each post is necessary.

6.6 Assuming that the basic events fail independently and all occur with probability 0.05, determine the probability of the top event for the New Waterway fault tree.

7

Fault trees – analysis

7.1 The MOCUS algorithm for finding minimal cut sets

We illustrate an algorithm for calculating the minimal cut sets of a fault tree by calculating the cut sets for the tree in Figure 6.3. In the algorithm we essentially calculate a formula for the top event T that has the form 6.2. This is done by substituting in an expression for each gate of the tree and applying the distributive law given in Table 6.1.

Note that the outcome of the first stage of this algorithm is actually a list of *cut sets* with the property that every minimal cut set is included as a subset of one of the cut sets. After the list of cut sets is generated, each cut set must be reduced to a minimal set and the multiple occurrences of minimal cut sets must be removed. A more efficient implementation of the algorithm can be made by searching for (and removing) multiple occurrences during each iteration of the algorithm.

7.1.1 Top down substitution

Label the gates of the tree by E_1 for the upper gate and E_2 for the lower gate. We interpret E_1 resp. E_2 as the events such that there is output from the upper resp. lower gate. We then have

$$T = E_1 = G \cdot E_2 = G \cdot (S + B)$$
$$= G \cdot S + G \cdot B.$$

The cut sets are $\{G, S\}$ and $\{G, B\}$. These cut sets are both *minimal* cut sets, and the algorithm is finished. Notice that the laws of Boolean algebra are applied to reduce the expression for the top event to an expression of the form 6.2. For the reactor protection example, the expression for the top event given in Equation 6.1 was obtained in this way.

T	G1	G2G3	P1G3 E1G3 G4G3	P1P2 P1E1 P1G4 E1P2 E1E1 E1G4 G4P2 G4E1 G4G4	P1P2 E1 G4	P1P2 E1 A G5	P1P2 E1 A G6G7	P1P2 E1 A S1G7 E2G7	P1P2 E1 A S1S2 S1E2 E2S2 E2E2	P1P2 E1 A S1S2 E2

Fig. 7.1. Cut set calculation for the reactor protection system

Another way to make this calculation which does not depend on writing out the expressions for the top event works as follows. We find the minimal cut sets by a branching process: working from the top of the fault tree at an OR gate we branch and at an AND gate we list those events (or gates) directly underneath. The final output is a list containing (possibly after reduction) the minimal cut sets. Using the reactor protection example again, we make the calculation shown in Figure 7.1. Note that we have applied the reduction rules several times to reduce the size of the list. For example $E1E1$ can be replaced by $E1$, and $P1E1$ can be removed because $E1$ is already in the list. This calculation gives the same answer for the minimal cut sets as we had determined previously.

To calculate the minimal paths for the reactor protection example we first write down the dual tree with top event T', see Figure 6.12, and then perform the above algorithm again. This gives the minimal paths shown in Figure 6.11.

7.1.2 Bottom up substitution

In a bottom up substitution procedure we calculate the minimal cut sets at gates lower in the tree before moving upwards and calculating minimal cut set expressions for higher gates. Eventually one has an expression for the minimal cut sets of the top event. This procedure seems more laborious than top down substitution, but one obtains minimal cut sets for each intermediate event instead of for just the top event. Another advantage is that there is often much repetition in complex trees. The bottom up procedure enables one to save on calculation time in such a situation.

7.1.3 Tree pruning

Definition 7.1 *A minimal cut set containing n basic events is called an n*-event cut set.

If all basic events occur independently (in the probabilistic sense) then one expects that large minimal cut sets will have a very low probability. For this reason many computer programs for finding cut sets are designed to generate only those cut sets with a small number of events. In order to further reduce the number of cut sets, the tree is often pruned of secondary events (secondary events are generally all assumed to be of low probability). Such simplifications are justifiable *if* the probabilities concerned are indeed low and independent.

7.2 Binary decision diagrams and new algorithms

Many systems are not coherent. Here is a simple example (taken from [Meeuwsen, 1998]).

Example 7.1 *An electrical network supplies current to two substations. The network consists of four lines from the supply generator to the substations, and a single line between the substations as shown in Figure 7.2a. The demand in substation 1 is 4 units, and that in substation 2 is 2 units (designated with −4 and −2 in the figure). If the capacity of lines 1 to 3 is 4 units, and the capacity of each of the lines 4 and 5 is 2 units then the current flow will be as follows: 4.29/3 over each of lines 1, 2 and 3, 1.71 on line 4 and 0.29 on line 5 (from substation 1 to substation 2). This is shown in Figure 7.2a. If lines 1 and 2 fall out (for example they might be lines mounted on the same pylon which then gets damaged in a storm) then the load flow changes to that shown in 7.2b. Note that line 4 is now overloaded. This is a situation that can only be accepted for a short period of time as it may lead to severe damage of the line. However, when line 5 either falls out of service or is manually removed from service, the load flow changes to that shown in Figure 7.2c. No lines are operating beyond capacity. Hence the failure of line 5 has changed the system so that it operates within the operating constraints between the towns.*

For many systems, when fault trees are constructed manually it is fairly easy to avoid making the tree non-coherent. In fact the 'no miracles' rule helps to force the analyst to make the tree coherent. Increasingly though, fault trees are being produced by software which is not capable of applying such rules. Such fault trees are typically non-coherent. The MOCUS algorithm, along with other common FT algorithms, assumes that a fault tree is coherent and it is therefore not generally applicable.

In this section we shall look at binary decision diagrams (BDDs) as a compact method of representing fault trees. BDDs can be manipulated very easily, so that computations (including probabilistic computations) can be carried out reasonably economically.

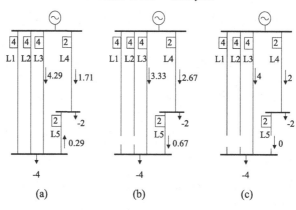

Fig. 7.2. A power system

BDDs are a popular way of representing Boolean functions, and were introduced by Bryant (see for example Bryant's survey paper [Bryant, 1992]). Since fault trees are simply a way of representing Boolean functions, we can always represent the same function with a BDD.

We introduce some notation to allow us to state the formal definitions. A *Boolean formula* is an expression built up inductively from a set of variables $\mathcal{X}$, the two constants 0 and 1, and the logical connectives · (and), + (or), ′ (negation), etc. (In our applications these variables represent statements such as 'basic event i occurs', 1 means 'true' and 0 means 'false'.) A *literal* is either a Boolean variable x, or its opposite *not x*, written x'. A *product* is a set of literals not containing both a literal and its opposite (for example $a \cdot b$ or $a \cdot b'$). An *assignment* over $\mathcal{X}$ is a map from $\mathcal{X}$ into $\{0, 1\}$ and tells us which variables are true and false.

We can now define the notion of an implicant, which for the case of coherent systems corresponds to a cut set. When f is a Boolean formula and σ is a product of literals, then we say that σ is an *implicant* of f if any assignment satisfying σ also satisfies f. When there is no subproduct γ of σ which is an implicant of σ then we say that σ is a *prime implicant*. In the case of coherent trees this notion is equivalent to a minimal cut set.

Example 7.2 *[Bryant, 1992]. Consider the simple coherent fault tree shown in Figure 7.3. Here $\mathcal{X} = \{a, b, c\}$. The fault tree is a representation of the Boolean formula $g_1 = (a + b) \cdot c$, and has five implicants,*

$$a \cdot c, \quad b \cdot c, \quad a \cdot b \cdot c,$$
$$a' \cdot b \cdot c, \quad a \cdot b' \cdot c.$$

7.2 *Binary decision diagrams and new algorithms* 125

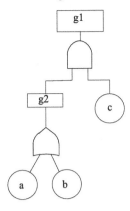

Fig. 7.3. The simple coherent fault tree from Example 7.2

Table 7.1. *Truth table for Example 7.2*

a	b	c	g1
0	0	0	0
0	0	1	0
0	1	0	0
0	1	1	1
1	0	0	0
1	0	1	1
1	1	0	0
1	1	1	1

Three of these are cut sets (cut sets do not contain negations). Of the five implicants there are two prime implicants (minimal cut sets),

$$a \cdot c, \quad b \cdot c.$$

The information represented in the fault tree can also be represented in a truth table. *A truth table gives the value of the Boolean function (in this case the occurrence or not of the top event), with 1 denoting failure and 0 denoting no failure. The truth table is given in Table 7.1.*

Example 7.3 *[Aralia, 1995] We consider the simple non-coherent fault tree shown in Figure 7.4. Here the set $\mathscr{X} = \{a, b, c\}$. The fault tree is a representation of the Boolean formula $g_1 = a \cdot b + c \cdot a'$, and has 7 implicants,*

$$a \cdot b, \quad a \cdot b \cdot c, \quad a \cdot b \cdot c', \quad a' \cdot b \cdot c,$$
$$a' \cdot b' \cdot c, \quad a' \cdot c, \quad b \cdot c.$$

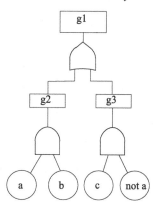

Fig. 7.4. *The simple non-coherent fault tree from Example 7.3*

Table 7.2. *Truth table for Example 7.3*

a	b	c	g1
0	0	0	0
0	0	1	1
0	1	0	0
0	1	1	1
1	0	0	0
1	0	1	0
1	1	0	1
1	1	1	1

Of these three implicants there are three prime implicants,

$$a \cdot b, \quad a' \cdot c, \quad b \cdot c.$$

The information represented in the fault tree can also be represented in a truth table. *A truth table gives the value of the Boolean function (in this case the occurrence or not of the top event), with* 1 *denoting failure and* 0 *denoting no failure. The truth table is given in Table 7.2.*

A binary decision diagram is just a way of representing the output of a Boolean function, truth table, or fault tree in a compact way. First of all, a truth table can be given in terms of a tree in which the nodes correspond to variables, and the two branches at each node correspond to the node taking the values 0 and 1 respectively. The nodes at the end of these branches will be called the 0-child and 1-child respectively. At the end of the tree, the value of the Boolean function is given corresponding to the whole path of node values going up to the top node. This is illustrated for Example 7.2

7.2 Binary decision diagrams and new algorithms

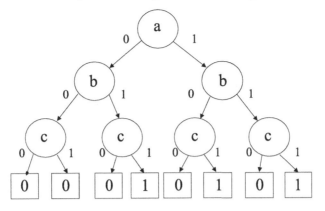

Fig. 7.5. A binary decision tree for Example 7.2

in Figure 7.5. The branching in the tree can be interpreted in terms of an *if then else* statement, also known as the Shannon decomposition. The first branching of the tree can be interpreted as 'If $a = 0$ then... else...', where the dots represent the logical structures given below in the tree. The same trick is often used in the calculation of structure functions in reliability theory [Barlow and Proschan, 1965].

Now, this is not a very compact representation, but Bryant [Bryant, 1986, Bryant, 1992] shows that there is a way to make a canonical compact representation if we impose an order on the variables. An *ordered BDD* is one in which there is a total ordering $<$ on the variables (for example $a < b < c$) and furthermore any node in the BDD is strictly lower in the ordering than its children.

There are three rules which can be then used to simplify the BDD (that is, reduce the number of nodes). When these rules have been applied a canonical representation of the BDD is obtained. The rules are

(i) Removal of duplicate terminal nodes: Remove all but one terminal nodes with a given label and redirect their input edges to the remaining node.
(ii) Remove duplicate non-terminal nodes: If two different nodes have the same label, their 0-children are identical, and their 1-children are identical, then one of the two nodes is removed and its incoming arcs are redirected to the other vertex. This rule must first be applied to the variable which is highest in the ordering.
(iii) Remove redundant nodes: If the 0-child and the 1-child of a non-terminal node have the same label then that node is removed and its incoming arcs are redirected to the child.

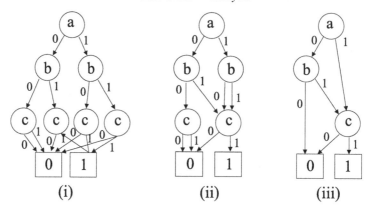

Fig. 7.6. Application of the simplification rules

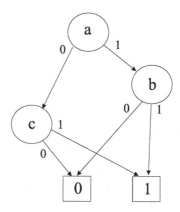

Fig. 7.7. BDD for Example 7.3

We illustrate the iterative use of these rules with the tree given in Figure 7.5. After applying the first rule, we obtain the BDD in Figure 7.6i. Application of the second rule merges two of the old c nodes and the result is shown in Figure 7.6ii. Finally, we can apply the last rule to remove two redundant nodes labeled b and c, giving the final result shown in Figure 7.6iii.

Note that the ordering given to the variables is extremely important in determining the size of the BDD. There are examples (see [Bryant, 1992]) with $2n$ variables in which one ordering gives a BDD with $2n + 2$ nodes, while another ordering of the same variables gives a BDD with 2^{n+1} nodes. We discuss the problem of finding a good ordering below.

The BDD associated with Example 7.3 (using the order $a < b < c$) is given in Figure 7.7.

7.2.1 Prime implicants calculation

The basis of the algorithms used to calculate prime implicants of a Boolean formula is the following theorem, which shows that an iterative calculation is possible. The calculation is however very slow. A faster alternative calculation will be discussed in the next subsection.

We denote the set of prime implicants of a Boolean formula f by $PI[f]$. Recall that at the top level of a BDD, the Boolean formula is decomposed by the Shannon decomposition 'If... then... else...'. It will be convenient to introduce the notation $ite(f, g, h)$ for 'If f then g else h', or formally,

$$ite(f, g, h) = (f \cdot g) + (f' \cdot h).$$

Theorem 7.1 *Let f be a Boolean formula, x a variable, and consider the Shannon decomposition $f = ite(x, f_1, f_0)$ for some Boolean formulae f_0 and f_1. Then*

$$PI[f] = PI_{10} \cup PI_1 \cup PI_0,$$

where PI_{10} is defined as $PI[f_1 \cdot f_0]$, PI_1 is defined as $x \cdot (PI[f_1] - PI_{10})$ and PI_0 is defined as $x' \cdot (PI[f_0] - PI_{10})$.

Returning briefly to Example 7.3 and the BDD shown in Figure 7.7, it is easy to see that the BDD is equivalent with

$$g1(a, b, c) = ite(a, ite(b, 1, 0), ite(c, 1, 0)).$$

The prime implicants can be calculated by means of the above theorem. Clearly $ite(b, 1, 0) = b$, $ite(c, 1, 0) = c$, and

$$ite(b, 1, 0) \cdot ite(c, 1, 0) = b \cdot c.$$

Hence the decomposition given in the theorem (taking $x = a$) says that $PI(g1) = PI_{10} \cup PI_1 \cup PI_0$, where

$$PI_{10} = \{b \cdot c\},$$

PI_1 is the set of products $a \cdot \sigma$ for σ in $\{b\}$ and not in $\{b \cdot c\}$, that is $PI_1 = \{a \cdot b\}$, and similarly $PI_0 = \{a' \cdot c\}$. We conclude that

$$PI(g1) = \{a \cdot b, a' \cdot c, b \cdot c\}.$$

Note that the implementation of the above theorem in a computer algorithm uses new variables to represent the presence and sign of the original variables in a prime implicant. For more details of the actual implementation the reader is referred to [Aralia, 1995], [Coudert and Madre, 1992].

7.2.2 Minimal p-cuts

Rauzy, Dutuit and Signoret introduced the notion of a minimal p-cut in [Rauzy et al., 1997], because often one is only interested in failures. In calculating probabilities of prime implicants we may often conservatively (and to a very good degree of approximation) assume that those components assumed to work in the prime implicant do so with probability 1. Hence only the positive part of the prime implicant (corresponding to the components that really fail) is really important.

To formalize this idea, consider a Boolean formula f and a product π containing only positive literals (that is, no negations). We denote by π_f^* the product obtained by conjoining all the negative literals to π obtained from variables that occur in f but not in π. (For example suppose that f is a formula in the variables a, b and c, and that $\pi = ab$. Then $\pi_f^* = abc'$.) We say that π is a *p-cut* of f if π_f^* is an implicant of f, and say that π is *minimal* if there is no product $\rho \subset \pi$ such that ρ_f^* is an implicant of f. The set of minimal p-cuts of f is denoted by $PC[f]$.

Returning again to Example 7.3, it is easy to see that

$$PC(g1) = \{a \cdot b, c\}.$$

It is straightforward to see that $PI[f] = PC[f]$ if and only if f is a monotone formula (that is, the fault tree is coherent). This shows that minimal p-cuts are a natural generalization of minimal cut sets.

In order to calculate minimal p-cuts it is necessary to have a decomposition result analogous to Theorem 7.1. The decomposition is the basis of an iterative scheme to calculate the p-cuts. The idea is that when $f = ite(x, f_1, f_0)$, then the set of minimal p-cuts can be divided up into two sets: the minimal p-cuts of f not containing x, and those which do contain x. The first collection is just the minimal p-cuts of f_0. The second collection is almost but not quite the same as the minimal p-cuts of f_1. Clearly x has to be added to each such minimal p-cut. The main problem is however that not all are minimal p-cuts for f because of the existence of a smaller p-cut for f_0. Two approaches have been suggested.

In order to state the first result a new set operation is needed:

Definition 7.2 *Suppose that S and T are two sets of products. Then*

$$S \div T = \{\pi \in S | \forall \rho \in T, \rho \not\subseteq \pi\}.$$

Thus $S \div T$ is the set of products in S that do not contain any product from T. In a moment we shall see how to compute with this set operation. But first note how we can get a decomposition result. Suppose that $f = ite(x, f_1, f_0)$.

7.2 Binary decision diagrams and new algorithms

The set PC_0 of minimal p-cuts of f that do not contain x is $PC[f_0]$. The set PC_1 of minimal p-cuts containing x is precisely that subset of $x \cdot PC[f_1]$ that do not include a product of $PC[f_0]$ (otherwise, they would not be minimal). This proves the first decomposition result,

Theorem 7.2 Suppose $f = ite(x, f_1, f_0)$, then

$$PC[f] = PC_1 \cup PC_0$$

where $PC_0 = PC[f_0]$ and $PC_1 = x \cdot (PC[f_1] \setminus PC_0)$.

The operation $\div$ has to be carried out inductively. This is done using the following lemma.

Lemma 7.3 Suppose that S, S_1, S_0, T, T_1, T_0 are sets of positive products not containing the variable x, and let

$$S = x \cdot S_1 \cup S_0, \quad T = x \cdot T_1 \cup T_0.$$

Then $S \div T = x \cdot (S_1 \div (T_1 \cup T_0)) \cup (S_0 \div T_0)$.

Proof First note that the following rules apply:

$$S \div (A \cup B) = (S \div A) \cap (S \div B),$$
$$(A \cup B) \div T = (A \div T) \cup (B \div T).$$

Note also that since x is not present in products from T_0 and T_1 we have $x \cdot S_1 \div T_0 = x \cdot (S_1 \div T_0)$ and $S_0 \div x \cdot T_1 = S_0$. This gives

$$\begin{aligned} S \div T &= [(x \cdot S_1 \div x \cdot T_1) \cap (x \cdot S_1 \div T_0)] \cup [(S_0 \div x \cdot T_1) \cap (S_0 \div T_0)] \\ &= [x \cdot (S_1 \div T_1) \cap x \cdot (S_1 \div T_0)] \cup [S_0 \cap (S_0 \div T_0)] \\ &= [x \cdot (S_1 \div (T_1 \cup T_0))] \cup [S_0 \div T_0], \end{aligned}$$

as claimed. $\square$

A second decomposition theorem gives another expression which avoids use of the $\div$ operator. Note that the set PC_1 is not the same as that defined in Theorem 7.2.

Theorem 7.4 Suppose $f = ite(x, f_1, f_0)$, then

$$PC[f] = PC_0 \cup PC_1$$

where $PC_0 = PC[f_0]$ and $PC_1 = x \cdot (PC[f_1 + f_0] \setminus PC_0)$.

Both of these decomposition results can be used inductively to determine the minimal p-cuts of a Boolean formula. In [Rauzy et al., 1997] some numerical evidence is given that the algorithm based on the first decomposition result works faster.

The algorithms based on the above decomposition theorems can easily be adapted to generate only those minimal p-cuts containing less than k variables for any given k. They can also be adapted to give all minimal p-cuts with probability greater than a given p. See [Rauzy et al., 1997] for details.

7.2.3 Probability calculations

Probabilities are quite simple to calculate in a BDD, because corresponding to the Shannon decomposition *ite*,

$$f = (x \cdot f_1) + (x' \cdot f_0),$$

is the probabilistic formula

$$P(f = 1) = P(x = 1)P(f_1 = 1) + P(x = 0)P(f_0 = 1). \qquad (7.1)$$

Using this formula iteratively through the BDD one obtains the exact probability of the top event.

7.2.4 Examples

Reactor protection system A BDD representation of the fault tree for the reactor protection example is given in Figure 7.8. Since the system is coherent, the concepts of prime implicant, minimal p-cut, and minimal cut sets are all identical.

Electrical power system The power system of Example 7.1 is not coherent. A BDD representation for the top event 'overload of power line' is given in Figure 7.9.

The probability of the top event is calculated by repeated application of the conditioning formula given in Equation 7.1. Applying the conditioning rules here we obtain

$$\begin{aligned}
P(T) &= P(T|L1=0)P(L1=0) + P(T|L1=1)P(L1=1) \\
&= P\{T|(L1,L2)=(0,1)\}P\{(L1,L2)=(0,1)\} \\
&\quad + P\{T|(L1,L2)=(1,0)\}P\{(L1,L2)=(1,0)\} \\
&\quad + P\{T|(L1,L2)=(1,1)\}P\{(L1,L2)=(1,1)\}
\end{aligned}$$

7.2 Binary decision diagrams and new algorithms

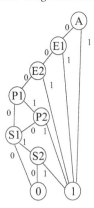

Fig. 7.8. BDD representation for the reactor protection example

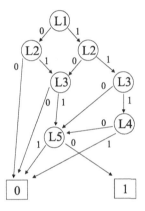

Fig. 7.9. BDD representation for the electrical power system example

$$\begin{aligned}
&= P\{T|(L1,L2,L3) = (0,1,1)\}P\{(L1,L2,L3) = (0,1,1)\} \\
&\quad + P\{T|(L1,L2,L3) = (1,0,1)\}P\{(L1,L2,L3) = (1,0,1)\} \\
&\quad + P\{T|(L1,L2,L3) = (1,1,0)\}P\{(L1,L2,L3) = (1,1,0)\} \\
&\quad + P\{T|(L1,L2,L3) = (1,1,1)\}P\{(L1,L2,L3) = (1,1,1)\} \\
&= P\{(L1,L2,L3,L5) = (0,1,1,0)\} \\
&\quad + P\{(L1,L2,L3,L5) = (1,0,1,0)\} \\
&\quad + P\{(L1,L2,L3,L5) = (1,1,0,0)\} \\
&\quad + P\{(L1,L2,L3,L4,L5) = (1,1,1,0,0)\}.
\end{aligned}$$

If we suppose that each line fails independently with a probability of 0.01, then we obtain a probabilty for the top event of $3 \times 0.01^2 \times 0.99^2 + 0.01^3 \times 0.99^2 = 2.95 \times 10^{-4}$.

134   7 Fault trees – analysis

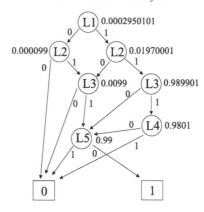

Fig. 7.10. Probability calculations for the electrical power system example

The probability calculations can be done quite simply in an iterative way using the nodes of the BDD to store intermediate calculations. The number given at a node is the probability of reaching the state '1' by a randomly chosen path starting at the current node (for the first node in the BDD this corresponds to the probability of the top event). The number is calculated by a bottom up procedure by applying the conditioning formula of Equation 7.1 and using the numbers already calculated for the child nodes. For example, the node $L5$ gets the number $0.99 = 0.01 \times 0 + 0.99 \times 1$. The calculations are shown in Figure 7.10. We return to this example to perform an uncertainty analysis in Chapter 17.

We demonstrate the calculation of minimal p-cuts using an algorithm based on the first theorem given above. The algorithm works inductively working up from the bottom of the BDD. We calculate a collection of p-cuts for each node. This collection is the set of p-cuts for the BDD obtained by deleting all nodes above the current one.

The calculation is performed in Table 7.3, and shows that two out of three failures of lines 1, 2, and 3 are the causes of overload.

7.2.5 The size of the BDD

The form of BDD is uniquely determined when the variables are put in some order. Changing the order, however, can change the size of the BDD quite dramatically. Good heurisitics for ordering the variables seem difficult and depend upon characteristics of the fault tree, see [Bouissou, 1997]. Even when a 'recommended' heuristic is used, the size of the fault tree and the corresponding BDD representation may bear little relation to one

Table 7.3. *The p-cut calculation for the electrical power system*

Node, f	PC_0	$PC[f_1]$	$PC[f_1] \div PC_0$	$PC[f]$
L5	$\emptyset$	$\emptyset$	$\emptyset$	$\emptyset$
L4	$\emptyset$	$\emptyset$	$\emptyset$	$\emptyset$
Left L3	$\emptyset$	$\emptyset$	$\emptyset$	$\{L3\}$
Right L3	$\emptyset$	$\emptyset$	$\emptyset$	$\{L3\}$
Left L2	$\emptyset$	$\{L3\}$	$\{L3\}$	$\{L2 \cdot L3\}$
Right L2	$\{L3\}$	$\{L3\}$	$\emptyset$	$\{L2, L3\}$
L1	$\{L2 \cdot L3\}$	$\{L2, L3\}$	$\{L2, L3\}$	$\{L1 \cdot L2, L1 \cdot L3, L2 \cdot L3\}$

Table 7.4. *Comparison of fault tree size and BDD size for various test cases*

Problem name	No. BE	No. gates	No. branches	BDD size
B5500	129	86	271	2949
DRESDEN3	57	60	123	303
ES200D	280	111	552	1577
ESS	67	57	124	100
EUROPEA1	61	84	219	5085
MDL10	585	248	1016	1467
STS012	495	734	1630	–

another. Table 7.4, from [Kirkegaard and Kongsø, 1999], compares fault tree (in terms of number of basic events, number of gates and number of branches) and BDD size (number of nodes) for a few industrial fault trees. We have only taken data from the fault trees with more than 10 basic events. In the STS012 case the computer code was unable to deal with the size of the BDD being created, and failed.

7.3 Importance

When making a quantitative analysis of a particular system one wants to rank the different contributors to system failure. A number of different definitions have been made to quantify basic event importance. Each is appropriate for a different kind of ranking problem.

Definition 7.3 *A basic event is an n-point failure event ($n \geq 1$) if the smallest minimal cut set containing it has n elements. When $n = 1$ the basic event is called a* single point failure.

This notion has traditionally been used as a qualitative tool at the design stage of a system in order to prioritize the development process. NASA just used critical item lists in the development of its systems until taking a more quantitative view in the aftermath of the Challenger disaster.

Definition 7.4 *A component C is* critical *if its failure causes system failure, i.e. if all other basic components in some minimal set have failed except for C.*

Note that depending on the state of other system components, a particular component may change status many times between critical and non-critical.

Definition 7.5 *The* (Birnbaum) structural importance *of a component C is the probability that C is critical.*

This definition, though simple, is not considered too useful as a component may contribute to a systems failure even when that component is not critical to the system.

Definition 7.6 *The* Fussell–Vesely importance *of a basic event A is the conditional probability of A given the top event T,*

$$I^{FV}(A) = P(A|T) = \frac{P(A \cap T)}{P(T)}.$$

There is a similar definition for the FV-importance of a minimal cut set. A good approximation to $I^{FV}(A)$ is given by

$$\frac{\sum_{C:A \in C} P(C)}{\sum_C P(C)}, \qquad (7.2)$$

where C denotes a cut set. We give two more important definitions of importance.

Definition 7.7 *The* Barlow–Proschan importance *of a component C is the average number of system failures up to time t caused by the failure of C.*

Definition 7.8 *The* Barlow–Proschan importance *of a minimal cut set M is the average number of system failures up to time t caused by the failure of M.*

Various authors have suggested using the Fussell–Vesely importance as a tool for optimizing the number of inspections required to identify the failed minimal cut set when a system has failed. However, during the process of inspection one obtains information about components which can dramatically affect the conditional failure probability of the remaining components.

7.3 Importance

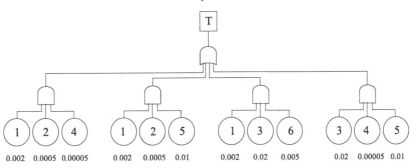

Fig. 7.11. Fault tree example

Table 7.5. *FV importance of basic events*

Basic event	FV importance
1	0.9546
2	0.0457
3	0.9543
4	0.0457
5	0.0909
6	0.9089

Consider the fault tree shown in Figure 7.11. The tree is given in minimal cut set form, and the probabilities of the basic events are given below them. It is assumed that the basic events are independent. Using the rare event approximation we find that the probability of the top event is $P(T) = 2.2005 \times 10^{-7}$. The FV importances are shown in Table 7.5. Given that the top event has occurred, the basic events are *not* independent. For example, if we observe that Component 1 has not failed, then it immediately follows that Components 3, 4, and 5 have failed.

In a similar way to that in which the rare event approximation can be derived, one can show that the probability of a basic event A occurring given that B and the top event T have occurred is approximately

$$\frac{\sum_{C:A,B\in C} P(C)}{\sum_{C:B\in C} P(C)}.$$

For example, if we inspect Component 2 and observe that it has failed, then the conditional probabilities of failure of the other components are as shown in Table 7.6. For Components 3 and 6 the approximation gives 0 because it assumes that only those minimal cut sets containing Component 2 could have failed. Exact calculations can be made easily to show that those probabilities

Table 7.6. *Approximate conditional probability of basic events given failure of Component 2 and T*

Basic event	Conditional probability
1	0.005
2	1
3	0
4	0.005
5	0.995
6	0

are not actually zero but are very, very small. In any case, it is clear that when determining conditional probability (and thus the 'importance' of a component) it is important to take account of all the information available.

An inspection strategy based on inspecting the component with the largest (conditional) FV importance is poor: often we will just confirm what we already are nearly certain of. The strategy of inspecting those components whose (conditional) FV importance is close to 1/2 would seem better [Xiaozhong and Cooke, 1991], although for the tree shown here it gives slightly worse results. However, [Norstøm et al., 1999] show that a better strategy is obtained by formulating the problem as a value of information problem using decision theory. They report that the strategy of maximal FV importance gives an expected number of inspections equal to 2.955 for the tree given above. The strategy of [Xiaozhong and Cooke, 1991] gives an expected number of inspections equal to 3.01. The decision theory strategy of [Norstøm et al., 1999] has expected number of inspections equal to 2.05.

7.4 Exercises

7.1 Apply the MOCUS algorithm to the non-coherent fault tree shown in Figure 7.4. Why is the result of the algorithm incorrect?

7.2 Change the ordering of the basic events for the reactor protection example, and determine the associated BDD.

7.3 Calculate the prime implicants and p-cuts for the non-coherent fault tree shown in Figure 7.4 using the methods of subsections 7.2.1 and 7.2.2.

7.4 Justify the approximation to the Fussell–Vesely importance given in Equation 7.2.

7.5 Choose your own probabilities for the basic event in the reactor protection and New Waterway examples and calculate the various sorts of importance for the basic events. Compare the rank orders given by the different notions of importance.

8

Dependent failures

8.1 Introduction

The subject of dependent failures is one of the more important issues affecting the credibility and validity of standard risk analysis methods; and it is an issue around which much confusion and misleading terminology exist. This treatment draws on procedures for dealing with common causes as issued by the US Nuclear Regulatory Commission, and the International Agency for Atomic Energy.

8.2 Component failure data versus incident reporting

Component data reliability banks typically collect individual component failure events and demands and/or operational times. From such data alone it is impossible to estimate the probabilities of dependent failures. For this we need information on the joint failures of components, which becomes available only when incidents involving multiple failures of components are recorded as such. Standard data banks do not collect data on incidents. There are, however, isolated exercises in incident reporting. There is also an ongoing program to analyze the so-called 'licensee event reports' in the American commercial nuclear power sector, and draw conclusions for probabilistic risk analysis. An issue of *Reliability Engineering and System Safety* (**27**, 1990) was devoted to 'accident sequence precursor analysis' and several contributions describe this program (see for example [Cooke and Goossens, 1990]). The use of incident reporting in probabilistic risk analysis deserves more attention than it has received to date, but will not be discussed further here. Recently an international common cause database for the nuclear sector has been established [Carlsson *et al.*, 1998].

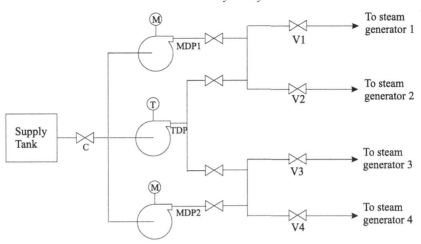

Fig. 8.1. Auxiliary Feedwater System

8.3 Preliminary analysis

The issue of common cause failure most frequently arises in connection with redundant components. This is perhaps due to the fact that such components are tested jointly, and it is relatively easy to gather joint failure data. An example which has been frequently studied [Fleming et al., 1986] is the 'Auxiliary Feedwater System' (AFWS) at pressurized water reactors. The schematic of a three train AFWS is shown in Figure 8.1. The system's mission is to provide water to extract heat from the primary cooling system and deliver steam to the steam generators, in the event that the Main Feedwater System should be lost. Failure to remove heat from the primary cooling system can lead to reactor core meltdown. Mission success is achieved when at least one of the three pumps delivers steam to at least two steam generators. The system depicted in Figure 8.1 consists of three mechanically identical pumps, two of which are motor driven (MDP) and one of which is steam-turbine driven (TDP). Four identical motor operated valves $(V1, \ldots, V4)$ are normally closed, but are required to open as one or more pumps are activated.

Let A and B denote failure events on demand, for example the events that two pumps in Figure 8.1 fail to start following a demand. Recall that

$$P(A \cap B) = P(A \mid B)P(B),$$

and that A and B are independent when

$$P(A \cap B) = P(A)P(B). \tag{8.1}$$

If 8.1 does not hold, then the failure events A and B are said to be dependent. For any event C, we may write

$$\begin{aligned} P(A \cap B) &= P(A \cap B \cap C) + P(A \cap B \cap C') \\ &= P(A \cap B \mid C)P(C) + P(A \cap B \mid C')P(C'). \end{aligned} \quad (8.2)$$

We could say that C 'is of influence upon' the probabilities of A and B if

$$P(A \cap B \mid C) \neq P(A \cap B) \text{ or } P(A \cap B \mid C') \neq P(A \cap B). \quad (8.3)$$

If these inequalities hold (one inequality implies the other – see Exercise 1), then some authors would call C a *common cause* for A and B. Note that 8.3 can hold while A and B are independent conditional on C and independent conditional on C',

$$P(A \cap B \mid C) = P(A \mid C)P(B \mid C), \quad (8.4)$$
$$P(A \cap B \mid C') = P(A \mid C')P(B \mid C'). \quad (8.5)$$

An example is given in Exercise 2.

In a typical situation A and B refer to identical, redundant systems. If common cause C has been identified, a common engineering approximation is to write

$$P(A \mid C') = P(B \mid C') \approx P(A),$$
$$P(A \cap B \mid C') \approx P(A \mid C')P(B \mid C') \approx P(A)^2,$$

and

$$P(A \cap B \mid C) \approx P(A \mid C).$$

These approximations then together imply

$$P(A \cap B) \approx P(A)^2(1 - P(C)) + P(A \mid C)P(C). \quad (8.6)$$

The approximation in (8.6) corresponds to a situation in which $P(C) \ll 1$, and in which the occurrence of C completely couples the occurrences of A and B, but does not cause A to occur. For example, C might be an event like 'fire', 'flooding', 'overload', which will destroy either both components or neither.

The importance of common cause is easily seen by filling in some representative values into (8.6). If $P(A) = 10^{-3}$, $P(A \mid C) = 1$, $P(C) = 10^{-5}$, then the 'common cause contributor' to (8.6) dominates by an order of magnitude, even though only 1% of occurrences are due to common cause.

8.4 Inter-system dependencies

The area of dependent failures is sometimes broken into two large sub-areas, namely, 'inter-system dependencies' and 'inter-component dependencies'. Large (sub)systems will exhibit dependencies when their proper functioning depends on common support systems. For example, a cooling system and a fire suppression system may both depend on the supply of service water, and on the electrical power system. Techniques for correctly modeling such dependencies involve extending the fault tree and event tree modeling so as to make these dependencies explicit. These techniques are not the subject of this chapter.

8.5 Inter-component dependencies – common cause failure

The techniques of fault tree and event tree modeling allow the failure of a system to be analyzed into combinations of failures of basic components. This can be done in such a way that all inter-system dependencies are properly taken into account. This does not mean that the issue of dependent failures has been removed. The component failure events may still exhibit dependencies. Table 8.1 from the *Procedures Guide* [NRC, 1983] lists a number of generic causes of dependent component failures. The rest of this chapter reviews techniques for analyzing incident failure data and modeling dependent component failure probabilities.

8.6 The square root bounding model

The Rasmussen report WASH-1400 [NRC, 1975] used a very simple method for dealing with dependent failures, which is mentioned briefly for historical interest. For a given system of multiple components, a probability of failure P_I assuming independence is computed, and a failure probability P_D assuming a maximal amount of dependence is computed. The overall system failure probability is then computed as

$$\sqrt{P_I P_D}.$$

The Lewis report [Lewis *et al.*, 1979], which reviewed WASH-1400, concluded that the 'degree of arbitrariness in this procedure boggles the mind', and it is no longer in use.

8.7 The Marshall–Olkin model

A very general model for common cause failure was developed by Marshall and Olkin [Marshall and Olkin, 1967]. We consider a system of m identical

Table 8.1. *Generic causes of dependent component failures*

	Extreme condition (generic cause)	Example of source	Environmental channel
1.	Impact	Pipe whip, water hammer, missiles, structural failure, earthquakes	Common location, hydraulic coupling, common structural base
2.	Vibration	Machinery in motion, earthquake	Common structural base
3.	Temperature	Fire, lightning, welding equipment, cooling system faults, electrical short circuits	Common location, ventilation ducts
4.	Moisture	Condensation, pipe rupture, rainwater, floods	Common location, ventilation ducts, hydraulic coupling
5.	Pressure	Explosion, out-of-tolerance system changes (pump overspeed), flow blockage	Common location, ventilation ducts, hydraulic coupling
6.	Grit	Airborne dust, metal fragments generated by moving parts with inadequate tolerances, crystallized boric acid from control system	Common location, ventilation ducts
7.	Electromagnetic interference	Welding equipment, rotating electric machinery, lightning, power supplies, transmission lines	Spatial proximity to source
8.	Radiation	Neutron sources and charged-particle radiation	Spatial proximity to source
9.	Corrosion or other chemical reaction	Acid, water, or chemical attack	Common location, ventilation ducts, hydraulic coupling
10.	Conductive medium	Conductive gases	Common location, ventilation ducts

components. Different kinds of shocks can occur to the system resulting in the failure of groups of components. A group of components is denoted by an m-dimensional vector $\underline{x}$ of 0s and 1s, containing at least one 1, where a 1 in the ith position indicates that component i is a member of the group. The Marshall–Olkin model assumes that shocks killing the different groups of components occur independently with a constant rate $\lambda_{\underline{x}}$. That is, the constant failure rate of group $\underline{x}$ is $\lambda_{\underline{x}}$. A given component i can thus fail

8.7 The Marshall–Olkin model

from any one of a number of different causes, depending on which shock occurs first. The overall (constant) failure rate of component i from all possible shocks is $\sum_{\underline{x}|x_i=1} \lambda_{\underline{x}}$.

Suppose that data is available to quantify this model. Since all the failure rates are constant, the number of observations $N_{\underline{x}}(t)$ of the failure of group $\underline{x}$ in time t has a Poisson distribution with parameter $\lambda_{\underline{x}} t$ (see Chapter 3). The maximum likelihood estimator of $\lambda_{\underline{x}}$ is then $N_{\underline{x}}(t)/t$ (see Chapter 4).

Because of its generality, the Marshall–Olkin model as such is not often used. The data requirements for quantifying the model are huge (there are $2^m - 1$ different parameters). This model has however given rise to a number of other models which make assumptions (e.g. symmetry assumptions) about the failure rates. These models have fewer parameters, and are therefore better able to use sparse data.

Moving from the Marshall–Olkin model to a less general model is typically done by clustering some of the groups of components together (for example all those with a given number of components). Since the Marshall–Olkin model assumes that the shock times of each group of components follow an independent Poisson process, a clustering of these groups corresponds to a *superposition* of the corresponding Poisson processes. Happily (see [Kingman, 1993], pp. 5–6), it is a well known result that the superposition of independent Poisson processes (with rates λ_i, $i = 1, \ldots, n$, say) is again a Poisson process whose rate is the sum of the rates of the superimposed processes, $\sum_{i=1}^{n} \lambda_i$. Furthermore, if we ask what the probability is that a particular failure event came originally from process i ($i = 1, \ldots, n$) then it can be shown that this probability is just

$$\frac{\lambda_i}{\sum_{j=1}^{n} \lambda_j}$$

and that the 'origin' of each point in the superimposed process is independent of those of the other points. In particular, if we observe s failure times then the probability that r_i of these have origin i ($i = 1, \ldots, n$) is given by the multinomial distribution,

$$\frac{s!}{r_1! \ldots r_n!} \left(\frac{\lambda_1}{\sum_{j=1}^{n} \lambda_j} \right)^{r_1} \ldots \left(\frac{\lambda_n}{\sum_{j=1}^{n} \lambda_j} \right)^{r_n}.$$

These facts are often used in determining estimators for the parameters of common cause failure models.

8.8 The beta-factor model

The *beta-factor* model, [Fleming, 1975], can be used to model dependencies between dissimilar and not necessarily redundant equipment. Of the current models it is the simplest, the most limited and not surprisingly, the most often applied.

In this model, there is only one kind of common cause shock, and this will fail all m of the components with certainty. This corresponds to the special case of the Marshall–Olkin model with $\lambda_x \neq 0$ only when the number of components in the group is 1 or m. Furthermore, the components are assumed identical so that λ_x is the same for all groups of one component.

There are thus just two parameters in the model, which we shall denote λ_i (for the rate of independent shocks killing single components), and λ_d (for the rate of independent shocks killing all components).

The overall failure rate λ of a particular component can be written as the sum of independent and dependent failure contributions:

$$\lambda = \lambda_i + \lambda_d. \tag{8.7}$$

The β-factor is defined as the fraction of the total failure rate attributable to dependent failures:

$$\beta = \lambda_d/\lambda. \tag{8.8}$$

The β-factor may be interpreted as the conditional probability that a component fails due to a common cause, given that it fails, $P(C|A)$.

Dissimilar components may have different failure rates and different beta-factors. We give an example in which the model is applied to a situation in which components fail on demand.

Example 8.1 *Suppose a parallel system consists of two identical components and is subject to demand once per month. A is the event that the first component fails on demand, and B the event that the second fails on demand. The system fails on demand when neither component is available, $A \cap B$. Suppose that each component has a failure probability of p per demand, and let β denote the fraction of single component failures due to a common cause event C. In the presence of the common cause, both components fail with certainty. In the absence of the common cause the components fail independently. The probability of system failure on demand is computed analogously to Formula 8.6. Take $P(A|C) = 1$, $P(A|C') = (1 - \beta)p$ and $P(C) = P(A)P(C|A) = p\beta$. We then get*

$$P(\text{system fails}) \approx [(1-\beta)p]^2(1-p\beta) + p\beta. \tag{8.9}$$

8.8 The beta-factor model

An example arising from an analysis of observed failures in auxiliary feedwater systems is illustrated below ([NRC, 1983], Chapter 3):

number of component months 4449, T,
number of single failure events 68, N_i,

number of component failures in
multiple failure occurrences 24, N_d.

We assume that the components have identical exponential distributions. The beta factor is estimated as

$$\beta = N_d/(N_d + N_i) = 24/(24 + 68) = 0.26. \tag{8.10}$$

This number is taken to estimate β_D in (8.9). The probability on demand of a single component failure is estimated as

$$(N_i + N_d)/T = 0.02.$$

A formula similar to (8.9) holds here and gives

Probability of failure on demand of a 1-out-of-2 AFWS ≈ 0.0054.

If the data collection effort had not distinguished between multiple and single failure events, but had simply recorded the number of single failures, then an estimation of β would not have been possible. The analyst who neglected the possibility of common cause failures would compute the system failure probability as

$$P \approx (0.02)^2 = 0.0004$$

which is a factor 14 too low.

8.8.1 Parameter estimation

It is clear from the above discussion that the unavailability will be highly sensitive to the value of β that was estimated. Since the β-factor is the conditional probability that a component fails due to a common cause, given that it fails, the maximum likelihood estimator of β is just $n_d/(n_d + n_i)$ where n_d is the number of components in the sample that have experienced common cause failures, and n_i is the number of single (necessarily independent) component failures. A Bayesian analysis is also possible [Fleming et al., 1983], for the likelihood of observing n_c component failures due to common cause in a total of n component failures is given by the binomial distribution

$$P(n_c|n, \beta) = \binom{n}{n_c} \beta^{n_c}(1 - \beta)^{n-n_c}.$$

Table 8.2. *Some β-factors*

Component (failure mode)	Beta factor (MLE)
High-pressure injection pump (run)	2.6×10^{-2}
High-pressure injection pump (start)	6.1×10^{-2}
MOV (open/close)	2.9×10^{-2}
Diesel generator (run)	2.72×10^{-2}
Diesel generator (start)	1.05×10^{-2}

Hence if we have a prior density $\pi_0(\beta)$ on β then the posterior density after observing n_c component failures due to common cause in a total of n component failures is proportional to

$$P(n_c|n, \beta)\pi_0(\beta)$$

by Bayes' theorem.

Some values of β (estimated by maximum likelihood) taken from [Fleming et al, 1983] are given in Table 8.2.

A disadvantage with the beta-factor model is that it does not distinguish between the multiplicities of multiple failure events. The common cause is assumed to fail *all* the redundant components. Of course it is easy to introduce a more sensitive model, the problem is to estimate its parameters from available data. A slightly more sensitive model is given in the next section.

8.9 The binomial failure rate model

The binomial failure rate model is an adaptation by Vesely [Vesely, 1977] of the Marshall–Olkin model [Marshall and Olkin, 1967].

The original BFR model assumes that there are two sorts of shocks – independent shocks killing individual components, and a common cause shock which might kill any group of components. Each component can fail independently with a constant failure rate λ. Common cause shocks are

8.9 The binomial failure rate model

assumed to impinge on the system according to a Poisson process with rate μ. Given that a shock has occurred, each component has a probability p of failure independently of the other components. The number of components failing following a shock is therefore binomially distributed.

The binomial failure rate model implies that the failure rate for any given group of components depends only on the number of components in that group.

The constant failure rate for any *given* component is

$$\lambda + \mu p q^{m-1},$$

where $q = 1 - p$, and the constant rate at which failures of *some* single component occur λ_1 is given by

$$\lambda_1 = m\lambda + \mu m p q^{m-1}.$$

Likewise, the rate at which a given group of i components will fail is

$$\mu p^i q^{m-i}, \quad i = 2, \ldots, m.$$

while the rate at which some group of i components fail is

$$\lambda_i = \mu \binom{m}{i} p^i q^{m-i}, i = 2, \ldots, m.$$

We now see how to estimate the parameters from the data.

The rate of multiple dependent failures λ_+ is defined as

$$\lambda_+ = \sum_{i=2}^{m} \lambda_i = \mu(1 - q^m - mpq^{m-1}) \tag{8.11}$$

and if N_i is the number of dependent failures of i components, then the number N_+ of dependent failure events is

$$N_+ = \sum_{i=2}^{m} N_i.$$

To use this model, $\lambda_i, i = 1, \ldots, m$, must be determined, and for this it is sufficient to estimate μ, p and λ_1 from data. We assume that data is available of the form

$$\{N_1 = n_1, N_2 = n_2, \ldots, N_m = n_m\},$$

that is, we have observed n_1 single component failures, n_2 failures of two components simultaneously, etc. Under the model $N_1, \ldots, N_m$ are independent

Poisson distributed random variables. So in particular, N_1 is independent of N_+, and we can use the maximum likelihood estimators

$$\lambda_1 = n_1/T, \lambda_+ = n_+/T, \tag{8.12}$$

where T is the time on test and $n_+ = n_2 + \ldots + n_m$.

Given N_+, the random vector $(N_2, \ldots, N_m)$ follows a multinomial distribution, for given a failure event of more than one component the probability that it involves precisely i components is

$$r_i = \frac{\lambda_i}{\lambda_+} = \binom{m}{i} \frac{p^i q^{m-i}}{1 - q^m - mpq^{m-1}}. \tag{8.13}$$

So

$$P(N_2 = n_2, \ldots, N_m = n_m | N_+ = n_+)$$
$$= \frac{n_+!}{n_2! \ldots n_m!} \prod_{i=2}^{m} r_i^{n_i}. \tag{8.14}$$

We can use this equation to determine the maximum likelihood estimator for p as follows.

Filling (8.13) into (8.14), taking the logarithm, then setting its derivative with respect to p equal to zero, we get

$$0 = \sum_{i=2}^{m} n_i \left(\frac{i}{p} - \frac{m-i}{q} - \frac{m(m-1)pq^{m-2}}{1 - q^m - mpq^{m-1}} \right)$$
$$= \frac{S - n_+ mp}{p} - \frac{m(m-1)n_+ pq^{m-1}}{1 - q^m - mpq^{m-1}}$$

where $S = \sum i n_i$ is the total number of components observed in common cause failures. Re-arranging terms

$$S = n_+ mp \left(1 + \frac{(m-1)pq^{m-1}}{1 - q^m - mpq^{m-1}} \right)$$
$$= n_+ mp(1 - q^{m-1})/(1 - q^m - mpq^{m-1}). \tag{8.15}$$

Now p can be determined from (8.15), and then μ can be retrieved from (8.12) and (8.11); λ_1 is determined directly from (8.12).

For the special case $m = 3$, (8.15) yields

$$p = 3(S - 2n_+)/(2S - 3n_+). \tag{8.16}$$

Atwood [Atwood, 1986] describes a variation on the above model in which there is also a lethal shock (as in the β-factor model) which occurs independently of the other shocks with a constant rate ω. Atwood shows how to perform a Bayesian analysis of operational data, and how to perform frequentist goodness-of-fit tests.

8.10 The α-factor model

This model (see for example [Siu and Mosleh, 1989]) assumes that the failure rate for a group of components just depends on the number of components in the group. Shock occurrence rates are not modeled explicitly, but all shocks are implicitly assumed to affect groups of components of the same size in the same way.

The α-factor model for a system of m components contains m parameters $\alpha_1, \ldots, \alpha_m$, summing to 1, where α_k is the conditional probability that a set of any k components fails given that there is a failure. Letting λ_k be the rate at which the simultaneous failure of k components is observed, and $\tilde{\lambda}_k$ the rate at which a particular group of k components is observed to fail, we have

$$\lambda_k = \binom{m}{k} \tilde{\lambda}_k,$$

and

$$\alpha_k = \lambda_k / (\lambda_1 + \cdots + \lambda_m).$$

Estimation of the α_k from operational data is straightforward since the maximum likelihood estimator of α_k is just $N_k/(N_1 + \cdots + N_m)$ where N_i is the number of failures of i components. A Bayesian analysis is also easy since the likelihood of data

$$\{N_1 = n_1, N_2 = n_2, \ldots, N_m = n_m\}$$

is given by the multinomial distribution

$$P(n_1, n_2, \ldots, n_m | (\alpha_1, \ldots, \alpha_m)) = \frac{(n_1 + \ldots + n_m)!}{n_1! \ldots n_m!} \alpha_1^{n_1} \ldots \alpha_m^{n_m}.$$

8.11 Other models

For the sake of completeness we mention briefly other models which can be found in the literature. Mosleh, Fleming and Deremer [Fleming et al., 1986] discuss a 'Basic Parameter Model' which occupies a position of intermediate generality, between the Marshall–Olkin and the binomial failure rate models. In the same article a 'multiple greek letter model' is similar to the beta-factor model, except that failures to start and failures to run are considered separately. An 'alpha factor model' recommended by the International

Agency for Atomic Energy is similar to the binomial failure rate model, but expresses the probability of k components failing dependently as a proportion of the total (independent and dependent) single component failure probability. A Markov model for common cause failure is introduced in [Kumamoto and Henley, 1996], in which a common cause state is modeled as a 'failure-repair process' with a constant occurrence rate and a constant disappearance rate. When the common cause state exists, common-mode failures occur simultaneously.

All of these models are intended to apply to more or less identical components in some redundant configuration. The common causes listed in Table 8.1 will also surely apply to dissimilar components. Effects of such contributions can only be assessed if an elaborate program of incident reporting is in place. Analyzes of failure incidents suggests that the dependency on dissimilar components is not negligible ([Ballard, 1985], [SRD, 1983], [Minarick and Kukielka, 1982], and later reports from the accident sequence precursor program).

8.12 Exercises

8.1 Give an example in which 8.3 and 8.4 hold. Show that if $0 < P(C) < 1$, then either both inequalities in 8.3 hold, or both fail.

8.2 A fair coin and a fair die have been painted strangely. The head side of the coin, and the '1' side of the die, are white, while the other sides are all black. The random variable X is defined as

$$X = \begin{cases} \text{coin} & \text{with probability } \frac{1}{2}, \\ \text{die} & \text{with probability } \frac{1}{2}. \end{cases}$$

Now, X is tossed twice. Define A and B by

$$A = [\text{outcome first toss is white}],$$
$$B = [\text{outcome second toss is white}].$$

Are A and B independent? Are A and B conditionally independent given X? Justify your answers by calculating $P(A \cap B)$, $P(A)$, $P(B)$, and $P(A \cap B | X = \text{die})$, $P(A | X = \text{die})$, $P(B | X = \text{die})$.

The outcome of a toss of X is an element of $S = \{H, T, 1, 2, 3, 4, 5, 6\}$. Determine the probability table showing each outcome $(i, j) \in S \times S$ for i the first and j the second toss of X.

9

Reliability data bases

9.1 Introduction

Reliability data is not simply 'there' waiting to be gathered. A failure rate is not an intrinsic property of a component like mass or charge. Rather, reliability parameters characterize populations that emerge from complex interactions of components, operating environments and maintenance regimes. This chapter presents mathematical tools for defining and analyzing populations from which reliability data is to be gathered. This chapter is long, the reason being that the mathematical sophistication required by a practicing risk/reliability analyst has increased significantly in the last years. Whereas in the past the choices of statistical populations and analytic methods were hard wired with the design of the data collection facility, today the analyst must play an increasingly active role in defining statistical populations relative to his/her particular needs.

The first step is to become clear about why we want reliability data. Modern reliability data banks (RDBs) are intended to serve at least three types of users: (1) the maintenance engineer interested in measuring and optimizing maintenance performance, (2) the component designer interested in optimizing component performance, and (3) the risk/reliability analyst wishing to predict reliability of complex systems in which the component operates.

To serve these users modern RDBs distinguish up to ten failure modes, often grouped into *critical failures*, *degraded failures* and *incipient failures*. Degraded and incipient failures are often associated with preventive maintenance. Whereas critical failures are of primary interest in risk and reliability calculations, a maintenance engineer is also interested in degraded and incipient failures. A component designer is interested in the particular component function that is lost and in the failure mechanisms. In addition to failure modes, modern RDBs also have data fields relating to component characteristics, component boundary, operating characteristic, and maintenance/repair policy. Since decision makers increasingly demand an assessment of uncertainty in risk and reliability calculations, modern RDBs also provide 'uncertainty bands' for failure rate estimates.

Representative modern RDBs in the public domain are IEEE-Std-500 [IEEE, 1984], OREDA [OREDA, 1984], EIREDA [EIREDA, 1991], the T-Book [ATV, 1987], and the *Guidelines* of the Center for Chemical Process Safety [CCPS, 1989]. An older tradition, represented by the Military Handbook [USAF, 1991], Green and Bourne [Green and Bourne, 1972], and going back to the Titan missile program, attempts to account for behavior under different environments via 'K-factors' which multiply a nominal failure rate. In the first RDB, the information communicated about one component was contained in one number; in current RDBs it is apt to consume two pages. Guidelines for the design of reliability databases are [Kelly and Seth, 1993], [IAEA, 1988], [Tomic, 1993], [Johanson and Fragola, 1982], and [Fragola, 1987]. Lannoy [Lannoy, 1996] and the ESReDA Working Group [ESReDA, 1999] represent current practice in Europe. A special issue of *Reliability Engineering and System Safety* (**51** no. 2) was devoted to the design of reliability data bases. For a historical perspective see [Fragola, 1996].

Modern RDBs assess competing failure rates by appealing to the theory of colored Poisson processes. This theory entails that the rate of occurrence of each of the competing risks would be unaffected by removing the others. For competing risks corresponding to critical failure and preventive maintenance, this means that the rate of occurrence of critical failures would be unaffected by stopping preventive maintenance activity altogether, a bizarre assumption to say the least. A colored Poisson process can be seen as a renewal process generated by an independent exponential competing risk model for interarrival times (see subsection 9.5.2). The general theory of competing risk, not necessarily independent and not necessarily exponential, is advanced here as the proper mathematical language for modeling reliability data.

A simple example illustrates the problem of competing failure risks. A reliability engineer buys a new car and collects reliability life data. He logs the times at which his car is taken in to repair a breakdown (failure). He maintains his car fastidiously: every Sunday morning he repairs anything which is not functioning normally even though no breakdown has occurred. These 'non-critical repairs' are logged as preventive maintenance. His neighbor, who never works on his car, asks the engineer how reliable the new car is.

If the neighbor uses the standard methods employed in RBD's, he will estimate the rate of occurrence of breakdowns (failure rate) via the 'total time

9.1 Introduction

on test statistic' (see subsection 9.6.1: the number of failures divided by the total time in operation). The total time on test statistic is the *observed failure rate*, i.e. the rate at which failures befall the fastidious engineer in spite of his preventive maintenance. It may also be something else, and herein resides much confusion. *If* the preventive maintenance were performed randomly, then the total time on test statistic would *also* estimate the rate of failures when no maintenance is performed. This latter is called the *naked failure rate*.

Now what does the neighbor want to know? He does not want to know the rate of failure if he spends every Sunday working on his car, as he will not spend every Sunday working on his car. He wants to know the rate of failures when no maintenance would be performed. Since the neighbor performs no maintenance, the observed failure rate of his car will be the naked failure rate of the reliability engineer's car. If these two rates were the same, then our reliability engineer would be a very poor reliability engineer indeed. His maintenance efforts would be totally ineffective; failures would be just as frequent if he did nothing. Preventive maintenance is supposed to *prevent* failures while losing as little useful service time as possible. This entails that preventive maintenance should be highly correlated to failure. Ideally, the car is preventively maintained at time t if and only if it *would* have failed shortly after t.

The modern RDB user is interested in not only *when* a component expires, but also *why* it expires. Such users distinguish several risks which are competing, as it were to kill the component. When the component expires, we observe which risk killed it, but we do not observe which other risks would have killed it a little later. Competing risk data resemble what statisticians call 'right censored data'. The language of censoring, however, is more appropriate to the analysis of cohort populations where we are interested in death due to a specific cause (e.g. the treated disease), and lump together as 'censoring' everything which prevents us from observing this cause (e.g. moving out of town, death by other causes, termination of observation period, etc.).

Standard RDB analysis techniques are based on the assumption that competing risks are independent. For independent competing risks, the observed rate of failure (not necessarily constant) and the naked rate of failure coincide (Section 9.5). This is a highly exceptional situation, and has masked the important distinction between observed and naked failure rates. If a reliability engineer analyzes the system that generated the data, then the observed rate of failure may be what he needs. However, if he contemplates changes in maintenance regime, changes with regard to other failure modes, or contemplates applying the results to other systems, then he requires the naked failure rate.

The body of this chapter begins with taxonomies for maintenance and failure. Next we describe the data structure for component socket histories and operations which may be performed on such data. Standard techniques for data analyses are based on the absence of competing risks. After discussing the standard tehniques we turn to the mathematics of competing risk and develop some competing risk model. Uncertainty in failure rate estimation may be caused either by sampling fluctuations or by 'non-identifiability' of competing risks. Both sources of uncertainty must be taken into account. The final substantive section illustrates the ideas in the chapter with data from pressure relief valves from a Swedish nuclear power plant. This chapter is based on [Cooke et al., 1993], [Paulsen et al., 1996], [Cooke and Bedford, 1995], [Cooke, 1996].

9.2 Maintenance and failure taxonomies

We first distinguish types of maintenance jobs (preventive and corrective), and ways in which maintenance is scheduled (calendar based, condition based, opportunity based and emergency).

9.2.1 Maintenance taxonomy

Maintenance types:

- **Preventive:** These are jobs which do not repair a fault or failure, but form part of regular servicing.
- **Corrective:** These are jobs to repair a defect, fault or failure.

Scheduling types:

- **Calendar based:** These are planned maintenance activities, not based on observed deterioration of component, but scheduled from calendar time, or some surrogate such as cycles, cumulative load, etc. Calendar based maintenance does not repair anything, and involves preventive maintenance jobs.
- **Condition based:** The component's state is observed to deviate from that which the manufacturer intended, though the component's functionality (as required by the system in which it serves) is still maintained. Hence it is possible to leave the component in socket until a suitable maintenance opportunity arises. Condition based maintenance is therefore planned and

Table 9.1. *Maintenance jobs and schedules*

Job	Schedule			
	Planned			Unplanned
	Calendar based	Condition based	Opportunity based	Emergency
Preventive	X		X	
Corrective		X	X	X

corrective. It is typically triggered by indicators of degradation such as leaking, dirty oil, abnormal power consumption, vibration, noise, heat, etc.
- **Opportunity based:** Maintenance is undertaken when a suitable opportunity presents itself. If the system is shut down for overhaul of one component, other components may be maintained as well. The decision to maintain a component at an opportunity may or may not be triggered by the condition of the component.[1]
- **Emergency:** These are actions to repair a component which is in a state that disables the system. The system cannot function until the repair is carried out, and the repair activity was not planned beforehand. Repair actions usually start almost immediately after component failure. Postponing repair is not an option. Maintenance crews typically would like to prevent emergency maintenance from occurring.

The way maintenance is scheduled may influence the type of maintenance which is performed. Although usage is not standardized in this regard, we believe there is some currency in using these concepts in such a way that Table 9.1 emerges.

9.2.2 Failure taxonomy

We first distinguish types of failure:

- **Degraded failure:** A component is not in the state which the manufacturer intended for performing its function, but the system function is still being fulfilled. The component is in a **non-critically degraded state**.
- **Critical failure:** The component is unable to perform its function due to **critical degradation** of component state. Critically degraded components

---
[1] There is considerable overlap between the notions of condition-based and opportunity-based maintenance. In this chapter we will not consider opportunity based maintenance further. It is included here for the sake of completeness.

Table 9.2. *Maintenance and failure type*

	Schedule			
	Planned			Unplanned
Job	Calendar based	Condition based	Opportunity based	Emergency
Preventive	no repair		no repair	
Corrective		repair degraded failures	repair degraded failures	repair critical failures

are always repaired or renewed, and the repair typically begins as soon as possible. Critical failure may be total or non-total.

- **Entrained functional unavailability:** The functionality of the component is lost due to the upstream failure of some supporting system. The component need not be repaired.

- **Total failure:** The component not only fails to meet specifications, but fails to meet specifications to any degree. Non-total critical failure occurs when the component fails to meet specifications but still has some residual functionality, as when a pump cannot deliver the required flow rate but can still deliver some flow, or a valve does not close in the required time, but still closes.

The terms 'functional and non-functional' are used to describe failures in a way which is complementary to the way they are used to describe component states. Thus a 'functional failure' arises when a component is in a non-functional state.

The previous table of maintenance can now be supplemented to include the types of repairs typically performed, as shown in Table 9.2. There will of course be exceptions. Some components may be allowed to remain in-line even though their functionality is totally lost. In this case the repair of a critical failure might be planned.

Figure 9.1 displays the various maintenance and failure concepts in a plot of state against time.

9.2.3 Operating modes; failure causes; failure mechanisms and failure modes

It is common nowadays to distinguish three *operating modes:*

9.2 Maintenance and failure taxonomies

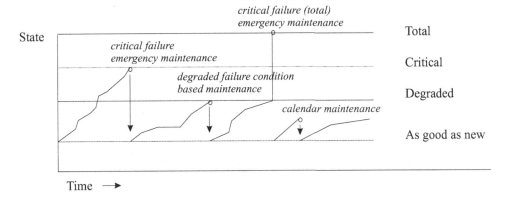

Fig. 9.1. Maintenance and failure in time

- **Continuous operation:** the component operates as long as the system operates.
- **Standby:** the component is dormant unless required to function on demand.
- **Alternating (or intermittent):** the component rotates sojourns of continuous operation with one or more spares.

Components in the alternating mode are planned to relieve each other according to a regular schedule, whereas components on standby are typically brought into service when the front line component is failed or in service.

The following terms are encountered in the literature. Though their usages are not standard, we suggest the following definitions:

Failure causes give the reasons why a component fails. RDBs often list only local causes, that is causes within the component itself (vibration, leak, crack, etc.). However, these local causes may themselves be caused by factors outside the frame of the RDB, as when a component fails due to over-stress caused by other failures upstream. These are sometimes called *root causes*. If root cause analysis has been performed, the results would typically enter an RDB in the form of free text. *Failure mechanisms* describe the actual physical processes leading to a failure. Thus, the cause of failure might be 'leak in the oil lines' and the mechanism of failure might be jamming of piston. The distinction of cause and mechanism is frequently discretionary.

Failures are sometimes distinguished according to whether they are *time related* or *demand-related*. This distinction usually coincides with operating mode. Thus, a failure occurring when the component is called into service from the standby mode is classified as demand-related. Failures occurring while the component is in continuous operation are classified as time related.

160   9 Reliability data bases

Fig. 9.2. Hierarchical categories

Table 9.3. *Component socket time histories, 314 pressure relief valves*

PLANT		Component socket		Failure fields					Repair fields				
					FAILURE DETECTION DATE	FAILURE TIME	CAUSE OF FAILURE	TYPE OF FAILURE	ACTION TAKEN	START IRREGULAR OPER.	START REPAIR	REPAIR AVAILABLE	TEAM	MAN HOURS
B	1	314 V001	M/I-UNIT	B	6/28/76	L	Z	Z	7/23/76	7/23/76	7/23/76	1	1
B	1	314 V001	SENSOR	B	10/6/77	L	N	B	10/6/77	10/6/77	10/6/77	3	9
B	1	314 V001	VALVE	C	9/17/80	K	F	C	9/17/80	9/18/80	9/18/80	2	18
B	1	314 V001	VALVE	C	8/1/82	L	E	C	8/2/82	8/2/82	8/2/82	2	8
B	1	314 V001	VALVE	C	7/11/83	L	U	C	7/12/83	7/12/83	7/22/83	2	24
B	1	314 V001	SENSOR	B	10/7/85	F	Y	B	7/31/86	7/31/86	7/31/86	1	1
B	1	314 V001	SENSOR	D	8/18/86	E	P	Z	7/20/87	7/20/87	7/20/87	2	4
B	1	314 V001	VALVE	E	9/14/92	K	F	C	9/14/92	9/14/92	9/15/92	2	8
B	1	314 V001	SENSOR	D	7/3/93	K	J	B	7/10/93	7/10/93	7/10/93	1	1
B	1	314 V001	VALVE	B	9/6/93	F	P	C	9/6/93	9/6/93	9/6/93	2	2
B	1	314 V001	VALVE	B	5/13/95	K	E	C	5/20/95	5/20/95	5/20/95	1	2
B	1	314 V002	M/I-UNIT	B	10/6/77	L	N	C	10/6/77	10/6/77	10/6/77	3	9
B	1	314 V002	VALVE	D	5/8/78	B	M	C	5/8/78	5/8/78	5/8/78	2	6
B	1	314 V002	VALVE	D	5/8/78	B	M	C	5/8/78	5/8/78	5/8/78	2	2
B	1	314 V002	VALVE	E	5/8/78	L	F	C	5/9/78	5/8/78	5/9/78	1	15
B	1	314 V002	VALVE	C	9/17/80	K	F	C	9/17/80	9/18/80	9/18/80	2	18
B	1	314 V002	PIPE	B	7/7/83	L	E	C	7/25/83	7/25/83	7/25/83	2	16
B	1	314 V002	VALVE	C	7/11/83	L	U	B	7/12/83	7/12/83	7/22/83	2	24
B	1	314 V002	SENSOR	B	8/8/85	L	J	Z	8/26/85	8/26/85	8/26/85	2	2
B	1	314 V002	SENSOR	D	7/17/86	E	Y	B	7/31/86	7/31/86	7/31/86	2	2
B	1	314 V002	VALVE	C	7/19/86	L	F	C	7/19/86	7/19/86	7/24/86	2	24
B	1	314 V002	M/I-UNIT	A	11/17/92	K	N	B	11/23/92	11/23/92	11/23/92	1	1
B	1	314 V003	VALVE	C	5/8/78	L	F	C	5/9/78	5/8/78	5/9/78	1	20
B	1	314 V003	PIPE	B	9/1/79	K	E	C	9/18/79	9/18/79	9/18/79	2	3
B	1	314 V003	VALVE	C	9/17/80	K	F	C	9/17/80	9/18/80	9/18/80	2	18
B	1	314 V003	VALVE	C	7/11/83	L	U	B	7/12/83	7/12/83	7/22/83	2	24
B	1	314 V003	VALVE	C	7/19/86	L	F	C	7/19/86	7/19/86	7/24/86	2	24
B	1	314 V003	VALVE	E	7/23/91	L	E	C	7/23/91	7/24/91	7/26/91	2	40
B	1	314 V003	SENSOR	D	8/27/94	L	P	H	9/12/94	9/12/94	9/12/94	1	1

Since the same component type may operate continuously or in standby, this component may have both demand-related and time related failures.

Failure modes describe the way in which a component fails, usually from a functional or sub-functional point of view (e.g. 'fails to open', 'fails to close'). It becomes necessary to distinguish failure modes when the consequences of a failure depend on the way in which a component fails.

9.3 Data structure

The data consists of component time histories, organized in hierarchical categories and incident data shown in Figure 9.2.

The subcomponent is the smallest maintainable part. When a subcomponent is in repair, the component is not available. The system may or may not be available. An example from the Swedish TUD RDB is shown in Table 9.3.

9.3.1 Operations on data

Mathematically, a time history is a colored point process. The 'points' are the event times logged in the data base. Each 'point' bears a number of properties describing the event. A grouping of properties into mutually exclusive and exhaustive sets is called a coloring. The notion of a colored point process resembles the 'marked point process' as found in the literature. However, in a marked point process the distribution over the possible marks, given that an event has occurred, is independent of the history of the process.

Competing risk processes produce colored point processes. After each event, risks (or colors) $1, \ldots, k$ are competing to cause the next time event. When the next time event occurs, we observe which risk caused it.

Three main types of operation may be performed on the data:

Superposition Time histories having the same begin and end points may be superposed. The set of event times of the superposition is the union of the event times of superposed processes. In general, superposing scrambles any structure in the individual processes. A notable exception occurs in the case of superposing independent (homogeneous or non-homogeneous) Poisson processes. Since the intensity function of these processes does not depend on the history of the process, the intensity function of the superposition is the sum of the intensity functions of the superposed processes, and is independent of the history of the process.

Pooling Pooling is applied to data generally. The pooled data is considered as multiple realizations of the same random variable or stochastic process. When time histories are pooled, these are considered as realizations of the same (colored) point process. The process determines the '$M(t)$' function, were $M(t)$ is the expected number of events up to time t. The expected number of events up to t can be estimated from the pooled data. It is not necessary that the individual processes all have the same calendar begin and end points.

In case inter-event times are pooled, these are considered as realizations of the same distribution. Typically, but not always, inter-event times are pooled when the individual processes are believed to be renewal processes.

When the pooled objects are realizations of different distributions or processes, then their pooled combination realizes a mixture of the distributions or processes. Any information contained in the ordering of the time histories or inter-event times is lost when these are pooled. When pooling inter-event times from different distributions, collected over the same calendar period, the distribution with the smallest expected inter-event time will contribute the greatest number of inter-event times. Suppose we pool processes 1 and

2 with expected inter-event times E_1 and E_2, over the same (large) calendar period. The proportion of inter-event times coming from process 1 will be approximately $(1/E_1)/(1/E_1 + 1/E_2)$.

Coloring/uncoloring A coloring of time events is a partition of the event properties into mutually exclusive and exhaustive classes (colors). A colored process can be uncolored by simply giving each event the same color.

In addition to these three main operations, two other operations may be defined:

Splicing Splicing involves the removal of intervals of time and adjoining the unremoved parts. It may be applied, for example, to remove functional unavailabilities if these are not considered relevant to the aging of a component.

Concatenation Concatenation applies to time histories. Whereas in pooling, these are considered as multiple realizations of one process, concatenation appends one history to the other, so as to create one new time history. Concatenation will not be applied further; it is mentioned only for the sake of completeness.

To illustrate superposition and pooling; suppose that outages of the steam pressure relief valve V001 of the 314 system are due to either the subcomponent 'sensor' or the subcomponent 'valve'. Then the outage times of the valve and sensor may be superposed to form the outage times of the V001 pressure relief valve. If the sensor may be regarded as renewed whenever the valve is repaired and vice versa, then the superposition *will* be a renewal process, and the individual valve and sensor histories *will not* be renewal processes. On the other hand, if the sensor is not renewed when the valve is repaired, then the superposition, i.e. the V001 valve, *will not* be a renewal process. The inter-outage times of the V001 and V002 valves in the 314 system may be pooled to form the set of inter-outage times for 314 valves. The outage events may be colored according to the subcomponent which failed the component, but they may also be colored according to method of detection or effect or repair type, or length of repair, etc.

In general, superposition is performed in order to obtain a renewal process. Discussions with the maintenance crew must determine at what hierarchical category they aim for as good as new. With regard to the 314 isolation valves, the maintenance crew strives to return the valves to service as good as new, in other words, if the valve fails, the sensor is checked and if necessary restored as well before returning to service. Thus, all subcomponents should be superposed. Pooling is performed on identical independent renewal processes in order to obtain better statistical estimates. In some cases non-identical

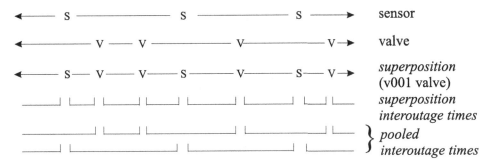

Fig. 9.3. Superposed and pooled time histories

renewal processes may be pooled. Thus, some maintenance indicators can be defined from pooling of non-identical renewal processes.

In general, the rule of thumb is

Superpose until 'as good as new', and then pool similar histories.

Figure 9.3 shows the operations of pooling and superposing. For a more detailed discussion of operations on component socket time histories and relevant statistical tests to support these operations, see [Paulsen *et al.*, 1996]. Statistical reference books are [Cox and Lewis, 1966], [Hoyland and Rausand, 1994] and [Aven and Jensen, 1999].

9.4 Data analysis without competing risks

We first consider the analysis of data not subject to competing risk. A component enters service and remains in service until it fails. There is no other cause of termination of a service sojourn. We then turn to competing risk analysis. After introducing the mathematical background for competing risk, we develop some models for analyzing dependent and independent competing risks.

9.4.1 Demand related failures: non-degradable components

When components subject to demand do not degrade while on standby, the statistical analysis of the failure data is quite simple. In this case we can model each demand as a flip with a coin; we assume for each component type that the probabilities of failure per demand are independent and identical. If the probability of failure per demand is p, then the probability of observing r failures in n demands is given by the binomial distribution:

$$P(r \text{ failures in } n \text{ trials } |p) = \binom{n}{r} p^r (1-p)^{(n-r)}.$$

As shown in 4.3.1.1 the maximum likelihood estimator (MLE) of p, given r failures in n trials, is r/n.

For r failures in n observations, the 90% statistical confidence bounds are given by two values p_l and p_u such that (see 4.3.3.1)

p_l = largest p such that : $P(r \text{ or more failures in } n \text{ trials}|p) \leq 0.05$,
p_u = least p suchthat : $P(r \text{ or fewer failures in } n \text{ trials}|p) \leq 0.05$.

Approximation Tables for the binomial distribution for n up to 20 are readily available. For larger values of n we can use the fact that r is approximately normally distributed with mean np and variance $np(1-p)$, provided that $np > 5$ (for $p < \frac{1}{2}$). In risk and reliability applications n is typically large and p typically very small, and in this case the Poisson approximation to the binomial is most useful:

$$\lim_{n \to \infty, np \to \mu} \binom{n}{r} p^r (1-p)^{(n-r)} = \frac{e^{-\mu} \mu^r}{r!}.$$

9.4.2 Demand related failures: degradable components

A component called to function on demand is in the standby mode. Some failure mechanisms, notably those associated with wear, are disengaged during standby, while others, notably those associated with corrosion, oxidation or embrittlement, continue during standby. Therefore components on standby are usually subjected to maintenance. We mention the two most popular maintenance models and indicate how they affect the probability of failure on demand.

For components on standby, failure on demand at time t is modeled as the *unavailability at time t*, that is, the probability that the component is in a failed state at t. When t is far removed from the component's birthday, it is customary to use the *equilibrium* (or *steady-state*) *unavailability*, which is the limit time-average unavailability.

Test-and-replace According to this maintenance regime, components are tested at a regular interval I, and if found in a failed state they are replaced immediately by new components of the same type. If λ is the standby failure rate, assumed constant, the failure on demand for uniformly distributed demand is (from subsection 3.3.1)

$$1 - (1 - e^{-\lambda I})/\lambda I \approx \lambda I/2 \quad \text{for } \lambda I < 0.1.$$

Fail-and-fix Suppose an exponential component with failure rate λ is allowed to fail and then taken off-line and repaired. During repair the component is unavailable and would lead to a demand failure if demanded during outage. Suppose that the repair process is exponential with repair rate μ, then as shown in subsection 3.3.2 the equilibrium unavailablity is

$$\frac{\lambda}{\lambda + \mu}.$$

This formula remains valid when the variables are not exponential, and when the rates above are interpreted as the inverses of expected lifetimes.

From these results it is apparent that identical components subject to standby degradation will not yield the same demand probabilities when they are maintained and repaired in different ways. For such components, the user cannot interpret a failure probability on demand, unless he is told the testing interval (when using test-and-replace) or the repair rate (when using fail-and-fix). Additional complications arise if a hybrid maintenance policy is pursued; components tested regularly and pulled off-line for repairs.

9.4.3 Time related failures; no competing risks

The statistical models underlying the treatment of time related failure data are less simple, and more problematic, than those for demand related failures. Let $X_1, \ldots, X_n$ be independent exponential variables with parameter λ, then the cumulative distribution function of each X_i is $F(x) = 1 - e^{-\lambda x}$ and the MLE for λ given observations $x_1, \ldots, x_n$ is (see 4.3.3.2)

$$\hat{\lambda} = \frac{n}{\sum_{i=1}^{n} x_i}.$$

To find statistical confidence bands, recall that $\sum X_i$ follows a gamma distribution with shape n and scale λ. Substituting $(z/2) = \lambda \sum x_i$ we find that the density of z is

$$f(z) = \frac{2^{-n} z^{n-1} e^{-z/2}}{\Gamma(n)},$$

which is the chi-squared density with $2n$ degrees of freedom. The 90% statistical confidence band is given by two values λ_l and λ_u:

$$\lambda_l = \text{largest } \lambda \text{ such that } P \text{ (less than } \sum x_i | \lambda) \leq 5\%,$$

$$\lambda_u = \text{least } \lambda \text{ such that } P \text{ (more than } \sum x_i | \lambda) \leq 5.$$

Instead of n identical components, suppose we have one exponential component with parameter λ which fails and is repaired to as good as new, or replaced with a new, identical component. The number of failures as a function of time is then a Poisson process with parameter λ. The expected number of failures in time t is λt (as discussed in Chapter 3). Given n failures in time t, the MLE of λ is n/t. The mean time between failures (MTBF) is just the mean of the exponential distribution, $1/\lambda$.

9.5 Competing risk concepts and methods

If we wish to assess the failure rates corresponding to different competing ways of ending a service sojourn, we are in a situation of *competing risk*. Different 'ways of dying' are competing as it were to terminate the component's service sojourn. As we have seen in Section 9.3, a failure event may be described by assigning values, or colors, to a number of failure fields. Figure 9.6 shows possible fields for the 314 valve.

In each failure event exactly one value in each field is realized. A coloring is simply a partition of the set of failure events into disjoint subsets. In practice, we will choose one field for coloring, and then assign different colors to the values in that field. Thus we may color the field 'method of detection' by grouping failure events into 'alarm or unintended discovery' and 'operator or test or revision'. In this section we introduce the basic mathematical formalism for describing competing risks. The goal is to extract information about the failure rates of competing failure modes, sometimes called the naked failure rates. This section reviews the theory of independent competing risks [Cox, 1959, Gail, 1975, David and Moeschberger, 1975, Crowder, 1991, Crowder, 1994].

Assume we have k competing risks, $X_1, \ldots, X_k$, and denote by $\wedge X_i$ the minimum of $X_1, \ldots, X_k$. In a competing risk context, we observe the minimum of the X_i, and observe which it is. In other words, we observe $Y = (\wedge X_i, 1_{\wedge X_i = X_j}, j = 1, \ldots, k)$. The failure rate of X_i, h_i, is the failure rate which *would* be observed if we could observe X_i without the observation being censored by earlier occurrences of X_j, $j \neq i$. We say that risk X_j is *cured*, if it is eliminated without disturbing the distributions of the other risks. *Mathematically*, curing risk j corresponds to observing $Y_{(j)} = (\wedge_{i \neq j} X_i, 1_{\wedge X_i = X_h}, h = 1, \ldots, k, h \neq j)$, where the distribution of $Y_{(j)}$ is gotten from that of Y by integrating out over variable X_j. If we could cure all risks other than i, then we should observe the failure rate h_i.

9.5 Competing risk concepts and methods

When the competing risks are not cured, we observe a different rate of failure for X_i. The observed failure rate for X_i is defined as

$$\begin{aligned} \operatorname{obr}_i(t) &= \lim_{\Delta \to 0} P(\wedge X_j = X_i, X_i \in (t, t+\Delta)| \wedge X_j > t)/\Delta \\ &= \frac{-dR_i^*(t)/dt}{\sum_{j \le k} R_j^*(t)}. \end{aligned}$$

If the competing risks are independent, then $P(\wedge X_i > t) = \prod R_i(t) = \sum R_i * (t)$. This is the basis of the identity of observed and naked failure rates for independent competing risks:

Theorem 9.1 *[Cooke, 1996] If competing risks $X_1, \ldots, X_k$ are independent, with differentiable survival functions, then $h_i(t) = \operatorname{obr}_i(t)$, $i = 1, \ldots, k$.*

The proof is a straightforward calculation using techniques introduced in the following section (see Exercise 5).

If we assume that the competing risks are independent, then the problem of estimating the naked failure rate simply does not arise, as the naked and observed failure rates coincide.

In many cases, we are primarily interested in one risk, so we may as well put $X = X_1$ and $Z = \min\{X_2, \ldots, X_k\}$. The observed variable is

$$Y = (\min\{X, Z\}, 1_{X<Z}).$$

In other words, we observe the lesser of X and Z, and observe which it is. In such contexts we often speak of X as the life variable and Z as the censoring variable.

We write '$X \perp Z$' to denote that X and Z are independent. Let Y_i, $i = 1, 2, \ldots$, be independent copies of Y. The process $\mathbf{Y} = Y_1, Y_2, \ldots$ is called the competing risk renewal process associated with Y. Suppose $X \perp Z$; re-arranging if necessary, we write the first N observations as

$$(y_1, \ldots, y_N) = (x_1, \ldots, x_n, z_1, \ldots z_m)$$

with $n + m = N$ and where $x_1, \ldots, x_n$ are inter-arrival times at which the variable X was observed, and $z_1, \ldots, z_m$ the inter-arrival times at which the variable Z was observed. If F_Z and f_Z denote the cumulative distribution and density of Z respectively, and F_Z and f_Z are defined similarly, the probability of $(y_1, \ldots, y_N)$ is

$$P(y_1, \ldots, y_N) = \prod_{i=1}^{n} f_X(x_i)(1 - F_Z(x_i)) \cdot \prod_{j=1}^{m} (1 - F_X(z_j)) f_Z(z_j). \tag{9.1}$$

If the distribution F_X is described by parameters θ which do not occur in

F_Z, then it is easy to show that the MLE of θ does not depend on F_Z (see Exercise 4).

9.5.1 Subsurvivor functions and identifiability

According to Theorem 9.1 above we could identify the naked failure rate from the competing risk data, assuming independence. This means that we can identify the distributions of independent competing risks from indefinitely repeated independent observations.

This is an example of a general result from the theory of competing risk. The notions underlying this result are of interest in themselves. If F_X is the cumulative distribution function of X, then $R_X = 1 - F_X$ is called the *survival* or *reliability function* of X. If F_X has a density (unless stated otherwise, we assume throughout that all distributions have densities) then the failure rate $h_X(t)$ of X is

$$h_X(t) = f_X(t)/R_X(t) = -(dR_X/dt)/R_X(t).$$

Since

$$d[\log(R_X)] = dR_X/R_X,$$

we have

$$R_X(t) = \exp\left\{-\int_0^t h_X(s)\,ds\right\}.$$

The function $R_X^*(t) = P(X > t, X < Z)$ is called the *subsurvival function* of X [Peterson, 1977]. Note that R_X^* depends on Z, though this fact is suppressed in the notation. If R_X^* is continuous at 0 then $R_X^*(0) = P(X < Z)$. If X and Z are independent we have

$$R_X*(t) + R_Z*(t) = P(X > t, Z > t) = P(X > t)P(Z > t) = R_X(t)R_Z(t).$$

Definition 9.1 *Real functions R_1^* and R_2^* on $[0,\infty)$ form a (continuous) subsurvival pair if*

(i) R_1^* and R_2^* are non-negative and non-increasing (continuous, continuous from the right at zero), $R_1^*(0) < 1$, $R_2^*(0) < 1$,
(ii) $\lim_{t\to\infty} R_1^*(t) = 0$, $\lim_{t\to\infty} R_2^*(t) = 0$,
(iii) $R_1^*(0) = 1 - R_2^*(0)$.

If $\lim_{t\to\infty} R_X(t) > 0$, then we say that X has an atom at infinity. This means that there is finite probability that a component with life distribution X never expires. Atoms at infinity are invoked in Theorem 9.2 below.

Clearly R_X^* and R_Z^* form a subsurvival pair. If we have data for Y then we can calculate the *empirical subsurvival functions*; these contain all the information in the data, that is, any parameter that can be estimated from the data can be written as a function of the empirical subsurvival functions. The *conditional subsurvival function* is the subsurvival function, conditioned on the event that the failure mode in question is manifested. Assuming continuity of R_X^* and R_Z^* at zero:

$$P(X > t, X < Z | X < Z) = R_X^*(t)/R_X^*(0),$$
$$P(Z > t, Z < X | Z < X) = R_Z^*(t)/R_Z^*(0).$$

Closely related to the notion of subsurvival functions is the probability of censoring beyond time t,

$$\Phi(t) := P(Z < X | Z \wedge X > t) = \frac{R_Z^*(t)}{R_Z^*(t) + R_X^*(t)}.$$

Note that for continuous subsurvival functions $\Phi(0) = P(Z < X)$.

The *subdistribution functions* for X and Z are

$$F_X^*(t) = P(X \leq t, X \leq Z) = R_X^*(0) - R_X^*(t),$$

and

$$F_Z^*(t) = P(Z \leq t, Z \leq X) = R_Z^*(0) - R_Z^*(t).$$

Peterson [Peterson, 1976] derived bounds on the survival function R_X by noting that

$$P(X \leq t, X \leq Z) \leq P(X \leq t) \leq P(X \wedge Z \leq t);$$

which entails

$$1 - F_X^*(t) \geq R_X(t) \geq R_X^*(t) + R_Z^*(t).$$

Note that the quantities on the left and right hand sides are observable.

We can now state the main result for independent competing risks. This theorem generalizes for n competing modes.

Theorem 9.2 *[Tsiatis, 1975, Peterson, 1977, van der Weide and Bedford, 1998]*
(i) Let X and Z be independent life variables, with F_X and F_Z continuous. Let X' and Z' be independent life variables such that $R_X^* = R_{X'}^*$ and $R_X^* = R_{Z'}^*$; then $F_X = F_{X'}$ and $F_Z = F_{Z'}$.
(ii) If R_1^* and R_2^* are a subsurvival pair and are continuous, then there exist independent life variables X and Z such that $R_X^* = R_1^*$ and $R_X^* = R_2^*$, and at most one of X, Z has an atom at infinity.

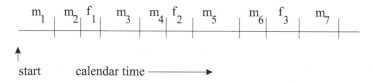

Fig. 9.4. Calendar time picture of censored data

By observing independent copies of $Y = (\min\{X,Z\}, 1_{X<Z})$ we can estimate the subsurvival functions. *Assuming* independence of X and Z we can determine uniquely the survival functions of X and Z. In this case the distributions of X and Z are said to be *identifiable* from the censored data. Of course, X and Z may not actually be independent, and in this case the survival functions gotten via Theorem 9.2 would NOT be correct. Moreover, the independence assumption can never be tested by the censored observations since, according to Theorem 9.2 (ii), any censored observations can be explained by an independent model.

9.5.2 Colored Poisson representation of competing risks

Some RDBs [EIREDA, 1991, Pörn, 1990] suggest a colored Poisson representation of censoring. This subsection shows that the colored Poisson representation is equivalent with independent exponentially distributed competing risks. It also affords additional insight into the model's assumptions. For purposes of visualization we change notation. The failure process is called F, to be colored fuchsia, and the maintenance (censoring) process is called M, to be colored magenta.

Let us first consider the process $\mathbf{Y} = Y_1, Y_2, \ldots$, where Y_i are independent copies of $Y = (\min\{F, M\}, 1_{F<M})$, and imagine data generated by instantly replenishing a component socket with as-good-as-new components whenever a component exits service. The components exit service either because of failure (F) or because of preventive maintenance (M). If we plot the set of observed inter-arrival times $\{f_1, f_2, \ldots; m_1, m_2, \ldots\}$ in calendar time, then the picture shown in Figure 9.4 emerges.

The process $\min\{F, M\}$ is just the process obtained by removing all labels. Think of $\min\{F, M\}$ as the 'uncolored' process, and think of the Ms as colored magenta and the Fs as colored fuchsia. The 'coloring theorem' for Poisson processes says (see Section 8.7 and [Kingman, 1993]):

Theorem 9.3 *If the uncolored process is a Poisson process with intensity v, and if the coloring of a point is determined by the outcome of an independent* *coin toss, heads for magenta, tails for fuchsia with* $P(\text{heads}) = p$, *then the magenta points are a Poisson process with intensity* vp, *the fuchsia points are a Poisson process with intensity* $v(1-p)$, *and the magenta and fuchsia processes are independent.*

A colored Poisson process starting at $t = 0$ may therefore be represented as $\mathcal{Y} = \{M_1, M_2, \ldots; \mu; F_1, F_2, \ldots; \phi\}$ where $\{M_i\}$ and $\{F_i\}$ are the inter-arrival times of two independent Poisson processes, starting at $t = 0$, with intensities μ and ϕ respectively; the uncolored (Poisson) process gotten by uncoloring the points has intensity $v = \mu + \phi$. The interleaving of the two processes is uniquely determined in every realization $\{f_1, f_2, \ldots; m_1, m_2, \ldots\}$: the jth failure occurs between the kth and the $(k+1)$th preventive maintenances if

$$\sum_{i \leq k} m_i < \sum_{i \leq j} f_i < \sum_{i \leq k+1} m_i.$$

The process $\mathcal{Y}$ may be associated with a subsurvival pair as follows: Letting $R_U(t) = \exp(-(\mu + \phi)t)$ denote the survival function for the uncolored inter-arrival times, $R_M^*(t) = R_U(t)\mu/(\mu+\phi)$, $R_F^*(t) = R_U(t)\phi/(\mu+\phi)$. Now, $\mu/(\mu+\phi)$ is the probability of magenta, $\phi/(\mu+\phi)$ is the probability of fuchsia.

Theorem 9.4 *[Cooke, 1996]* (i) *Let* $\mathcal{Y} = M_1, M_2, \ldots; \mu; F_1, F_2, \ldots; \phi$ *be a colored Poisson process. Then there is a unique independent competing risk process,* $\mathbf{Y} = (\min\{F, M\}, 1_{F<M})_i, i = 1, 2, \ldots$, *associated with* $\mathcal{Y}$. *Moreover, M and F are exponentially distributed with survival functions* $\exp(-\mu t)$ *and* $\exp(-\phi t)$ *respectively.*
(ii) *Let* Y_i *be independent copies of* $(\min\{F, M\}, 1_{F<M})$ $(i = 1, 2, \ldots)$, *where M and F are independent and exponentially distributed with survival functions* $\exp(-\mu t)$ *and* $\exp(-\phi t)$ *respectively, and let* $\mathbf{Y} = Y_1, Y_2, \ldots$ *be the competing risk renewal process associated with* Y. *Then* $\mathbf{Y}$ *is a colored Poisson process with intensities* μ *and* ϕ *for the M and F processes respectively.*

Proof (i) The subsurvival functions associated with $\mathcal{Y}$ determine a unique independent competing risk process by Theorem 9.2 (see below); filling the subsurvival functions into Theorem 9.5 concludes the proof.
(ii) Since $\mathbf{Y}$ is a renewal process, it suffices to verify that the fuchsia and magenta inter-arrival times have survival functions $\exp(-\phi t)$ and $\exp(-\mu t)$ respectively. This calculation uses the remark following the proof of Theorem 9.5, the fact that $F \wedge M$ is exponential with rate $\mu + \phi$, and the fact that the sum of k independent exponentials with the same failure rate follows a gamma distribution with shape k. We may start the process at $t = 0 = Y_0$,

numbering as in Figure 9.5, and putting $\gamma = \mu + \phi$. Let T^* be the time at which the first F is observed, then

$$P(T^* > t) = \sum_{k=0}^{\infty} P \left\{ (F_i \wedge M_i) = M_i, i = 1, \ldots, k; \right.$$
$$\left. \sum_{i=1}^{k} M_i < t < \sum_{i=1}^{k} M_i + (F_{k+1} \wedge M_{k+1}) \right\}$$
$$= e^{-\gamma t} + \sum_{k=1}^{\infty} \left(\frac{\mu}{\gamma} \right)^k \int_0^t \frac{\gamma^k z^{k-1} e^{-\gamma z} e^{-\gamma(t-z)}}{\Gamma(k)} \, dz$$
$$= e^{-\phi t} \left\{ \sum_{k=0}^{\infty} e^{-\mu t} (\mu t)^k / k! \right\} = e^{-\phi t}.$$

$\square$

Consider the distance between an uncolored point π_i, and the previous uncolored point π_{i-1}. This distance follows an exponential distribution with failure rate $\mu+\nu$, since $\mu+\nu$ is the failure rate of $\min\{F, M\}$. If π_i is colored by flipping an independent coin, this does not affect the distribution of $\pi_i - \pi_{i-1}$. Hence, the distance between a maintenance point and its nearest predecessor has the same distribution as the distance between a failure point and its nearest predecessor. In other words, given that a service sojourn terminates in preventive maintenance, the distribution for the length of that sojourn is the same as the distribution for the length of sojourn given termination in failure. This is illustrated in a different way in Remark (i) after the proof of Theorem 9.5.

9.6 Competing risk models
9.6.1 Independent exponential competing risk

We asume that X and Z are independent and exponentially distributed with failure rates λ and γ. Suppose we observe N independent copies of $Y = (\min\{X, Z\}, 1_{X<Z})$. Arrange the observations so as to distinguish the Xs and Zs, as follows:

$$y_1, \ldots, y_N = x_1, \ldots, x_n, z_1, \ldots, z_m.$$

Substituting the exponential distribution and density functions into Equation 9.1 and setting the derivative of the logarithm equal to zero, one finds that

9.6 Competing risk models

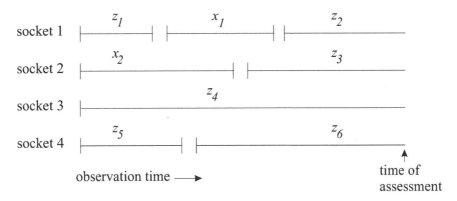

Fig. 9.5. Censored failure data from four plants

the maximum likelihood estimator is (Exercise 4)

$$\hat{\lambda} = \frac{n}{\sum_{i=1}^{n} x_i + \sum_{j=1}^{m} z_j}. \tag{9.2}$$

The quantity on the right hand side of Equation 9.2 is called the *total time on test statistic (TTT)*. Note that this result does not depend on the distribution of Z, so long as it is independent of X.

Neglecting the censored lifetimes (the z_is) in a reliability context could produce a significant overestimate for λ. For reliable components, the sum of the censored lifetimes is typically much larger than the sum of the observed failure times, and the overestimation can be large. Suppose our data comes from different sockets, and we estimate λ as $1/MTTF$ (mean time to failure), then Figure 9.5 shows how wrong this estimate can be.

Here the x_is are the failure times and the z_is are the censoring times, and $1/MTTF$ would be estimated as $2/(x_1 + x_2)$; whereas the MLE of λ is $2/(\sum x_i + \sum z_i)$.

Statistical confidence intervals cannot be readily computed for the MLE of λ under random right censoring, as the sampling distribution of the TTT is not available. We can however appeal to the asymptotic normality of the MLE to derive bounds, but we shall not pursue this here (for a discussion see [Bain, 1978] or [Nelson, 1990]).

Although competing risk data can always be explained by an independent model (Theorem 9.2) this does not mean that any censored observations can be explained by a model with *exponential life variables*. Rather, we can derive a very sharp criterion for exponentiality in terms of the subsurvival functions [Cooke, 1993]:

Theorem 9.5 *Let X and Z be independent life variables, then any two of the following imply the others:*

(i) $R_X(t) = \exp(-\lambda t)$,
(ii) $R_Z(t) = \exp(-\gamma t)$,
(iii) $R_X^*(t) = \lambda \exp(-(\lambda + \gamma)t)/(\lambda + \gamma)$,
(iv) $R_Z^*(t) = \gamma \exp(-(\lambda + \gamma)t)/(\lambda + \gamma)$

Proof Using independence, we have $R_X R_Z = R_X^* + R_Z^*$, and $dR_X^* = R_Z dR_X$; thus

$$R_X(t) = \exp\left\{-\int_0^t \frac{\frac{dR_X^*(u)}{du}}{R_X^*(u) + R_Z^*(u)} du\right\} \qquad (9.3)$$

and similarly for R_Z. We sketch the proof (see Exercise 6).

That (i) and (ii) imply (iii) and that (i) and (iii) imply (ii) is shown by substituting the information into Equation 9.3. To show that (i) and (ii) imply (iv), put $f_Z(u) = -dR_Z(u)/du$, then

$$R_Z^*(t) = \int_0^t e^{-\lambda u} f_Z(u)\, du = \frac{-\gamma e^{-(\lambda+\gamma)t} \lambda + \gamma}{\lambda + \gamma}. \qquad (9.4)$$

Take derivatives of both sides. The other statements are straightforward computations. □

Remarks (*i*) If X and Z are independent exponential life variables with failure rates λ and γ, then

$$R_X^*(t)/R_X^*(0) = R_Z^*(t)/R_Z^*(0) = \exp(-(\lambda + \gamma)t)$$

and

$$\Phi(t) = \frac{\gamma}{\lambda + \gamma}.$$

In other words, the conditional subsurvival functions are equal and exponential, and the conditional probability of $Z < X$ given survival up to t (the function Φ), is constant.

(*ii*) From the above theorems, we see that for independent exponential competing risks, $R_X^*(t) = P(X > t, X < Z)$ can be computed as $P(X \wedge Z > t)P(X < Z)$.

(*iii*) The first equality in Equation 9.4 is true generally, namely

$$dR_Z^*(t)/dt = R_X(t)dR_Z(t)/dt.$$

Assuming independence, standard methods may be used to test the hypothesis that both empirical conditional subsurvival functions follow the same exponential distribution.

Theorems 9.1, 9.4 and 9.5 yield the following characterization of independent exponential competing risks:

If competing risks are independent and exponential, then the rate of occurrence of each risk is unaffected by curing other risks.

In particular, if preventive maintenance and critical failure are independent exponential competing risks, then abandoning all preventive maintenance would not affect the rate of occurrence of critical failure.

9.6.2 Random clipping

Perhaps the simplest model of interaction between an exponential life process X and a warning process Z is obtained by assuming that X is always censored by a random amount $X - Z$. More specifically, we assume that X is exponential and for some positive random variable W independent of X, we observe $X - W$. W may be thought of as a warning which a component emits prior to expiring at time X. Of course W may be greater than X, which we interpret as censoring at birth. Let us suppose that censors at birth are simply not recorded. Suppose, in other words, that components emitting warnings at birth are simply repaired until the warning disappears, and that the false start is not recorded as an incipient failure at time 0. Indeed, this is what usually happens. We call the variable $X - W$ given $X - W > 0$ a random clipping (RC) of X. The following result entails that in this case

$$\lambda = \lim_{n \to \infty} \frac{n}{\sum_{i=1}^{n} z_i}.$$

Recall, X is exponential with parameter λ.

Theorem 9.6 *Let $W > 0$ be a random variable independent of X, and $U = X - W$. Then conditional on $U > 0$, U has the same distribution as X.*

Proof For $u \in (-\infty, \infty)$ and $v = \max\{0, -u\}$ we have

$$dF_U(u) = \int_{w > v} \lambda e^{-\lambda(u+w)} \, dF_W(w).$$

If $u > 0$, then $v = 0$ and $dF_U(u) \propto \lambda e^{-\lambda u}$. □

9.6.3 Random signs

Perhaps the simplest model for dependent competing risk is *random signs competing risk* (called *age-dependent censoring* in [Cooke, 1993]). Consider a component subject to right censoring, where X denotes the time at which a component would expire if not censored. Suppose that the event that the component expire due to competing risk Z is independent of the age X at which the component would expire, but *given* that the component expires from Z, the time at which it expires may depend on X. For example, suppose that a component emits some warning of deterioration (leakage, vibration, noise) before expiring (life variable X). If these warnings are perceived by the maintenance personnel, then the component will be preventively maintained and we observe competing risk Z. If the process of perceiving the warnings is independent of the component's time in service, then the random signs model is appropriate. This situation is captured in the following definition.

Definition 9.2 *Let X and Z be life variables with $Z = X - \xi$, where ξ is a random variable, $\xi \leq X$, $P(\xi = 0) = 0$, whose sign is independent of X. The variable $Y = (\min\{X,Z\}, 1_{X<Z})$ is called a random signs censoring of X by Z.*

Theorem 9.7 *[Cooke, 1993] Let (R_1^*, R_2^*) be a pair of continuous strictly monotonic subsurvival functions; then the following are equivalent:*

(1) *there exist random variables ξ and X, $X \perp \text{sgn}(\xi)$, such that $R_1^*(t) = P(X > t, \xi < 0)$ and $R_2^*(t) = P(X - \xi > t, \xi > 0)$.*
(2) *for all $t > 0$, $R_1^*(t)/R_1^*(0) > R_2^*(t)/R_2^*(0)$.*

Proof (1) $\Longrightarrow$ (2) Since X is independent of $\text{sgn}(\xi)$,

$$R_1^*(t) = P(X > t | \xi < 0) P(\xi < 0) = P(X > t | \xi > 0) P(\xi < 0)$$
$$= P(X > t | \xi > 0) R_1^*(0),$$

and

$$R_2^*(t) = P(X - \xi > t | \xi > 0) P(\xi > 0) = P(X - \xi > t | \xi > 0) R_2^*(0).$$

Hence for $t > 0$,

$$\frac{R_1^*(t)}{R_1^*(0)} = P(X > t | \xi > 0) > P(X - \xi > t | \xi > 0) = \frac{R_2^*(t)}{R_2^*(0)}.$$

(2) $\Longrightarrow$ (1) Let X and Z be random variables with survival functions $R_X(t) = R_1^*(t)/R_1^*(0)$ and $R_Z(t) = R_2^*(t)/R_2^*(0)$. Then $R_Z(t) < R_X(t)$ for all $t > 0$, and R_X^{-1}, R_Z^{-1} exist. Choose a random variable δ independent of X with $P(\delta = 1) = 1 - P(\delta = 0) = R_2^*(0)$, and define

$$\xi = \delta(X - R_Z^{-1}(R_X(X))) - (1 - \delta).$$

We have $\xi > 0$ if and only if $\delta = 1$, and $\xi = -1$ if and only if $\delta = 0$. Hence X is independent of sgn(ξ) and

$$P(X - \xi > t, \xi > 0) = P(X - \xi > t|\xi > 0)R_2^*(0)$$
$$= P(R_Z^{-1}(R_X(X)) > t)R_2^*(0) = P(X > R_X^{-1}(R_Z(t)))R_2^*(0)$$
$$= R_X R_X^{-1}(R_Z(t))R_2^*(0) = R_Z(t)R_2^*(0) = R_2^*(t).$$

Also $P(X > t, \xi < 0) = R_X(t)P(\xi < 0) = R_X(t)R_1(0) = R_1^*(t)$. □

Condition (1) of Theorem 9.7 says that the subsurvival functions R_1^* and R_2^* are consistent with a random signs censoring model. For random signs censoring under the conditions of Theorem 9.7 the conditional probability of censoring is maximal at the origin:

$$\Phi(t) = 1/(1 + R_X^*(t)/R_Z^*(t)) < \Phi(0).$$

Not every set of censored observations is consistent with a random signs censoring model. Assuming continuity and strict monotonicity, Theorem 9.7 says that a random signs censoring model exists if and only if for all $t > 0$ the conditional subsurvival function for failure dominates that of incipient failure. Under random signs censoring the population of observed failures statistically equivalent to the uncensored population, hence: $\lambda \approx n/\sum x_i$. Recalling the estimate under independent right censoring, $\lambda \approx n/(\sum x_i + \sum z_i)$, and noting that typically $\sum z_i \gg \sum x_i$ for reliable systems, it is evident that the independence assumption can lead to gross underestimates of the critical failure rate in the case of random signs competing risk (see example subsection 9.8.2).

9.6.4 Conditionally independent competing risks

A somewhat more complex model views the competing risk variables, X and Z, as sharing a common quantity Y, and as being independent given Y:

$$X = Y + W, \quad Z = Y + U,$$

where Y, U, W are mutually independent.

Explicit expressions can be derived for the case that Y, U, and Z are exponential [Hokstadt and Jensen, 1998, Paulsen et al., 1996]:

Theorem 9.8 *Let Y, U, W be independent with $R_Y(t) = e^{-\lambda_Y t}$, $R_U(t) = e^{-\lambda_U t}$ and $R_W(t) = e^{-\lambda_W t}$, then*

(i) $R_X^*(t) = \frac{\lambda_Y \lambda_W e^{-(\lambda_U + \lambda_W)t}}{(\lambda_U + \lambda_W)(\lambda_Y - \lambda_W - \lambda_U)} - \frac{\lambda_W e^{-\lambda_Y t}}{\lambda_Y - \lambda_W - \lambda_U}$,

(ii) $R_Z^*(t) = \frac{\lambda_Y \lambda_U e^{-(\lambda_U + \lambda_W)t}}{(\lambda_U + \lambda_W)(\lambda_Y - \lambda_W - \lambda_U)} - \frac{\lambda_U e^{-\lambda_Y t}}{\lambda_Y - \lambda_W - \lambda_U}$,

(iii) $R_X^*(t) + R_Z^*(t) = \frac{\lambda_Y e^{-(\lambda_U + \lambda_W)t}}{\lambda_Y - \lambda_W - \lambda_U} - \frac{(\lambda_W - \lambda_U) e^{-\lambda_Y t}}{\lambda_Y - \lambda_W - \lambda_U}$,

(iv) $\frac{R_X^*(t)}{R_X^*(0)} = \frac{R_Z^*(t)}{R_Z^*(0)} = R_X^*(t) + R_Z^*(t)$,

(v) $\frac{R_X^*(0)}{E(X \wedge Z)} = \frac{\lambda_W}{\lambda_U} \frac{R_Z^*(0)}{E(X \wedge Z)}$,

(vi) *if Y has an arbitrary distribution such that $P(Y \geq 0) = 1$, and Y is independent of U and W, then*

$$\frac{R_X^*(t)}{R_X^*(0)} = \frac{R_Z^*(t)}{R_Z^*(0)}.$$

Proof Writing

$R_X^*(t) = P(X > t, Z > X)$

$= P(Y > t)P(U > W) + \int_{y=0}^{t} \int_{u=t-y}^{\infty} P(t - y < W < u) f_U(u) f_Y(y) \, du \, dy,$

we fill in the exponential distributions and densities to obtain (i). Statement (ii) is similar; and from these (iii), (iv) and (v) are straightforward calculations. To prove (vi) we write

$R_X^*(t) = P(X > t, X < Z) = \int P(X > t, X < Z | y) dF(y)$

$= \int_0^t P(W > t - y, W < U | y) dF(y) + P(W < U) \int_t^\infty dF_Y(y)$

using the independence of Y with (U, W). Since the conditional probability under the integral is just the subsurvival function for W starting at $t - y$, we may write the last expression as

$$\frac{\lambda_W}{\lambda_W + \lambda_U} \left[e^{-(\lambda_W + \lambda_U)t} \int_0^t e^{(\lambda_W + \lambda_U)y} dF_Y(y) \right] + R_X^*(0) R_Y(t).$$

statement (vi) follows by substitution. $\square$

Further, it is easy to see that $X \wedge Z$ is the sum of the Y and $U \wedge W$. Hence the expectation of $X \wedge Z$ is

$$\frac{1}{\lambda_Y} + \frac{1}{\lambda_U + \lambda_W}.$$

A calculation gives the variance,

$$\text{var}(X \wedge Z) = \frac{1}{\lambda_Y^2} + \frac{1}{(\lambda_U + \lambda_W)^2}.$$

The ratio of the naked over the observed failure rates for X is found to be

$$\frac{\lambda_Y + \lambda_U + \lambda_W}{\lambda_Y + \lambda_W}.$$

Together with $R_X^*(0)$, these give three equations for estimating the three parameters λ_Y, λ_U and λ_W, from which we see that this model is identifiable from the subsurvival functions (of course, maximum likelihood provides a better means of estimating these parameters).

9.6.5 Time window censoring

In analyzing data, when the period of observation is terminated, there may be components still in service. This type of censoring does not affect each service sojourn in the same way, as sojourns beginning near the end of the time window have a larger chance of being censored. Techniques for dealing with time window censoring and competing risk are not well developed, but we describe a crude procedure to assess the impact of this type of censoring. Let X denote the minimum of all competing risks, and let C denote the time window censoring variable, and put $C = \infty$ for each sojourn which is not time window censored. If $C < \infty$, imagine that we could look into the future and record the values of X for the time window censored variables. In this way, for each sojourn, we can actually record the values of X and C. Imagine that our data is now generated by placing all these sojourns in an urn and drawing them out with replacement, recording $Y = (\min\{X, C\}, 1_{X<C})$. Let us assume that X is exponential with failure rate λ and that the observations are recorded as $x_1, \ldots, x_n, c_1, \ldots, c_{k_s}$. The MLE of λ_X satisfies

$$\frac{1}{\lambda_X} = \frac{\sum_{i=1}^n x_i}{n} + \frac{\sum_{j=1}^k c_j}{n}.$$

Let $\lambda_u = n/(\sum_{i=1,\ldots,n} x_i)$ denote the estimate of the failure rate of X which we would make if the censoring times were simply ignored. If μ_C is the average of the cs, then rearranging the terms above we find

$$\frac{\lambda_u}{\lambda_X} \approx 1 + \frac{\sum_{j=1}^k c_j}{\sum_{i=1}^n x_i}.$$

This enables us to give a rough estimate of the effects of time window censoring. Note that $\lambda_u \geq \lambda_X$, and equality holds if there are no time window censors.

9.7 Uncertainty

In the Introduction it was observed that most modern reliability data bases provide some indication of the uncertainty of estimates derived from the failure data.

Uncertainty bounds convey the restrictions on the possible choices of reliability parameters arising from the observed life data. It is convenient to distinguish uncertainty due to non-identifiability from uncertainty due to sampling fluctuations. To exclude the effect of sampling fluctuations it is useful to consider how we should proceed if we actually had infinitely many observations of the censored life process.

9.7.1 Uncertainty due to non-identifiability: bounds in the absence of sampling fluctuations

Let F^* be the subcumulative distribution of the life process X of interest,

$$F^*(t) = P(X \leq t, X < Z),$$

and let $F_{\min}$ be the cumulative distribution of the minimum of X and Z,

$$F_{\min}(t) = P(X \wedge Z \leq t).$$

In [Peterson, 1976] it is observed that the survival function for X, R_X, satisfies

$$1 - F^*(t) \geq R_X(t) \geq 1 - F_{\min}(t)$$

and he showed that these bounds are sharp[1] in the following sense. For all t and all $\epsilon > 0$ there are joint distributions with survival functions R_1 and R_2, depending on t and ϵ, satisfying

$$R_1^* = R_2^* = R_X^*$$

and

$$1 - F^*(t) - \epsilon < R_1(t), \quad 1 - F_{\min}(t) + \epsilon > R_2(t).$$

If F is the cumulative distribution function for X, these bounds can also be written as

$$F^*(t) \leq F(t) \leq F_{\min}(t).$$

Through any point between the functions $F_{\min}$ and F^* there passes a (non-unique) distribution function for X which is consistent with the censored data. Since $X \geq X \wedge Z$, we have

$$E(X) \geq E(X \wedge Z).$$

If X follows an exponential distribution with failure rate λ, we therefore have

$$\lambda = 1/E(X) \leq 1/E(X \wedge Z).$$

Hence, the observed data yields an upper bound on λ.

---
[1] In [Bedford and Meilijson, 1997] it is noted that the bounds are only true under certain continuity conditions.

9.7 Uncertainty

The above cited result of Peterson does not say that any distribution function between F^* and $F_{\min}$ is a possible distribution of X. [Bedford and Meilijson, 1997] give a complete characterization of the possible marginals consistent with the 'observable' F^* and $F_{\min}$, and the following result is a simplified version of their result.

Theorem 9.9 *If F is a cumulative distribution function satisfying*

$$F^*(t) < F(t) < F_{\min}(t),$$

then there is a joint distribution for (X, Z) with F as marginal distribution for X if and only if for all t_1, t_2, with $t_1 < t_2$,

$$F(t_1) - F^*(t_1) \leq F(t_2) - F^*(t_2).$$

In other words as noted also by [Crowder 1991, 1994], the distance between $F(t)$ and $F^*(t)$ must be increasing in t.

Although there may be a wide band between $F_{\min}(t)$ and $F^*(t)$ the band of exponential cumulative distributions passing through this band may be quite narrow. A better visual appreciation of the data is afforded by considering the time-average failure rates. Recall that the failure rate $h_X(t)$ for X is given by

$$h_X(t) = -dR_X(t)/R_X(t) = -d(\log(R_X(t))/dt$$

so that the time average failure rate is

$$-\log(R_X(t))/t = (1/t) \int_0^t h_X(u)\, du.$$

Of course, if $R_X(t) = \exp(-\lambda t)$, then the time-average failure rate is just the (constant) failure rate λ. Applying this transformation the Peterson bounds become

$$\mathrm{lmin}(t) = -\log(1-F^*(t))/t \leq (1/t) \int_0^t h_X(u)\, du \leq -\log(1-F_{\min}(t))/t = \mathrm{lmax}(t).$$

At each time t, the set of numbers between $\mathrm{lmax}(t)$ and $\mathrm{lmin}(t)$ corresponds to the time-average failure rates at time t which are consistent with the data *up to time t*. As $t \to \infty$, $F^*(t) \to P(X \leq Z) < 1$ so that $\mathrm{lmin}(t) \to 0$. Hence the lower bound on the admissible values of the time-average failure rate decreases as time becomes large. The Peterson bounds are the end points of the intersection for all t of the admissible time-average failure rates:

$$\text{Peterson bounds} = \bigcap_{t>0} [\mathrm{lmax}(t), \mathrm{lmin}(t)] = [\lambda_l, \lambda_u].$$

9.7.2 Accounting for sampling fluctuations

The bounds developed in the previous paragraph reflect a lack of knowledge in λ due to non-identifiability of the distribution for X caused by censoring. This lack of knowledge cannot be reduced by observations unless the censoring is suspended. In practice we also have to deal with another lack of knowledge, namely that caused by a limited number of observations. The goal of this section is to give classical confidence bounds for λ_l and λ_u.

Estimate based on $E_{\min}$ The quantity $E_{\min} = E(X \wedge Z)$ can be estimated from the data. If X is exponential with failure rate λ we have $E(X) = 1/\lambda > E_{\min}$. This leads to an upper bound on λ. If μ and σ are the empirical mean and standard deviation of N independent observations of $X \wedge Z$, then $\sqrt{N}(\mu - E_{\min})/\sigma$ follows approximately a standard normal distribution. Hence

$$[\mu - 1.65\sigma/\sqrt{N}, \mu + 1.65\sigma/\sqrt{N}]$$

is an approximate classical 90% confidence interval for $E_{\min}$ and by inverting these we obtain a 90% classical confidence interval for λ.

Though simple and transparent, this procedure yields only confidence bounds for λ_u; it says nothing about λ_l.

9.7.3 Sampling fluctuations of Peterson bounds

One possibility for obtaining confidence bounds on the end points of the intervals $[\lambda_l, \lambda_u]$ is to use the Peterson bounds. A rather obvious idea gives insight into the interpretation of the Peterson bounds, but does not yield confidence bounds. Bounds based on the chi-squared test are sketched briefly. [Bedford and Meilijson, 1996] discusses a non-parametric test. The two tests are compared in [Cooke and Bedford, 1995]. We consider only the case where X follows an exponential distribution. We write the Peterson bounds as

$$\mathrm{lmin}(t) = -\log(1 - F^*(t))/t \leq (1/t)\int_0^t h_X(u)du$$
$$\leq -\log(1 - F_{\min}(t))/t = \mathrm{lmax}(t).$$

For each t, the probabilities $F^*(t)$ and $F_{\min}(t)$ can be estimated from the data. Classical confidence bounds on these estimates can be substituted into the above expression to yield classical confidence bounds for the time-average failure rate of X, for each time t. If X is exponential with failure rate λ then

$$\max_t \mathrm{lmin}(t) \leq \lambda \leq \min_t \mathrm{lmax}(t).$$

The sampling distributions of $\max_t \mathrm{lmin}(t)$ and $\min_t \mathrm{lmax}(t)$ are not known,

9.7 Uncertainty

and even if they were known they wouldn't help. Suppose, for example, that lmax(t) is the constant h (corresponding to $X \wedge Z$ exponential with failure rate h). Let us draw 100 samples of size N and compute the empirical lmax(t) in each N-sample. We will find in most cases, say 95 of the 100, that the minimum over t of the empirical lmax(t) is less than h. It may thus arise that the 95% quantile of the empirical $\min_t$ lmax(t) is less than the true limiting value of $\min_t$ lmax(t), namely h.

In spite of this it is useful to have a picture of the 'time-wise sampling fluctuations' of the Peterson bounds. If in $n(t)$ of N independent observations of $(X \wedge Z, 1_{X<Z})$ the event $(X < t, X < Z)$ is observed to occur, then the quantity

$$\sqrt{N}(n(t)/N - F^*(t))/\sigma$$

is approximately standard normal where $\sigma^2 = F^*(t)(1 - F^*(t))$ may be estimated as $(nN - n^2)/N^2$. Hence the classical 5% lower confidence bound for $-\log(1 - F^*(t))/t$ can be written as

$$\lambda_{l(t)} = -\log[1 - (n(t)/N - 1.65\sigma(t)/\sqrt{N})]/t. \tag{9.5}$$

Similarly, if $m(t)$ is the number of observations of the event $(X \wedge Z < t)$ in N independent observations of $(X \wedge Z, 1_{X<Z})$ then an upper 95% classical confidence bound for $-\log(1 - F_{\min}(t))/t$ is

$$\lambda_{u(t)} = -\log[1 - (m(t)/n + 1.65\sigma'(t)/N)]/t \tag{9.6}$$

where now σ'^2 is estimated as $(mN - m^2)/N^2$. The curves $\lambda_{l(t)}$ and $\lambda_{u(t)}$ have the following interpretation. If we repeatedly draw samples of size N from the distribution of $(X \wedge Z, 1_{X<Z})$, then for each t, in 95% of the N-samples the empirical version of lmin(t) is greater than $\lambda_{l(t)}$ and in 95% of the N-samples the empirical version of lmax(t) is less than $\lambda_{u(t)}$. This does not mean that 95% of the N-samples lie above $\lambda_{l(t)}$ (below $\lambda_{u(t)}$) for all t. Graphs of $\lambda_{l(t)}$ and $\lambda_{u(t)}$ are given in Section 9.8.

Chi-squared bounds for λ can be given by the following method, which we illustrate for the lower Peterson bound. Let $p_i = P(i - 1 < X < i, X < Z)$, $i = 1, \ldots, m$. The lower Peterson bound may be written as a set of inequality constraints on $p_1, \ldots, p_m$ forming a constraint set C:

$$\begin{aligned} p_1 &< 1 - e^{-\lambda}, \\ p_1 + p_2 &< 1 - e^{-2\lambda}, \\ p_1 + p_2 + p_3 &< 1 - e^{-3\lambda}, \\ &\vdots \quad \vdots \\ p_1 + p_2 + \cdots + p_m &< 1 - e^{-m\lambda}. \end{aligned}$$

Put $\mathbf{p} = (p_1,\ldots,p_{m+1})$, $p_{m+1} = 1 - p_1 - \cdots - p_m$. Suppose we have N independent observations of the variable $Y = (X \wedge Z, 1_{X<Z})$. Let $\mathbf{s} = (s_1,\ldots,s_{m+1})$, where $s_i = \#\{X|i-1 < X \leq i, X < Z\}/N$, $i = 1,\ldots,s_m$ and $s_{m+1} = 1 - s_1 - \cdots - s_m$. Fix a value λ and ask which probability vector $\mathbf{p}$ satisfying constraint set C has the greatest likelihood given the data $(\mathbf{s}, N)$? If the sample $\mathbf{s}$ itself satisfies C, then $\mathbf{s}$ is the vector of maximal likelihood satisfying C. In this case λ is certainly not 'too small'. As λ gets smaller, eventually the sample will violate one or more of the constraints. This does not mean that λ is 'too, small', as the violation may be explainable as a chance fluctuation. To determine whether λ is 'too small' we find a probability vector $\mathbf{p}$ which solves a constrained optimization problem:

$$\text{minimize } Q(\mathbf{p}) = 2N \sum s_i \log(s_i/p_i) \text{ subject to constraint set } C.$$

The quantity $Q(\mathbf{p})$ is the log likelihood ratio for the hypothesis that the data has been generated by independent samples from p; its asymptotic distribution is chi-squared with m degrees of freedom. If the optimal p, for fixed λ, is such that $Q(\mathbf{p})$ is so large as to fall in the upper critical region, then the hypothesis that the data was generated by independent samples from p would be rejected, hence every vector p satisfying C would also be rejected and accordingly λ would also be rejected. In numerical experiments performed to date this method gives good results, but is somewhat sensitive to the number of cells m and to the cell size.

9.8 Examples of dependent competing risk models

Pressure relief valves are designed to open when pressure exceeds a certain limit. A typical boiling water reactor has 20 pressure relief valves inside the containment. These are tested once a month. The data discussed below comes from one Swedish nuclear station operating two identical reactors, from the period 1/1/78 up to 1/9/95. This yields a total $17.5 \times 12 \times 20 \times 2 = 8400$ socket-months. Censoring due to termination of observation period accounts for 1995 months, leaving 6405 socket-months. The analyses below are based on these 6405 sojourns. The data fields which can be colored are shown in Figure 9.6.

Four figures illustrate different analyses of these component socket histories. The user chooses a field for coloring and applies two colors. Thus in Figure 9.6 we have chosen the field 'effect of failure' and colored the first nine effects dark, and the last effect, 'non-functional', light. The survival, and subsurvival, functions are shown at the top. 'Type 1' always corresponds to the dark coloring. The left bottom graph in Figure 9.7 shows the time-wise

9.8 Examples of dependent competing risk models 185

Failure detection			Failure Cause	
A	alarm		A	crack
B	operator supervision		D	internal pipe leak
H	accidental discovery		E	external leak in packing
C	preventive maintenance		F	internal leak in packing
D	test		G	deformation
E	inspection		H	vibration / suspicious noise
G	revision		I	fouling
Failure Effect			J	nick
			L	Glapp contact
A	fail to close/stop/break (contacts)		M	loss of power
B	fail to open/start/connect (contacts)		N	ground/isolation failure
C	spurious close/stop/disconnect		O	short circuit
D	spurious open/start/connect		P	deviation from calibration
E	signal /alarm failed		R	
F	spurious signal		S	
G	wrong measure value		T	
H	wrong control		U	corrosion/erosion/wear
K	other functional disturbance		W	
L	non-functional failure		X	other elect. power failure
Action taken			Y	control system - other
			Z	other
H	change type			
B	replace			
C	repair adjustment			
D				
Z	other (cleaning lubrication etc.)			

Fig. 9.6. Data fields for coloring

average failure rate bounds, as given in Equations 9.5 and 9.6, and the observed failure rate (the heavy line).

9.8.1 Failure effect

In Figure 9.7, the field 'failure effect' has been selected for coloring. The functional failures have been darkened, and the 'non-functional failure' is colored light. There are 144 functional and 148 non-functional failures with total sojourn times 2938 and 3467 months respectively. The functional failures include 'fail to open' and 'fail to close'. These are both essential functions of a valve, but from the risk point of view only 'fail to open' would be significant.

Figure 9.7 shows the subsurvival and conditional subsurvivor functions. The value $R_X^*(0)$ is the probability of a service sojourn terminating due to color X. We see from the subsurvival plots that the probability is about one half that a sojourn ends in functional failure. The conditional subsurvival functions are more or less equal. This is consistent with the independent exponential model (Theorem 9.5), and with the conditional independence model (Theorem 9.9). The independent exponential model also predicts that the function Φ is constant. Figure 9.7 is not wholly convincing in that respect, and the conditional independent model might be better. Nonetheless, we proceed here with the independent exponential model. The lower two plots

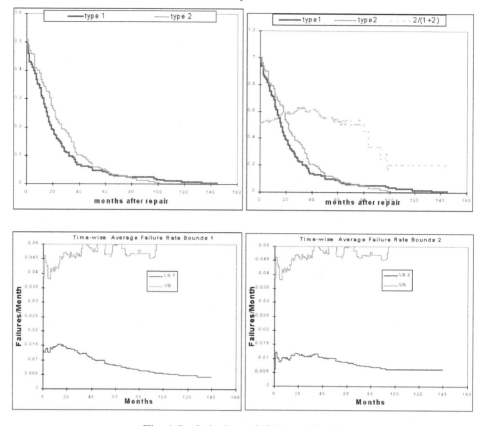

Fig. 9.7. Coloring of 'failure effect'

show the time-wise bounds on the time-average failure rate. The estimated failure rate with this model is $144/6405 = 0.022$ failures per month. The conditional independence model yields 0.02. Apparently the difference in these models is not great.

To gauge the effect of time window censoring we must consider the minimum of the dark and light variables, and must assume this follows an exponential distribution with parameter λ_m. Ignoring the time window censors, λ_m is estimated as $\lambda_u = (148+144)/6405$. According to subsection 9.6.5 we have

$$\frac{\lambda_u}{\lambda_m} = 1 + 1995/6405 = 1.3.$$

9.8.2 Action taken

We now color the field 'action taken'. The dark color corresponds to the action 'replace', the light color to the other possible actions. There are 57 (dark) sojourns terminating in 'replace' and 195 other (light) sojourns. The total

9.8 Examples of dependent competing risk models

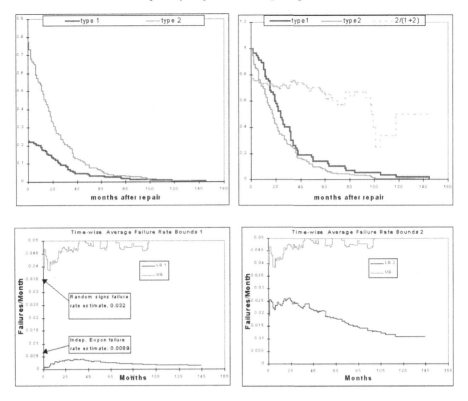

Fig. 9.8. Coloring of 'action taken'

sojourn times for these two colors are 1812 and 4593 months respectively. It is not unreasonable to suppose that the other actions are undertaken with the goal of prohibiting or intercepting the action 'replace'. In this case the maintenance personnel might plausibly behave in accordance with the random signs model. That model predicts the conditional subsurvivor function for 'other' should lie below that for 'replace', and that the function Φ should take its maximum value at the origin. Figure 9.8 confirms these predictions. These patterns are not at all consistent with the independent exponential or conditional independent model. The random signs estimate of the dark failure rate is $57/1812 = 0.031$. Had we used the independent exponential model, we should have estimated this as $57/(1812 + 4593) = 0.00089$.

If the maintenance crew is trying to prohibit 'replace' actions while losing as little useful service time as possible, then we should hope that the number of 'replace's is small relative to the number of 'other's and that the expected sojourn time is ending in 'other' only a little shorter than the expected sojourn time ending in 'replace' [Paulsen and Cooke, 1994]. When these maintenance

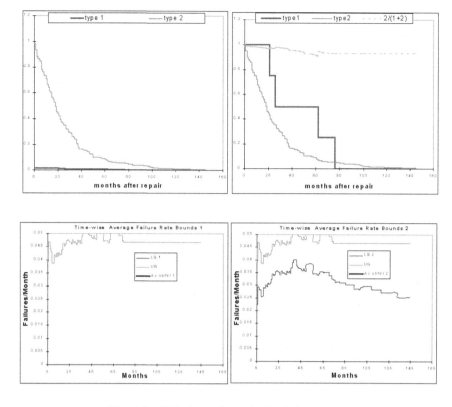

Fig. 9.9. Coloring of 'method of detection'

indicators are applied to this data, the maintenance crew appears to be doing a good job. The lower two graphs in Figure 9.8 show the time-wise average failure rate bounds. For 'replace', the random signs and independent exponential models are shown. Both lie within the bounds, and the latter is a factor 3 lower than the former.

9.8.3 Method of detection

Figure 9.9 shows the results of applying coloring to the field 'method of detection'. The dark color corresponds to 'alarm' or 'unintended discovery'. Discussions with maintenance personnel indicated that these were indeed events which they would try to avoid. There are only 4 such events (dark), and 248 other (light) events. The total sojourn times are 185 (dark) and 6220 (light). As in the previous example, the random signs model seems appropriate. The random signs estimate of the dark faliure rate is $4/185 = 0.022$.

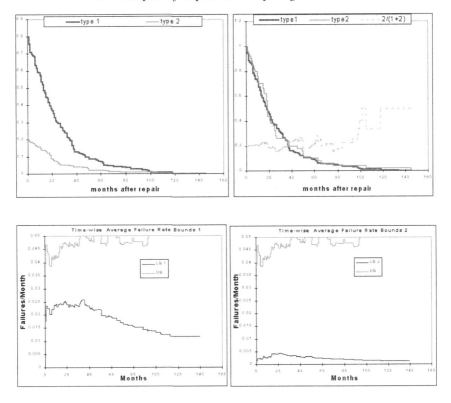

Fig. 9.10. Coloring of 'subcomponent'

9.8.4 Subcomponent

The most important (and expensive) subcomponents in the pressure relief valves are 'piston' and 'valve'. Figure 9.10 shows the coloring of 'subcomponent', with 'valve' and 'piston' colored dark, the others are light. There are 202 dark events with a total sojourn time of 5055 months; there are 50 light events with a total sojourn time of 1350 months. Here, the function Φ appears to be constant and the conditional subsurvivor functions appear to be equal. This is the signature for the independent exponential model. The naked failure rate would be estimated on this model as $202/(5505 + 1350) = 0.032$.

9.8.5 Conclusions

These examples show how competing risk models may be applied to estimate failure rates. It is clear that the independent exponential model is not always appropriate. When the failure mode of interest is competing with modes related to preventive maintenance, then the random signs model may be more appropriate, as in the examples 'action taken' and 'method of detection'. In such cases the independent exponential model would underestimate the failure rate of interest. If this error were repeated for a large number of different components, the errors could cumulate and produce a significant overall error.

9.9 Exercises

9.1 Verify that the maximum likelihood estimator of the probability p of failure per demand, given r failures in n trials, is r/n.

9.2 Let X and Z be independent competing risks as in Section 9.5, and let $F_X = F_{X,\theta}$ be described by a vector of parameters q, define the likelihood function

$$L(q|y_1,\ldots,y_N) = P(q|y_1,\ldots,y_N)$$
$$= \prod_{i=1}^{n} f_{X,\theta}(x_i)(1 - F_Z(x_i)) \times \prod_{j=1}^{m}(1 - F_{X,\theta}(z_j))f_Z(z_j).$$

Show that $\partial F_{X,\theta}/\partial \theta_k = 0$ is equivalent to

$$\sum_{i=1}^{n} \frac{(\partial f_{X,\theta}(x_i)/\partial \theta_k)}{f_{X,\theta}(x_i)} = \sum_{i=1}^{n} \frac{(\partial F_{X,\theta}(x_i)/\partial \theta_k)}{(1 - F_{X,\theta}(x_i))}.$$

9.3 Write the likelihood for k independent competing risks.

9.4 Derive the MLE of the failure rate of an exponentially distributed life variable subject to independent censoring.

9.5 A straightforward calculation [Peterson, 1977] shows that

$$\frac{dR_i^*}{dt}(t) = \lim_{\Delta \to 0} \frac{P(X_i \in (t, t+\Delta), X_j > t+\Delta, j \neq i)}{\Delta}$$
$$= \frac{dR_i}{dt}(t) \prod_{j \neq i} R_j(t).$$

Use this to prove Theorem 9.1.

9.6 Complete the proof of Theorem 9.5.

9.7 Use the heuristic for time window censoring to assess the effects of time window censoring on the data analyzed in subsection 9.6.5.

10

Expert opinion

10.1 Introduction

Probabilistic risk analysis treats events with a low intrinsic rate of occurrence, and large amounts of data are seldom available. Since its inception, expert opinion in the form of subjective probabilities has been a dominant source of data for failure probabilities. Two sources for data in risk analysis, WASH-1400 [NRC, 1975] and IEEE Standard 500 [IEEE, 1977], are based on unstructured surveys of expert opinions. The IAEA data base, the COVO study [COVO, 1982], and the Canvey Island reports [HSE, 1978, HSE, 1981] reference most data to these two sources. The Swedish *T-Book* uses Bayesian methods in combination with operational data to derive uncertainty bounds for failure rate assessments. The USNRC made extensive use of structured expert opinion surveys to quantify uncertainties in the NUREG 1150 study [NRC, 1989].

This chapter focuses on the mathematical background for utilizing expert subjective probabilities. It draws on Chapter 11 of [Cooke, 1991], to which source the reader is referred for more detail on the use of expert opinion in general.

Expert judgement techniques are useful for quantifying models in situations in which, because of either cost, technical difficulties or the uniqueness of the situation under study, it has been impossible to make enough observations to quantify the model with 'real data'. Expert judgement data may also be used to refine estimates from 'real data' when it turns out that the categorization of the data was not as fine as the models require. Finally, expert judgement techniques have been used to estimate model parameter uncertainties. An example of this type of application will be given at the end of this chapter, and be further developed in the context of uncertainty analysis in Chapter 17.

10.2 Generic issues in the use of expert opinion

Choosing experts can be a major issue in an expert opinion study. On the one hand, it is advantageous to use a group of experts wide enough to encompass all facets of scientific thought on a particular topic. On the other hand, there may be pressure to exclude representatives of some schools as being incompetent (not experts). All methods for combination of expert opinion are sensitive to this problem. The model discussed in Section 10.6 is probably the least sensitive.

Biases can occur at many levels. In [Otway and von Winterfeldt, 1992] mindsets, structural biases, motivational biases, and cognitive biases such as overconfidence, anchoring and availability are discussed. Mindsets are unstated assumptions that the expert (but not necessarily the analyst) uses. Structural biases can occur for example through the level of detail in one part of a study, or the choice of background scales on which the quantification is made. Motivational biases can occur when an expert has a stake in the outcome of the study (this kind of bias can occur without any feeling of impropriety on the part of the expert). Anchoring occurs when an expert has been given some estimate (possibly from another context) and subconsciously bases his opinion on that estimate. Availability refers to the cognitive bias in which those events that are easily or vividly recalled are likely to be overestimated, while other events which are difficult to recall may be underestimated.

There is no magic way to avoid these problems. If the analyst is aware of these difficulties, it is possible to guard against them to some extent. A so-called 'dry-run', a trial exercise with experts who are not part of the expert panel, can be useful to reveal many problems of this type.

10.3 Bayesian combinations of expert assessments

Many Bayesian models for combining expert assessments have been proposed in the literature, but few have been applied, and fewer still have been applied more than once (see [Cooke, 1991]). An exception is the model proposed in [Mosleh and Apostolakis, 1986], and independently in [Winkler, 1981]. The problem here is to combine experts' point assessments for an unknown quantity.

Let X denote an unknown quantity of interest, and let $x_1, \ldots, x_e$ denote estimates of X from experts $1, \ldots, e$. The decision maker starts with a *prior* density $p(x)$ over X, and updates this density with the information $x_1, \ldots, x_e$, from his/her expert advisors. Bayes' theorem reads

$$p(x \mid x_1, \ldots, x_e) \propto p(x_1, \ldots, x_e \mid x) p(x).$$

10.3 Bayesian combinations of expert assessments

In the simplest case, the experts are independent and the likelihood term $p(x_1,\ldots,x_e|x)$ factors:

$$p(x_1,\ldots,x_e|x) = \prod p(x_i|x).$$

The problem reduces to determining the terms $p(x_i \mid x)$. Two models, one with *additive* and one with *multiplicative* errors, are suggested in [Mosleh and Apostolakis, 1986]. On the additive error model the expert's assessment x_i is modeled as the sum of two terms:

$$x_i = x + \xi_i$$

where x denotes the true unknown value of X, and ξ_i is an additive error term. The model assumes that the errors are normally distributed with mean μ_i and standard deviation σ_i. The choice of these parameters rests with the decision maker and reflects his/her appraisal of expert i's bias and accuracy. Under these assumptions the likelihood $p(x_i \mid x)$ of getting estimate x_i given that the true value is x is simply the value of the normal density with mean $x + \mu_i$ and standard deviation σ_i.

Proposition 10.1 *Let the decision maker's prior $p(x)$ be normal with mean x_{e+1} and standard deviation σ_{e+1}; then $p(x \mid x_1,\ldots x_e)$ is normal with mean and variance given by*

$$E(x \mid x_1,\ldots,x_e) = \sum_{i=1}^{e+1} w_i(x_i - \mu_i),$$

$$V(x|x_1,\ldots,x_e) = \frac{1}{\left(\sum_{i=1}^{e+1} \sigma_i^{-2}\right)}$$

where

$$w_i = \frac{\sigma_i^{-2}}{\left(\sum_{j=1}^{e+1} \sigma_j^{-2}\right)} \quad \text{and } \mu_{e+1} = 0.$$

Proof $p(x_1,\ldots,x_e|x) = \prod_{i=1}^{e} p(x_i|x)$ so

$$p(x|x_1,\ldots,x_e) = Cp(x)\prod_{i=1}^{e} p(x_i|x)$$

$$= C\exp\left\{-\frac{1}{2}\sum_{i=1}^{e+1}((x_i - (x + \mu_i))/\sigma_i)^2\right\}$$

$$= C\exp\left\{-\frac{1}{2}\sum_{i=1}^{e+1}((x - (x_i - \mu_i))/\sigma_i)^2\right\}.$$

If we write the exponent as $(Kx - L)^2$ + terms without x, then these latter terms can be absorbed into the constant C and $p(x \mid x_1, \ldots, x_e)$ is normal with mean L/K and variance $1/K^2$. It may be verified that this is accomplished by setting

$$K^2 = \sum_{i=1}^{e+1} \sigma_i^{-2} \text{ and } KL = \sum_{i=1}^{e+1} ((x_i - \mu_i)/\sigma_i)^2.$$

□

The multiplicative model models the expert's assessment as $x_i = x\xi_i$ and assumes log-normality. It is therefore essentially the same as the additive model after the assessments have been logged. These models can be readily generalized to allow for dependence between the experts' assessments by choosing a joint (log-)normal likelihood. However, the assessment burden on the analyst is significantly increased, as he/she must then assess the covariance matrix for the experts' assessments.

These models treat the decision maker as the $(e+1)$-th expert, a role in which the decision maker is not always comfortable.

10.4 Non-Bayesian combinations of expert distributions

There are myriad models for combining opinions, expert or otherwise. Older methods, like the Delphi method or the 'nominal group technique', operate on experts' point assessments of unknown quantities. More recent methods concentrate on the combination of expert subjective probability distributions. In this section we examine some of the mathematical background for combining distributions.

We denote by $\mathscr{P}$ the set of probability measures over some fixed but unspecified probability space. If $P_1, \ldots, P_e \in \mathscr{P}$, then a combination rule which depends only on $P_1, \ldots, P_e$ is simply a function $G : \mathscr{P}^e \to \mathscr{P}$,

$$G(P_1, \ldots, P_e) = P.$$

We need to know the form of G, and so one seeks to restrict G by introducing 'natural' axioms. General results are available which show how constraints on G determine its form. Simple, but sufficiently rich, results can be obtained by restricting attention to a manageable subclass of G.

We assume that the probability space is generated by a finite number of atoms or elementary exclusive events $a_1, \ldots, a_n$. Then we write p_{ij} for the probability that expert i gives for event a_j, so that expert i gives a vector

10.4 Non-Bayesian combinations of expert distributions

of probabilities

$$P_i = (p_{i1}, \ldots, p_{in}) \quad p_{ij} \geq 0, \quad \sum_{j=1}^{n} p_{ij} = 1 \quad i = 1, \ldots, e. \tag{10.1}$$

A fairly natural class of functions is the class of *norm probabilities*. Let w_i ($i = 1, \ldots, e$) be a set of non-negative normalized *weights*

$$w_i \geq 0, \quad \sum w_i = 1. \tag{10.2}$$

For $r \in \mathbb{R}$, the *elementary r-norm weighted-mean probability* for outcome a_j is defined as

$$\mathcal{M}_r(j) = \left(\sum_{i=1}^{e} w_i p_{ij}^r \right)^{1/r}. \tag{10.3}$$

The *r-norm probability* for outcome a_j is defined as

$$\mathcal{P}_r(j) = \frac{\mathcal{M}_r(j)}{\sum_{k=1}^{n} \mathcal{M}_r(k)}. \tag{10.4}$$

The interpretation of $\mathcal{P}_r$ is facilitated by the following result. This says that for $r = 1$, $\mathcal{P}_r$ is the *weighted arithmetic mean* of the distributions P_i; for $r = 0$, $\mathcal{P}_r$ is the *weighted geometric mean*, and for $r = -1$, $\mathcal{P}_r$ is the *weighted harmonic mean*. For $r = \infty$, $\mathcal{P}_r$ takes the largest probability assessment for each atom a_i and renormalizes, for $r = -\infty$, $\mathcal{P}_r$ does likewise with the smallest probability assessment. These are sometimes called the *upper* and *lower probability distributions* generated by $P_1, \ldots, P_e$.

Proposition 10.2 (i) $\mathcal{M}_r(j) \to \prod_{i=1}^{e} p_{ij}^{w_i}$ as $r \to 0$.
(ii) $\mathcal{M}_r(j) \to \max_{i=1,\ldots e}\{p_{ij}\}$ as $r \to \infty$.
(iii) $\mathcal{M}_r(j) \to \min_{i=1,\ldots e}\{p_{ij}\}$ as $r \to -\infty$.
(iv) If $r < s$ then $\mathcal{M}_r(j) \leq \mathcal{M}_s(j)$ with equality if and only if $p_{ij} = p_{kj}$, $1 \leq i, k \leq e$.
(v) Define $\mathcal{M}_r(j+k) = (\sum_{i=1}^{e} w_i (p_{ij} + p_{ik})^r)^{1/r}$; and assume that p_{ij}/p_{ik} is not constant in i. Then the following (in)equalities hold:

if $r > 1$ then $\mathcal{M}_r(j+k) < \mathcal{M}_r(j) + \mathcal{M}_r(k)$,
if $r = 1$ then $\mathcal{M}_r(j+k) = \mathcal{M}_r(j) + \mathcal{M}_r(k)$,
if $r < 1$ then $\mathcal{M}_r(j+k) > \mathcal{M}_r(j) + \mathcal{M}_r(k)$.

Proof The proof uses Hölder's inequality: if $0 < \alpha < 1$, then for $a, b \in \mathbb{R}_+^e$, where $\mathbb{R}_+ = [0, \infty)$,

$$\sum a_i^\alpha b_i^{1-\alpha} \leq \left(\sum a_i \right)^\alpha \left(\sum b_i \right)^{1-\alpha} \tag{10.5}$$

with equality holding if and only if a and b are proportional vectors. An equivalent form is as follows: let $(1/k) + (1/k') = 1$, then unless a and b are proportional,

$$\left. \begin{array}{l} \sum a_i b_i < \left(\sum a_i^k\right)^{1/k} \left(\sum b_i^{k'}\right)^{1/k'} \quad \text{if } k > 1, \\ \sum a_i b_i > \left(\sum a_i^k\right)^{1/k} \left(\sum b_i^{k'}\right)^{1/k'} \quad \text{if } k < 1 \end{array} \right\} \quad (10.6)$$

([Hardy et al., 1983], pp. 22–24).

(i)
$$\log(\mathcal{M}_r(j)) = (1/r)\log\left(\sum w_i p_{ij}^r\right).$$

Write $p_{ij}^r = \exp\{r \log p_{ij}\}$ and take the Taylor expansion of $\sum w_i p_{ij}^r$ about $r = 0$. The right hand side becomes

$$(1/r) \log [1 + r \sum w_i \log(p_{ij}) + o(r)]$$

where $o(r)/r \to 0$ as $r \to 0$. As $r \to 0$, this quantity becomes

$$\sum w_i \log(p_{ij})$$

from which (i) follows.

(ii) Let
$$p_{kj} = \max_{i=1,\ldots,e} \{p_{ij}\}.$$

If $r > 0$ then

$$(w_k^{1/r} p_{kj})^r \le \left(\sum_{i=1}^{e} w_i p_{ij}^r\right) \le p_{kj}^r,$$

from which the result follows.

(iii) This is in Exercise 2.

(iv) Assume $r < s$ and that the proportionality condition does not hold. Put $r = s\alpha, 0 < \alpha < 1$, then by Hölder's inequality

$$\mathcal{M}_r(j)^r = \sum_{i=1}^{e} w_i p_{ij}^{\alpha s} = \sum_{i=1}^{e} (w_i p_{ij}^s)^\alpha w_i^{1-\alpha} \le \left(\sum_{i=1}^{e} w_i p_{ij}^s\right)^\alpha \left(\sum_{i=1}^{e} w_i\right)^{1-\alpha}$$

$$= \left(\sum_{i=1}^{e} w_i p_{ij}^s\right)^\alpha = \mathcal{M}_s(j)^{s\alpha}.$$

Raising both sides to the power $1/(\alpha s)$ yields the result.

(v) The result is trivial for $r = 1$. Assume $r > 1$ and write $s_i = p_{ij} + p_{ik}$.

$$\mathcal{M}_r(j+k)^r = \sum w_i s_i^r = \sum (w_i^{1/r} p_{ij})(w_i^{1/r} s_i)^{r-1} + \sum (w_i^{1/r} p_{ik})(w_i^{1/r} s_i)^{r-1}.$$

Since $r > 1$, we may apply Hölder's inequality to the first term on the right hand side ($k = r; k' = r/(r-1)$):

$$\sum (w_i^{1/r} p_{ij})(w_i^{1/r} s_i)^{r-1} = \sum (w_i p_{ij}^r)^{1/r} (w_i s_i^r)^{(r-1)/r}$$
$$< \left(\sum (w_i p_{ij}^r)\right)^{1/r} \left(\sum (w_i s_i^r)\right)^{(r-1)/r}$$
$$= \mathscr{M}_r(j) \mathscr{M}_r(j+k)^{r-1}.$$

Applying the same reasoning to the second term, we find

$$\mathscr{M}_r(j+k)^r < (\mathscr{M}_r(j) + \mathscr{M}_r(k)) \mathscr{M}_r(j+k)^{r-1}$$

from which the result follows. For $r < 1$ this is in Exercise 2. □

Now the question is: which r should we take? Clearly we should choose an r for which our function possesses some desirable properties.

The combination function $G(P_1, \ldots, P_e)$ is said to possess the *strong set-wise function property* if for every subset $A \subseteq \{a_1, \ldots, a_n\}$, the decision maker's probability of A only depends on the experts' judgements of the probability of A, that is, if $Q_i = (P_i(A), 1 - P_i(A))$ then

$$G(P_1, \ldots, P_e)(A) = G(Q_1, \ldots, Q_e)(A).$$

In fact, $\mathscr{P}_1$ is the only r-norm probability to possess the strong set-wise function property (see Exercises).

Another desirable property is the *marginalization property*. Intuitively this says that the assessed probabilities should not depend on the way in which events have been grouped together. Let $a'_{n-1} = a_n \cup a_{n-1}$, and for $i = 1, \ldots, e$ let P'_i be defined on $\{a_1, a_2, \ldots, a'_{n-1}\}$ by setting

$$P'_i(a_j) = P_i(a_j), j = 1, \ldots n-2;$$
$$P'_i(a'_{n-1}) = P_i(a_{n-1}) + P_i(a_n).$$

P'_i is called the *marginal distribution of P_i with respect to the sub-field of events generated by* $\{a_1, \ldots, a'_{n-1}\}$. The function $G(P_1, \ldots P_e)$ is said to possess the *marginalization property* if

$$G(P_1, \ldots, P_e)(a_j) = G(P'_1, \ldots, P'_e)(a_j) \text{ for } j = 1, \ldots, n-2,$$

and the same holds for any ordering of the $a_1, \ldots, a_n$.

It can be shown (see Exercises) that $\mathscr{P}_1$ is the only r-norm probability which possesses the marginalization property.

To illustrate the practical significance of the marginalization property consider the following flashlight example. Two experts give probabilistic assessments which are then combined with equal weights using the $\mathscr{P}_0$ rule.

Both experts give probability 0.8 that the light will not work. Combining these assessments gives a probability of 0.8 that the light does not work. Now, suppose (to keep the example simple) that the light only fails if the contacts fail (C) or if the battery fails (B), and that these two events are exclusive. Expert 1 assesses the probability of C as 0.7, that of B as 0.1 and that of the event 'light works' (W) as 0.2. Expert 2 assesses $P(C) = 0.1$, $P(B) = 0.7$ and $P(W) = 0.2$. Combining these probabilities according to $\mathscr{P}_0$ gives $\mathscr{M}_0(C) = \sqrt{(0.7 \times 0.1)} = 0.264$, $\mathscr{M}_0(B) = 0.264$, and $\mathscr{M}_0(W) = 0.2$, so that the probability of failure is now estimated as

$$(0.264 + 0.264)/(0.264 + 0.264 + 0.2) = 0.726.$$

Furthermore, even though both experts agree that the probability of no failure is 0.2, the combination rule now gives a probability of 0.274 for this event. The marginalization property requires that the probability of no failure remain at 0.2 whatever collection of failure modes are identified (assuming of course that the experts themselves remain consistent in the assignment of 0.2).

The final desirable property that we may require is that of independence preservation. A combination rule G is said to possess the *independence preservation property* if for all $A, B \subseteq \{a_1, \ldots, a_n\}$ such that $P_i(A \cap B) = P_i(A)P_i(B), i = 1, \ldots, e$

$$G(P_1, \ldots, P_e)(A \cap B) = G(P_1, \ldots, P_e)(A)G(P_1, \ldots, P_e)(B).$$

It can be shown (exercise!) that $\mathscr{P}_0$ has the independence preservation property, and that $\mathscr{P}_1$ does *not* have this property. This property is attractive as we tend to think of independence as a fundamental property. Hence if experts agree that two events are independent even if they do not agree on the underlying probabilities of the events, one might imagine that it would be reasonable to require that the events should be independent.

In general, i.e. not restricting to r-norm probabilities, it can be shown that $\mathscr{P}_1$ is the only rule which satisfies the strong set-wise function property, if $n > 2$. In the class of r-norm probabilities, marginalization is equivalent to the strong set-wise function property. In general this is not quite true, but the latter is equivalent to marginalization and the *zero preservation property*: if all experts assign probability zero to an event, then the combination must also assign probability zero to this event [McConway, 1981].

We have seen that it is not possible to have all the desirable properties in one function G. This means that we shall have to choose which 'desirable' property we are prepared to drop.

10.5 Linear opinion pools

Although some authors consider the independence preservation property important, there are strong arguments *against* combination rules which have this property. Consider outcomes of a coin tossing experiment with a coin of unknown composition. Two experts may agree that the outcomes on different tosses are independent and identically distributed, but may disagree sharply as to what the probability of heads is. Since each expert's assessment of the probability of heads on the n-th toss is independent of the outcomes of previous tosses, neither expert will 'learn from experience'. Rather they will both hold fast to their original assessments regardless of what relative frequencies of heads emerge from repeated tossings. Since the experts disagree, it seems unreasonable that the decision maker should also be unwilling to learn from experience, for experience might show one of the experts to be right and the other wrong, yet this would be the result if his combination rule satisfied the independence preservation property.

The results in the previous section, and the above argument, constitute strong motivation for restricting the admissible combinations of probability distributions to the class $\mathscr{P}_1$. These are known in the literature as 'linear opinion pools'.

Once it is decided to use a linear opinion pool, the rule is fixed by deciding on a set of weights. The simplest and easiest choice of weights is simply $w_i = 1/e$; i.e. simple arithmetic averaging. This method is widely used (for example in [NRC, 1989]), but has the disadvantage that it does not allow a judgement of expert quality.

10.6 Performance based weighting – the classical model

Cooke [Cooke, 1991] describes methods based on experts' performance on assessing variables whose true values become known post hoc. These 'performance based weights' are finding increasing application, and experience indicates that they usually outperform simple arithmetic averaging (see for example [Goossens *et al.*, 1998]). The main aim of this method is to provide the basis for achieving *rational consensus*. Recall from Chapter 2 that the scientific foundation for subjective probability comes from the theory of rational decision making. Since every rational individual has his own subjective probability (and in particular, experts usually do have different opinions) it is necessary to find a way of building consensus. The fundamental principle of performance based weighting is that the weights used to combine expert distributions are chosen according to the performance of the experts on so-called 'calibration questions', questions for which the answers are known to the analyst but not to the experts. We give an informal discussion of the principal concepts used in performance based weights.

Input for performance based weights are the quantile assessments from experts for query variables, both variables of interest and calibration variables. Calibration variables are variables whose true values or realizations are known post hoc. Experts' assessments are scored with respect to *calibration* or *statistical likelihood*, and *informativeness*. These scores are used to compute weights which satisfy an 'asymptotic strictly proper scoring rule': that is, an expert achieves his/her maximal expected weight, in the long run, by and only by stating assessments corresponding to his/her true beliefs. The result of linear pooling using performance based weights is a 'decision maker' who himself can be scored with respect to calibration and informativeness. The input and output quantiles can be chosen arbitrarily, but in risk analysis applications the 5%, 50% and 95% quantiles are most common. The analyst has to determine the so-called *intrinsic range*, i.e. upper and lower bounds that (to a very good approximation) contain the whole of the distribution.

10.6.1 Calibration

Suppose that we have asked for the experts' uncertainties over a number of calibration variables. The expert gives quantile information for each of his/her uncertainty distributions so that for each calibration variable we are given four intervals – 0% to 5%, 5% to 50%, 50% to 95%, 95% to 100%. Now intuitively if the expert is well calibrated, he/she should have given intervals such that 5% of the realizations of calibration variables fall into the corresponding 0% to 5% intervals, 45% of the realizations fall into the corresponding 5% to 50% intervals, etc. Hence we can measure the quality of calibration of an expert by looking at how far the empirical distribution given by the calibration variables differs from that given by the expert. This leads to the following definition of calibration:

calibration is the likelihood of a statistical hypothesis which is defined for each expert as:

The realizations may be regarded as independent samples from a distribution corresponding to the expert's quantile assessments.

The decision maker *wants* experts for whom the corresponding statistical hypothesis is well supported by the data gleaned from the calibration variables. This is sometimes expressed as 'the decision maker wants probabilistic assessments which correspond to reality'. We may sketch the matter very crudely as follows. If an expert gives 90% confidence bands for a large number of variables, then we might expect that 10% of all variables will actually fall outside his bands. If the expert has assessed 20 variables for which the realizations are known post hoc, then 3 or 4 of the 20 variables falling outside these bands would be no cause for alarm, as this can be interpreted as sampling fluctuations. The above hypothesis would still be reasonably supported by the data. If 10 of the 20 variables fell outside the bands, we should be worried, as it is difficult to believe that so many outliers should result from fluctuations; we should rather suspect that the expert chooses his bands too narrowly. Statistical likelihood measures the degree to which data supports the corresponding statistical hypothesis.

Suppose that we observe on the basis of N calibration variables that $s_1 N$ realizations fell into the 0% to 5% interval, $s_2 N$ realizations fell into the 5% to 50% interval, etc. Then the empirical density is $(s_1, \ldots, s_4)$, and we wish to measure how close this is to the hypothesized density of $(p_1, \ldots, p_4) = (0.05, 0.45, 0.45, 0.05)$.

A way of measuring this is the *relative information* of s with respect to p,

$$I(s;p) = \sum_1^4 s_i \log\left(\frac{s_i}{p_i}\right),$$

which is always non-negative and takes its minimal value of 0 if and only if $s = p$. A good expert should give an empirical density $(s_1, \ldots, s_4)$ close to $(p_1, \ldots, p_4)$, and hence have a relative information score close to 0. It is well known that for large N the distribution of $2N$ times the relative information is approximately chi-squared distributed with three degrees of freedom,

$$P(2N\, I(s;p) \leq x) \approx \chi_3^2(x),$$

where χ_3^2 is the distribution function of the chi-squared variable with three degrees of freedom. The *calibration* of expert e is defined as the probability of getting an information score worse than (greater than or equal to) that actually obtained under the assumption that the expert's true distribution is $(p_1, \ldots, p_4)$,

$$C(e) = 1 - \chi_3^2(2N\, I(s;p)).$$

Hence, an empirical distribution s equal to the hypothesized distribution p gives the best possible calibration score of 1.

However, calibration is not the only way we should measure the quality of an expert opinion. Another criterion is *informativeness*.

10.6.2 Information

To measure informativeness, a *background measure* is assigned to each query variable. In the examples described below, the background measure is uniform or log-uniform over an 'intrinsic range' for each variable. The intrinsic range is obtained by adding a $k\%$ overshoot to the smallest interval containing all quantiles and realizations, where k is selected by the analyst (the default is $k = 10$). Probability densities are associated with the assessments of each expert for each query variable in such a way that (i) the densities agree with the expert's quantile assessments, and (ii) the densities are minimally informative with respect to the background measure, given the quantile constraints. (Informativeness is scored per variable per expert by computing the relative information of the expert's density for that variable with respect to the background measure.) When the background measure is uniform, this means that an expert's 'interpolated' distribution on a query question is uniform between the 0% and 5% quantiles, uniform between the 5% and 50% quantiles, etc., see Figure 10.1. The relative information of expert e on a given query variable is thus

$$I(e) = \sum_{i=1}^{4} p_i \log \left(\frac{p_i}{r_i} \right)$$

where $p = (0.05, 0.45, 0.45, 0.05)$ is the expert's probability and the values r_i are the background measures of the corresponding intervals. *Overall informativeness per expert is the average of the information scores over all variables.* This average is proportional to the relative information in the expert's joint distribution over all variables under the assumption that the variables are independent (see Exercise 7).[1]

Information scores are always positive and, *other things being equal*, experts with high information scores are preferred.

Note that an arbitrary choice has been made in the background measure and the intrinsic range. Changing the range by a large amount usually has a negligible effect on the decision maker. The choice of background measure usually has a small effect, but in certain cases a modest effect has been found.

The information score is a positive number, with increasing values indicating greater information relative to the background measure. Since the intrinsic range depends on the expert assessments, this range can change as

---

[1] Subjective probability distributions are seldom independent, as independence of subjective distributions indicates unwillingness to learn from experience. The decision maker is not interested in experts' joint distributions since these have not been queried. Moreover, the decision maker wants experts whose marginal distributions may be regarded as (nearly) independent as expert learning is not the primary reason for performing the elicitations.

10.6 Performance based weighting – the classical model

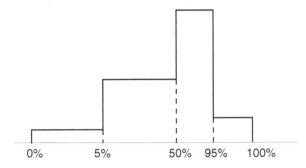

Fig. 10.1. The expert's interpolated density

experts are added or deleted, and this can exert a small influence on the information scores of the remaining experts. Information is a 'slow' function; that is, large changes in the quantile assessments produce only modest changes in the information score. On a data set with 20 realizations and 10 experts, calibration scores typically vary over four orders of magnitude, but information scores seldom vary by more than a factor 3.

10.6.3 Determining the weights

In order to determine the performance based weight that an individual expert gets we combine the information and the calibration scores. If we take the weights proportional to a product of calibration and information scores, the calibration score is most important and the information score serves to modulate between more or less equally well calibrated experts (this prevents the eventuality that very informative distributions – very narrow confidence bands – should compensate for very poor calibration).

In fact there is another property that we shall also insist upon for our expert weights. This property, defined below, of being *weakly asymptotically proper*, ensures that if an expert wishes to maximize his/her long run expected weight then he/she should do this by simply giving his/her true beliefs as answer to the query questions. This property is achieved by introducing a cut-off level α so that experts with a calibration score under α are automatically given a score of zero. This will be explained further below.

The ways in which weights are determined on the basis of information and calibration is shown in Figure 10.2. The distributions of three experts on four different items are given, together with realizations of each of the items. Experts 1 and 2 are equally well calibrated, and are both better calibrated than Expert 3. Experts 1 and 3 are equally informative, and are both more

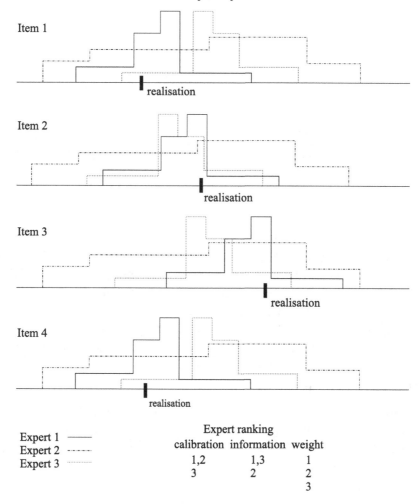

Fig. 10.2. Expert ranking

informative than Expert 2. Expert 1 gets the highest weight, followed by Expert 2 and finally Expert 3.

It is important to realize that the weight w_i of expert i used in the combination of expert opinions does *not* have the interpretation of 'the probability that expert i is correct'. The fact that the experts' weights sum to 1 does not imply that precisely one expert can be correct. It is perfectly possible for two experts to give different answers, but to both have precisely the right proportions of realizations in their quantile 'bins'. Hence it is not possible to consider the correctness of the experts as 'exclusive'. Instead, the experts' weights may be interpreted as *scores*.

10.6 Performance based weighting – the classical model

Definition 10.1 *A scoring rule R for a single uncertain quantity taking values $1,\ldots,n$ is a function giving reward $R(\underline{p},i)$ for a probability forecast of $\underline{p}$ when there is a realization i. The expected reward for a subjective probability $\underline{p}$ when the expert believes that the true distribution $\underline{q}$ is $E_q R(\underline{p}|i) = \sum_1^n q_i R(\underline{p},i)$. A scoring rule is proper if, for all $\underline{p}$ and is $\underline{q}$, $E_q R(\underline{p}|i)$ is maximized uniquely when $\underline{q} = \underline{p}$.*

Thus, when a proper scoring rule is used, an expert maximizes his expected score by stating the probabilities he believes to be correct. An example of a proper scoring rule is given by taking $R(\underline{q},i) = \log q_i$. Then the expected reward given a subjective probability $\underline{p}$ is $\sum_i p_i \log(q_i)$, which we recognize as the relative information. In the classical model more than one calibration variable is used, and so a generalization of the notion of a proper scoring rule is used to give a score on the basis of a set of assessments and realizations.

Suppose that the expert believes that the collection M of uncertain quantities $X_1, \ldots, X_m$, taking values $1,\ldots,n$, have a joint distribution Q. The expected relative frequency of outcome i is

$$q_i = E\frac{\#\{X_j = i\}}{m} = \frac{1}{m} E\left(\sum_j 1_{X_j=i}\right) = \frac{1}{m}\sum_j Q(X_j = i).$$

Suppose that we have a scoring rule giving reward $R(\underline{p}, M, \underline{s})$ when the expert states expected relative outcome frequency $\underline{p}$ to the set of M variables while the observed relative outcome frequency is $\underline{s}$. The expert's expected score is

$$E_Q(R(\underline{p}, M, \underline{s})).$$

Definition 10.2 *The rule $R(\underline{p}, M, \underline{s})$ is a strictly proper scoring rule for expected relative frequencies if*

$$\operatorname{argmax}_{\underline{p}} E_Q(R(\underline{p}, M, \underline{s})) = \underline{q}.$$

A rule is asymptotically strictly proper if for all $\underline{p} \neq \underline{q}$, for sufficiently large M (depending on $\underline{p}$),

$$E_Q(R(\underline{q}, M, \underline{s})) > E_Q(R(\underline{p}, M, \underline{s})).$$

To explain how the classical model gives weights that are asymptotically proper, we first have to explain how a set of weights for experts is used to make a combined distribution function.

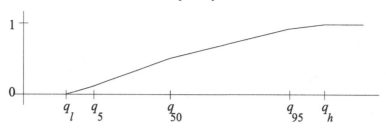

Fig. 10.3. Interpolation of expert quantiles

10.6.4 Approximation of expert distributions

In the implementation of the classical model, the experts are asked for a limited number of quantiles – typically 5%, 50%, and 95% quantiles – for each query variable. The analyst is asked to make a choice of scale for each query variable (logarithmic or uniform). If the scale is logarithmic then all the quantiles are logged before applying the same procedure as for uniform query variables. We therefore assume that the scale is uniform to explain the rest of the procedure.

Since we only have say 5%, 50%, and 95% quantiles we have to interpolate the rest of the distribution on a particular query variable. Let $q_i(e)$ be the i% quantile of expert e. The intrinsic range is obtained as follows with the k% overshoot rule. First we find the lowest and highest values named,

$$\ell = \min\{q_5(e), r|e\}, \quad h = \max\{q_{95}(e), r|e\},$$

where r is the value of the realization (if there was one on this query variable). Then we set

$$q_l = \ell - 0.1 \times [h - \ell],$$

and similarly

$$q_h = h + 0.1 \times [h - \ell].$$

The *intrinsic range* is thus $[q_l, q_h]$. The distribution of expert e is then approximated by linear interpolating the quantile information $(q_l, 0)$, $(q_5, 0.5)$, $(q_{50}, 0.5)$, $(q_{95}, 0.95)$, and $(q_h, 1)$. This is the distribution with minimum information (with respect to the uniform distribution on the intrinsic range) that satisfies the expert's quantiles. See Figure 10.3.

Although the above choice for interpolating the experts' quantiles is to some extent arbitrary, it generally makes relatively little difference to the weights given to the experts. This is because the calibration scores which usually drive the weighting *only* depend on the quantiles, and not on the interpolation. The information score depends only on the quantiles and the

10.6 Performance based weighting – the classical model

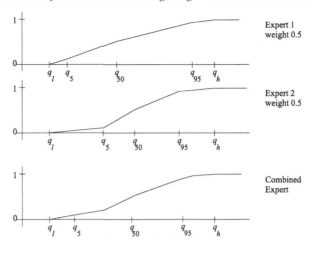

Fig. 10.4. Combination of expert distributions

choice of q_l and q_h. The interpolation does however make a difference to the estimate of the combined distribution, and hence influences the cut-off level for the weights.

The above procedure gives us a distribution function $F_e(t)$ for each expert on the current query variable. Given weights for each expert w_e, the combined distribution function is now $\sum_e w_e F_e(t)$. This is shown for two distributions and equal weights in Figure 10.4.

10.6.4.1 Calculation of the DM distribution

We now have all the elements needed to determine the output distribution for a given query variable. The calibration and information scores have been determined for each expert. In order to determine the weights we have to make a choice of the cut-off level α. For each choice of cut-off level we can determine weights (the weights depend on α because for higher values of α more experts are excluded, and the weights are more concentrated on the remaining experts), and hence also a combined expert (also depending on α). For this combined expert we can also calculate the 5%, 50%, and 95% quantiles for each query variable. Hence we are able to calculate calibration and information scores for the combined expert. In the classical model the choice of α is made by looking at the weight that the combined expert would get if he were to be scored within the pool of experts, and choosing that value of α that maximizes this weight.

10.7 Case study – uncertainty in dispersion modeling

In the joint NRC–EU study on uncertainty analysis for the consequences of a nuclear accident [Harper *et al.*, 1994] expert judgement methods were used to quantify the uncertainty in the parameters of dispersion models.

Experts are not asked to directly quantify their uncertainty over model parameters. This is because model parameters are often abstract, without a clear physical interpretation. Furthermore experts often have 'favorite' models and may be unwilling to work with another model when it has been chosen for programmatic reasons. Experts are therefore only asked to give uncertainties over physically realizable quantities. These uncertainties are then processed into uncertainties over model parameters by a procedure called post-processing that is explained further in Chapter 16.

Before the expert elicitation takes place it is necessary to determine what important information should be given to the experts, and for which physical variables no information is given. The expert is then asked to take account of this lack of information in determining his quantiles in answer to the elicitation questions.

For the dispersion problem it was decided to give information about the meteorological conditions by specifying temperature lapse rate (the change in temperature as one moves upwards), the average wind speed, the standard deviation of wind direction at 10 m measured over 10 minutes, the Monin–Obukhov length L (a measure of atmospheric stability), the surface roughness, the release height, and the sampling time (that is, the time after the release relevant for the elicitation questions).

The questions and data given here are taken from Appendix F of [Harper *et al.*, 1994], but the number of questions and the number of experts have been reduced to make the case more suitable for illustrative purposes. There were originally 8 experts, 23 calibration questions and 78 other questions. Each variable is given an item name in brackets, for example '(B-3-300)', which is used in the range graphs given below to illustrate the quantiles given the experts for each question.

The answers to these questions are given in Tables 10.1–10.5.

The expert data is shown for each item, together with the combined expert (the decision maker DM) produced by the model in Figures 10.5 and 10.6. These figures show, for each question (except Question 5), the 5, 50 and 95% intervals given by each expert and a realization for each case. It is straightforward to see from these graphs that some experts are better at capturing the realizations than others. The better experts are rewarded with a higher calibration score. Those experts with smaller uncertainty bands

10.7 Case study – uncertainty in dispersion modeling

Question 1

Meteorological conditions	
Temperature lapse rate	unknown
Average wind speed	8 m/s
Standard deviation of wind direction at 10 m measured over 10 min	15°
Monin–Obukhov length $1/L$	-0.02/m
Surface roughness	flat
Release height	22 m
Sampling time	60 min

Elicitation questions			
Variable	0.05 Quantile	Median	0.95 Quantile
(B-3-300) ground level concentration (χ_g/Q) at 300 m downwind			
(B-3-600) ground level concentration (χ_g/Q) at 600 m downwind			

Question 2

Meteorological conditions	
Temperature lapse rate	-1 K/100 m
Average wind speed	3 m/s
Standard deviation of wind direction at 10 m measured over 10 min	10°
Monin–Obukhov length $1/L$	unknown
Surface roughness	flat
Release height	22 m
Sampling time	60 min

Elicitation questions			
variable	0.05 Quantile	Median	0.95 Quantile
(B-4-300) ground level concentration (χ_g/Q) at 300 m downwind			
(B-4-600) ground level concentration (χ_g/Q) at 600 m downwind			

Question 3

Meteorological conditions	
Temperature lapse rate	stable
Average wind speed	1.9 m/s
Standard deviation of wind direction at 10 m measured over 10 min	6°
Monin–Obukhov length $1/L$	unknown
Surface roughness	urban and rural
Release height	45 m
Sampling time	60 min

Elicitation questions for release duration 60 min			
Variable	0.05 Quantile	Median	0.95 Quantile
(C-60-1) Ground level concentration (χ_g/Q) at 360 m downwind			
(C-60-2) Ground level concentration (χ_g/Q) at 970 m downwind			
(C-60-3) Ground level concentration (χ_g/Q) at 1970 m downwind			

Question 4

Meteorological conditions	
Temperature lapse rate	stable
Average wind speed	3 m/s
Standard deviation of wind direction at 10 m measured over 10 min	unknown
Monin–Obukhov length $1/L$	unknown
Surface roughness	flat
Release height	12 m
Sampling time	1 min

Elicitation questions for downwind distance 60 m			
Variable	0.05 Quantile	Median	0.95 Quantile
(D-60 sig-z) Standard deviation of vertical concentration (σ_z)			
(D-60 sig-y) Standard deviation of crosswind locations at release height (σ_y)			

Question 5

Meteorological conditions	
Temperature lapse rate	$-2\,\text{K}/100\,\text{m}$
Average wind speed	$2\,\text{m/s}$
Standard deviation of wind direction at 10 m measured over 10 min	$25°$
Monin–Obukhov length $1/L$	unknown
Surface roughness	urban and rural
Release height	10 m
Sampling time	60 min

Elicitation questions for standard deviation of crosswind locations at release height, σ_y			
Variable	0.05 Quantile	Median	0.95 Quantile
(A1-1-0.5) downwind distance 500 m σ_y Centerline concentration χ_c/Q			
(A1-1-1) downwind distance 1000 m σ_y Centerline concentration χ_c/Q			
(A1-1-3) downwind distance 3000 m σ_y Centerline concentration χ_c/Q			
(A1-1-10) downwind distance 10000 m σ_y Centerline concentration χ_c/Q			

are more informative and have a higher information score. We show how to perform the exact calculations in the exercises at the end of the chapter (the calculations are simple enough to be done on a spreadsheet, but were performed here with a computer code EXCALIBR). The information and calibration scores of the experts are shown in Table 10.6.

A number of remarks are in order here. The weights given in the column 'normalized weight, no DM' are those used to combine the various experts to form the synthetic expert (called here the DM, decision maker). The calibration scores of the experts differ more strongly than the information scores. This does not have to occur but usually does. Hence the calibration scores tend to be the more important factor driving the weighting. This is as it should be, for an expert who has a good information score but a poor calibration score will be surprised by almost all the realizations. We want

Table 10.1. *Expert 1 data*

Item	5%	50%	95%
A-1-0.5 sig-y	1.10E+2	2.15E+2	3.20E+2
A-1-1 sig-y	1.94E+2	3.81E+2	5.68E+2
A-1-3 sig-y	4.54E+2	8.90E+2	1.326E+3
A-1-10 sig-y	1.056E+3	2.071E+3	3.086E+3
A-1-0.5 chi/Q	1.54E-6	7E-6	2.39E-5
A-1-1 chi/Q	2.01E-7	9.19E-7	3.15E-6
A-1-3 chi/Q	3.63E-8	1.40E-7	4.04E-7
A-1-10 chi/Q	1.56E-8	6.02E-8	1.74E-7
B-3-300	2.92E-6	1.05E-5	2.81E-5
B-3-600	1.22E-6	4.48E-6	1.24E-5
B-4-300	2.60E-6	2.52E-5	1.83E-4
B-4-600	6.50E-6	2.39E-5	6.59E-5
C-60-1	1.00E-12	9.05E-9	6.69E-5
C-60-2	2.44E-7	4.66E-6	6.68E-5
C-60-3	1.28E-6	5.91E-6	2.07E-5
D-60 sig-z	3.40E-1	7.00E-1	1.00E+0
D-60 sig-y	7.00E-1	1.50E+0	2.20E+0

Table 10.2. *Expert 3 data*

Item	5%	50%	95%
A-1-0.5 sig-y	1.20E+2	1.70E+2	2.50E+2
A-1-1 sig-y	2.50E+2	3.50E+2	5.00E+2
A-1-3 sig-y	5.00E+2	7.00E+2	1.00E+3
A-1-10 sig-y	1.40E+3	2.20E+3	3.00E+3
A-1-0.5 chi/Q	2E-6	5E-6	1E-5
A-1-1 chi/Q	5E-7	1E-6	4E-6
A-1-3 chi/Q	8E-8	2E-7	8E-7
A-1-10 chi/Q	7E-9	6E-8	1.4E-7
B-3-300	2.00E-6	1.00E-5	4.00E-5
B-3-600	1.00E-6	5.00E-6	2.00E-5
B-4-300	8.00E-6	3.00E-5	1.00E-4
C-60-1	3.00E-8	7.00E-6	3.00E-5
C-60-2	5.00E-7	2.00E-5	5.00E-5
C-60-3	1.00E-6	5.00E-6	2.00E-5
D-60 sig-z	2.00E+0	3.00E+0	5.00E+0
D-60 sig-y	2.00E+0	4.00E+0	8.00E+0

to have experts who are not surprised by reality. Several experts have been given weight zero, so that they do not contribute to the final combined decision maker. This does not always occur, but frequently does. An expert with weight zero is not necessarily a bad expert, but may just be less good than another expert who is saying roughly the same in a slightly better way. On the other hand, an expert with weight zero may simply be overconfident and inaccurate.

Another way of fixing expert weights is simply to assign equal weight to

Table 10.3. *Expert 4 data*

Item	5%	50%	95%
A-1-0.5 sig-y	7.70E+1	1.55E+2	3.10E+2
A-1-1 sig-y	1.50E+2	3.00E+2	6.00E+2
A-1-3 sig-y	4.15E+2	8.30E+2	1.66E+3
A-1-10 sig-y	4.46E+2	2.23E+3	8.92E+3
A-1-0.5 chi/Q	7.2E-7	7.2E-6	7.2E-5
A-1-1 chi/Q	1.9E-7	1.9E-6	1.9E-5
A-1-3 chi/Q	2.2E-8	2.2E-7	2.2E-7
A-1-10 chi/Q	1.6E-9	3.2E-8	6.4E-7
B-3-300	4.20E-6	2.10E-5	1.00E-4
B-3-600	1.30E-6	6.30E-6	3.20E-5
B-4-300	1.30E-5	8.60E-5	6.90E-4
B-4-600	5.40E-6	3.60E-5	2.90E-4
C-60-1	2.10E-7	2.10E-6	2.10E-5
C-60-2	2.10E-6	2.10E-5	2.10E-4
C-60-3	1.00E-6	1.00E-5	1.00E-4
D-60 sig-z	5.00E-1	1.70E+0	3.00E+0
D-60 sig-y	1.30E+0	2.60E+0	1.50E+1

Table 10.4. *Expert 5 data*

Item	5%	50%	95%
A-1-0.5 sig-y	1.40E+2	2.05E+2	3.00E+2
A-1-1 sig-y	2.50E+2	3.65E+2	5.50E+2
A-1-3 sig-y	5.50E+2	8.50E+2	1.30E+3
A-1-10 sig-y	1.50E+3	3.90E+3	9.00E+3
A-1-0.5 chi/Q	4E-6	8E-6	2.5E-5
A-1-1 chi/Q	1E-6	2.7E-6	6E-6
A-1-3 chi/Q	2E-7	5.2E-7	1.3E-6
A-1-10 chi/Q	1.5E-8	3.4E-8	1E-7
B-3-300	5.00E-6	1.40E-5	4.00E-5
B-3-600	1.50E-6	4.60E-6	1.30E-5
B-4-300	2.80E-5	8.40E-5	2.50E-4
B-4-600	1.30E-5	3.80E-5	1.10E-4
C-60-1	1.00E-6	4.00E-6	1.20E-5
C-60-2	6.00E-6	1.70E-5	5.00E-5
C-60-3	4.00E-6	1.20E-5	4.00E-5
D-60 sig-z	5.00E-1	2.00E+0	6.00E+0
D-60 sig-y	1.00E+0	3.00E+0	8.00E+0

all the experts involved in the study. This is the route taken by NUREG 1150 (see for example [Hora and Iman, 1989]). Arguments in favor of equal weights given in [Harper *et al.*, 1994] are: simplicity; the weights are fixed and cannot be tampered with; outliers provided by one expert are not lost by giving that expert weight zero. Furthermore, a number of studies are cited that used equal weight schemes. The equal weight scheme could be applied to the expert data given above and gives the calibration and information

Table 10.5. *Expert 8 data*

Item	5%	50%	95%
A-1-0.5 sig-y	3.90E+1	7.80E+1	1.56E+2
A-1-1 sig-y	7.60E+1	1.53E+2	3.06E+2
A-1-3 sig-y	2.11E+2	4.21E+2	8.42E+2
A-1-10 sig-y	5.65E+2	1.13E+3	2.26E+3
A-1-0.5 chi/Q	8.5E-6	3.4E-5	1.36E-4
A-1-1 chi/Q	2.17E-6	8.66E-6	3.46E-5
A-1-3 chi/Q	2.63E-7	1.50E-6	4.2E-6
A-1-10 chi/Q	4.4E-8	1.76E-7	7.04E-7
B-3-300	9.53E-6	3.81E-5	1.524E-4
B-3-600	5.75E-6	2.30E-5	9.20E-5
B-4-300	2.53E-5	1.01E-4	4.04E-4
B-4-600	1.525E-5	6.10E-5	2.44E-4
C-60-1	7.50E-10	6.10E-9	4.90E-8
C-60-2	7.00E-7	5.40E-6	4.40E-5
C-60-3	1.35E-6	5.40E-6	2.00E-5
D-60 sig-z	8.90E-1	1.77E+0	3.54E+0
D-60 sig-y	1.14E+0	2.27E+0	4.54E+0

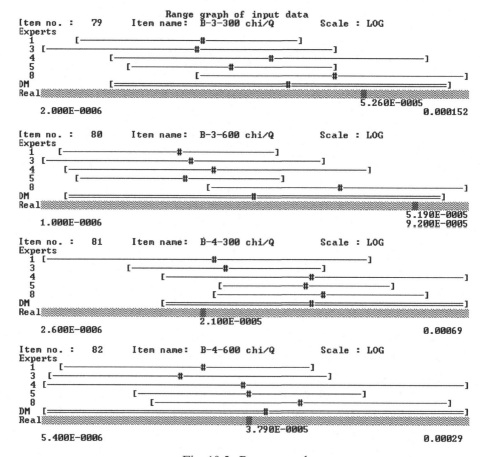

Fig. 10.5. Range graphs

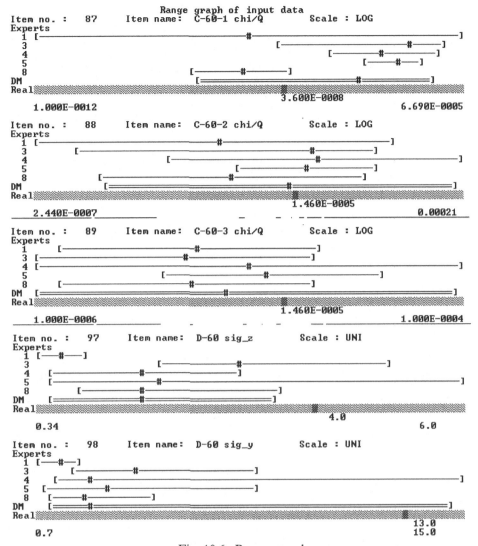

Fig. 10.6. Range graphs

results for the combined decision maker shown in Table 10.7. The results show that the equal weight decision maker has a lower calibration score than the performance-based decision maker. This does not have to happen, but typically does. The potential advantage of performance-based weighting above equal weight is that it makes the study less sensitive to the particular selection of experts. Equal weight seem fair pre adhoc, but beg the question of how the experts were selected. Clearly, any pressure group taking part in the process should maximize its influence by maximizing 'its' number of experts. The performance-based weighting involves a pre-hoc commitment

Table 10.6. *Global weights for the five experts*

Expert	Calibration	Mean rel. inform. total	Mean rel. inform. realiz.	number realiz.	unnorm. weight	Norm. weight no DM	Norm. weight with DM
1	0.001	1.146	1.435	9	0	0	0
3	0.050	1.136	1.265	9	0	0	0
4	0.200	0.870	1.178	9	0.23556	0.70661	0.24746
5	0.005	1.141	1.480	9	0	0	0
8	0.070	1.245	1.397	9	0.09781	0.29339	0.10275
DM	0.590	0.770	1.048	9	0.61855		0.64980

Table 10.7. *Equal weights for the five experts*

Expert	Calibration	Mean rel. inform. total	Mean rel. inform. realiz.	number realiz.	unnorm. weight	Norm. weight no DM	Norm. weight with DM
DM	0.320	0.690	0.934	9			0.64980

to the results of the study including the weights as they come out of the calibration exercise. A group that accepts the principles pre adhoc must have excellent technical reasons for rejecting the results of the study post adhoc. It is for this reason that performance-based weights give a better opportunity to build rational consensus around subjective uncertainties than equal weights.

The output of this study is a combined decision maker distribution over the four uncertain quantities for which no realizations were available. We shall see in Chapter 16 how the uncertainty distributions over these four (potentially observable) variables can be translated into uncertainty distributions over the (unobservable) model parameters.

10.8 Exercises

10.1 Assume that the errors ξ_i in the Apostolakis and Mosleh model are lognormally distributed and that $x_i = x\xi_i$, and show

$$E(x \mid x_1, \ldots x_e) = \prod_{i=1}^{e+1} \frac{x_i^{w_i}}{e^{w_i \mu_i}}. \qquad (10.7)$$

10.2 Do the exercises in the proof of Proposition 10.2.

10.3 Prove that $\mathscr{P}_1$ is the only r-norm probability to possess the strong set-wise function property.

10.4 Prove that $\mathscr{P}_1$ is the only r-norm probability which possesses the marginalization property.

10.5 Prove that $\mathscr{P}_0$ has the independence preservation property, and give an example to show that $\mathscr{P}_1$ does not have this property.

10.6 Experts assess 5%, 50%, and 95% quantiles for 10 variables. Their empirical distributions are

$$\begin{aligned} s(1) &= (0.1, 0.4, 0.4, 0.1), \\ s(2) &= (0.2, 0.4, 0.4, 0), \\ s(3) &= (0.2, 0.8, 0, 0), \\ s(4) &= (0.1, 0.7, 0.2, 0.1). \end{aligned}$$

Compute the relative information

$$I(s;p) = \sum_1^4 s_i \log\left(\frac{s_i}{p_i}\right),$$

for each expert.

10.7 The relative information for densities f and $g > 0$ is defined as $I(f;g) = \int f(x) \log\left(\frac{f(x)}{g(x)}\right) dx$. If f_i and $g_i > 0$ are densities on an interval A_i ($i = 1, \ldots, n$), show that for the product densities (that is, the independent densities) $\prod f_i$ and $\prod g_i$ on $A_1 \times A_2 \times \cdots \times A_n$ we have

$$I\left(\prod f_i \mid \prod g_i\right) = \sum_{i=1}^n I(f_i; g_i).$$

10.8 Check the results given in Table 10.6 (for example by programming the calculations in a spreadsheet). (Note that EXCALIBR rounds off the calculations at several places, so you will not get exactly the same results.)

11

Human reliability

11.1 Introduction

In many complex systems involving interaction between humans and machines, the largest contribution to the probability of system failure comes from basic failures or initiating events caused by humans. Kirwan ([Kirwan, 1994], Appendix 1) reviews twelve accidents and one incident occurring between 1966 and 1986, including the Space Shuttle accident and Three Mile Island, all of which were largely caused by human error. The realization of the extent of human involvement in major accidents has, in the Netherlands, led to the choice of a completely automated decision system for closing and opening that country's newest storm surge barrier.

Since humans can both *initiate* and *mitigate* accidents, it is clear that the influence of humans on total system reliability must be considered in any complete probabilistic risk analysis.

The first human reliability assessment was made as part of the final version of the WASH-1400 study. At that time the methodology was largely restricted to studies on the failure probability for elementary tasks. A human error probability, HEP, is the probability that an error occurs when carrying out a given task. In many situations in which human reliability is an important factor the operator has to interpret (possibly incorrect) instrumentation data, make deductions about the problems at hand, and take decisions involving billion dollar trade-offs under conditions of high uncertainty. Human reliability assessment attempts to indicate the probability of human error for these situations as well.

Some generic HEPs from WASH-1400 are shown in Table 11.1. Note that the term 'human error rate' here is used to mean human error probability, and should not be confused with the frequency of errors per unit time.

In the 1980s a number of different systems for estimating human error

Table 11.1. *Some generic HEPs used in WASH-1400*

Estimated rates	Activity
10^{-4}	Selection of a key-operated switch rather than a non-key switch (this value does not include the error of decision where the operator misinterprets situation and believes key switch is correct choice).
10^{-3}	Selection of a switch (or pair of switches) dissimilar in shape or location to the desired switch (or pair of switches), assuming no decision error. For example, operator actuates large handled switch rather than small switch.
3×10^{-3}	General human error of commission, e.g., misreading label and therefore selecting wrong switch.
10^{-2}	General human error of omission where there is no display in the control room of the status of the item omitted, e.g., failure to return manually operated test valve to proper configuration after maintenance.
3×10^{-3}	Errors of omission, where the items being omitted are embedded in a procedure rather than at the end, as above.
3×10^{-2}	Simple arithmetic errors with self-checking but without repeating the calculation by redoing it on another piece of paper.
$1/x$	Given that an operator is reaching for an incorrect switch (or pair of switches), he selects a particular similar appearing switch (or pair of switches), where $x =$ the number of incorrect switches (or pair of switches) adjacent to the desired switch (or pair of switches)). The $1/x$ applies up to five or six items. After that point, the error rate would be lower because the operator would take more time to search. With up to five or six items, he doesn't expect to be wrong and therefore is more likely to do less deliberate searching.

Notes. Modification of these underlying (basic) probabilities was made on the basis of individual factors pertaining to the tasks evaluated.

Unless otherwise indicated, estimates of error rates assume no undue time pressures or stresses related to accidents.

probabilities were constructed. Each has disadvantages and advantages. No model has been more than loosely based on our knowledge of the human cognitive process because that knowledge is not yet advanced enough to make it applicable to the determination of human error probabilities. Rather, the models take into account as many factors as the model builders thought desirable, and use a 'black box' to determine probabilities.

Throughout the late 1980s and 1990s a number of benchmark studies have been carried out with the general conclusion that none of the well known models work terribly well, but that some models work less well than others. Moreover, Kirwan ([Kirwan, 1994], p. 386) points out that the conclusions of two major assessments, Swain [Swain, 1989] and the *Human Reliability Assessor's Guide* [HFRG, 1995], differ in a significant number of cases.

One of the problems in making quantitative assessments of HEPs is that there is very little actual data. Even when data is available there are major discussions about its relevance. The human cognitive reliability (HCR) method for example was fitted to control room simulator data

[Hannaman and Worledge, 1988]. Critics hastened to point out that in the simulators, the operators know that it is a simulation. Hence they never have to make the billion dollar trade-offs that would be required in a real reactor emergency.

In this chapter we review a few of the models available, and discuss some issues such as parameter determination, dependency and uncertainty. An underlying assumption of the models is that one assumes that all humans involved act with good intentions. Deliberate acts of sabotage are not considered.

Important references for HRA methods are [Swain and Guttmann, 1983], [Dougherty and Fragola, 1988], and [Sayers, 1988]. An extensive recent survey of methods is given by Kirwan [Kirwan, 1994].

11.2 Generic aspects of a human reliability analysis

11.2.1 Human error probabilities

Most practitioners of HRA define a human error probability (HEP) as

$$\frac{\text{number of failures}}{\text{number of opportunities for failure}}.$$

This frequency definition is unsatisfactory on a number of counts. Formally, the probability should be defined in terms of 'limiting relative frequency'. More importantly, the frequency definition requires a categorization of 'failure opportunities'. We should be able to identify when two distinct failure opportunities were instances of the same event. Unfortunately this seems difficult for humans, because, unlike pieces of machinery, humans can learn from their mistakes (if they survive them). Hence even if the external environment could be kept constant, no individual would face the same situation twice with the same level of experience, and no two humans will face the same situation with the same level of experience. It is maybe because of this that the judgemental methods (such as APJ and SLIM – both discussed later) have acquired a degree of popularity at least equal to the methods based on simulator data.

It would seem that – even from a practical point of view – the subjective definition of probability is the only really workable notion on which HEPs can be based.

11.2.2 Task analysis

A human reliability analysis is usually part of a larger project. The models that we discuss here are applied once a great deal of the system analysis

11.2 Generic aspects of a human reliability analysis

Table 11.2. *Stages in a human reliability analysis*

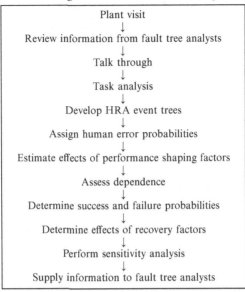

has been done. The system analysis is outside the scope of this book. However, we show the stages of a human reliability assessment given in [Swain and Guttmann, 1983] in Table 11.2. For more information on task analysis we refer to [Kirwan and Ainsworth, 1992].

11.2.3 Performance and error taxonomy

A number of HRA methods attempt to assess failure probabilities through a classification of different types of error. The error type is of course closely related to the way in which the human is performing.

The best known taxonomy of performance is the *skill-based, rule-based, knowledge-based* classification of Rasmussen [Rasmussen *et al.*, 1981]:

(i) *skill-based* refers to predetermined actions carried out in a certain order;
(ii) *rule-based* refers to actions following well defined rules such as *if...then...*
(iii) *knowledge-based* performance is required in novel situations in which knowledge of analogous situations is used together with analytic knowledge to determine and carry out a course of action.

With skill-based performance, errors will be mistaken selection of an instrument, incorrect activation of an instrument or incorrect ordering of activities.

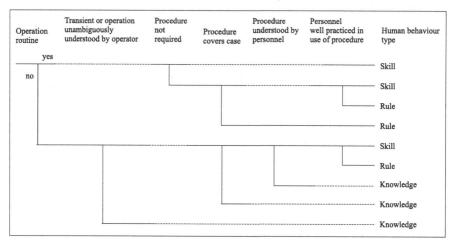

Fig. 11.1. Classification of expected cognitive performance

With rule-based performance, the errors might be in recall of the correct procedure, or in incorrectly assessing which of the possible states is pertaining. Errors during knowledge-based performance arise through incorrect or incomplete knowledge. In order to aid the analyst in the correct classification of skill, rule or knowledge based cognitive performance (as required to apply the HCR methodology that will be discussed later), Hannaman and Worledge [Hannaman and Worledge, 1988] gave a logic tree which is shown in Figure 11.1.

The work of Rasmussen has been built on by Reason [Reason, 1990], who describes his generic error-modeling system (GEMS). He distinguishes between monitoring failures and problem-solving failures. Slips and lapses tend to occur during monitoring failures due to inattention (that is, not making a necessary check), but may also occur due to over-attention (that is, making an additional check at an inappropriate moment). An important assumption in GEMS is that during problem-solving, the human operator will begin at the rule-based level and (possibly after a number of cycles through different rule-based action) after will progress to the knowledge-based when it seems that the problem cannot be solved. A switch from the KB level to the SB level is possible when an acceptable method of problem-solution has been determined and has to be implemented. The dynamics of GEMS is illustrated in Figure 11.2.

Reason lists a number of failure modes for the three performance levels. These are summarized in Table 11.3. For a detailed description of each of these failure modes, together with numerous examples, see [Reason, 1990].

11.2 Generic aspects of a human reliability analysis

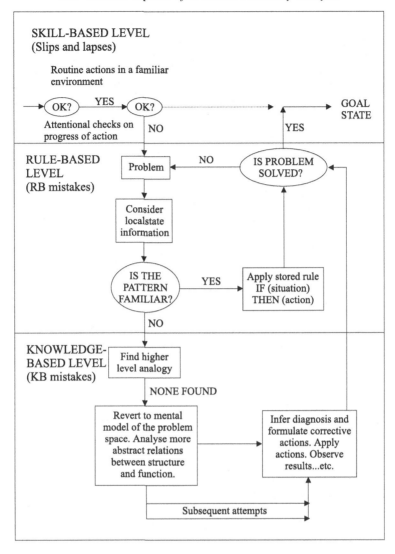

Fig. 11.2. The dynamics of GEMS

11.2.4 Performance shaping factors

Performance shaping factors (PSFs) are those factors in the environment of the operator that help determine the probability of failure. Examples of PSFs are time pressure, workload, adequacy of training, adequacy of procedures, population stereotypes, and complexity of task. Almost all quantitative methods for human reliability assessment attempt to take account of PSFs by making the HEP a function of the 'values' taken by the PSFs, but do this in different ways.

Table 11.3. *Failure modes for the three performance levels*

Skill-based performance	
Inattention	*Overattention*
Double-capture slips	Omissions
Omissions following interuptions	Repetitions
Reduced intentionality	Reversals
Perceptual confusions	
Interference errors	

Rule-based performance	
Misapplication of good rules	*Application of bad rules*
First exceptions	Encoding deficiencies
Countersigns and non-signs	Action deficiencies
Informational overload	Wrong rules
Rule strength	Inelegant rules
General rules	Inadvisable rules
Redundancy	
Rigidity	

Knowledge-based performance
Selectivity
Workspace limitations
Out of sight out of mind
Confirmation bias
Overconfidence
Biased reviewing
Illusory correlation
Halo effects
Problems with causality
Problems with complexity
Problems with delayed feedback
Insufficient consideration of processes in time
Difficulties with exponential developments
Thinking in causal series not causal nets
Thematic vagabonding
Encysting

11.3 THERP – technique for human error rate prediction

The best-known study of human reliability is the *Handbook of Human Reliability Analysis with Emphasis on Nuclear Power Plant Applications* by A.D. Swain and H.E. Guttmann [Swain and Guttmann, 1983]. This part of this chapter draws on that report. A methodology is described in the report which enables the assessment of possible human errors and estimation of

11.3 THERP – technique for human error rate prediction

the likely error probabilities. This methodology is called the technique for human error prediction (THERP).

THERP arose originally from the need to quantify the probability of human error in a bomb-assembly plant in the USA [Dougherty and Fragola, 1988], and was used in an early form in the WASH-1400 study. The methodology is primarily directed to nuclear power plant applications, but has frequently been applied in other contexts.

The THERP methodology is based on the use of event trees. A task analysis is used to determine the paths of operator actions that can lead to failure, and these paths are modeled in an event tree. Each node of the event tree corresponds to a particular action, and the branches correspond to operator failure or success in carrying out that action. Using the THERP database, the probabilities of failure and success for each action are determined. Using PSF factors and the THERP dependency model, these probabilities are adjusted to take account of the relevant PSFs and possible dependencies on previous actions. Finally, THERP contains an uncertainty model for human error probabilities and a diagnostic model to model the time required by the operator to make a correct diagnosis.

The THERP database uses a classification of error types as follows:

- Errors of omission
  - Omits step in task
  - Omits entire task
- Errors of commission
  - Selection error
    - Selects wrong control
    - Mispositions correct control
    - Issues wrong command (via voice or writing)
  - Error of sequence
  - Time error
    - Too early
    - Too late
  - Quantitative error
    - Too little
    - Too much

A *human error* is defined as a human output that has the potential for degrading the system in some way.

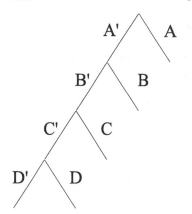

A: Control room operator omits ordering the following tasks

B: Operator omits verifying the position of MU-13

C: Operator omits verifying/opening the DH valves

D: Operator omits isolating the DH pump room

Fig. 11.3. Example human error event tree

11.3.1 Human error event trees

The human error event tree is the basic tool within THERP. An event tree indicating the sequence of actions carried out by the operator is given, and (unless recovery actions are specifically modeled) failure in carrying out any one step leads to human error.

A small example taken from [NRC, 1983] is shown in Figure 11.3. The event tree represents possible actions of an auxiliary operator outside a control room. The operator is first supposed to correctly order the tasks in hand (administrative error possible). Then he should verify that equipment MU-13 (a valve) is closed (error of omission possible). Thirdly he should verify that a group of DH valves are open (error of omission possible) and fourthly, he should isolate the DH pump room by locking the watertight doors (error of omission possible).

The first task is seen by the analyst as a failure to carry out plant policy, and is assigned an HEP of 0.01, based on selecting the appropriate table from the THERP *Handbook*. Furthermore an error range is given, in this case with a range of 0.005 to 0.05, whose meaning will be discussed in subsection 11.3.5. For the other errors shown in the event tree, the analyst assumes that the operator is working not from a set of written procedures but from oral instructions, and that the tasks are seen as three distinct unit tasks. Based on the appropriate table of HEPs in the THERP *Handbook*, the HEP is estimated as 0.01 with a range of 0.005 to 0.05. These HEPs may then be adjusted to take account of dependencies and performance shaping factors.

Table 11.4. *HEP modification factors, Table 20-16 from [Swain and Guttmann, 1983]*

Stress level	Task class	Skilled	Novice
Very low		2	2
Optimum	SBS	1	1
	DYN	1	2
Moderately high	SBS	2	4
	DYN	5	10
Extremely high	SBS	5	10
	DYN	0.25 ef=5*	0.5 ef=5*

* These are the actual HEPs, not the modifiers

11.3.2 Performance shaping factors

All baseline probabilities in THERP are changed by factors which represent the different circumstances under which a human operator is working. One example is poor lighting in the control room. If this PSF is present then the HEPs should be taken higher than the nominal values given.

For various PSFs, THERP gives modification factors which should be used to multiply the HEP. As an example, in Table 11.4 (taken from Table 20-16 of [Swain and Guttmann, 1983]) the modifiers for nominal HEPs are given, divided up into numbers for 'skilled' and 'novice' workers (a skilled person is one with more than six months experience in the tasks being assessed). The tasks are classified as *step-by-step (SBS) tasks*, which are routine, procedurally guided tasks and *dynamic tasks (DYN)* which include decision making, keeping track of several functions, controlling several functions, etc.

11.3.3 Dependence

It is easy to see that probabilistic dependence will be important in modeling human error. Dependence can occur both within and between people. For example, for one person activating two adjacent switches, the failure probabilities are likely to be highly dependent. If one person performs an action (such as throwing a switch) and another person checks that the action has been performed correctly, then the performances of the two are not likely to be independent.

Within THERP, a classification is made of pairs of events describing the level of dependency. This level can be zero (in which case the events are independent), low, moderate, high or complete (in which case there is complete determination). Then a linear model of the following kind is applied

Table 11.5. *Levels of dependency*

Level of dependency	a	b
Zero	0	1
Low	0.05	0.95
Moderate	0.14	0.86
High	0.5	0.5
Complete	1	0

to determine the probability $P(X|Y)$ in terms of the probability $P(X)$:

$$P(X|Y) = a + bP(X),$$

where a and b are given in Table 11.5.

This method of modeling dependence does not have a clear mathematical justification.

11.3.4 Time dependence and recovery

A part of the THERP methodology allows the use of probability distributions for *time to diagnosis*. Such distributions are now called TRC distributions (time reliability correlation).

Since such distributions vary over the population, one should estimate the median probability and upper and lower estimates. Such estimates have been made at the National Reliability Evaluation Program (NREP) Reliability Data Workshop in 1982, by group consensus. Lognormality for the time to diagnosis was assumed. An example graph from the THERP *Handbook* is shown in Figure 11.4.

11.3.5 Distributions for HEPs

An underlying assumption in human reliability analysis is that, given a particular task, each individual has a certain probability of failure. In order to model the use of a whole 'population' of individuals one chooses a probability distribution on $[0, 1]$, the space of failure probabilities.

The distribution taken here is the lognormal (X is lognormally distributed if $\log X$ is normally distributed). See Figure 3.4 for the density function of a lognormal distribution. Since we are modeling a probability, which must always lie in the interval $[0, 1]$, the distribution has to be truncated. Swain and Guttmann [Swain and Guttmann, 1983] suggest use of the lognormal distribution on the grounds that

11.3 THERP – technique for human error rate prediction

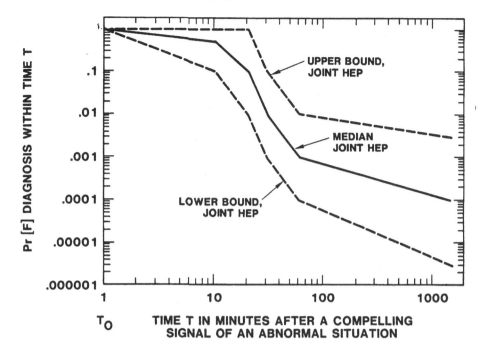

Fig. 11.4. Example human time reliability correlation

(i) '...the performance of skilled persons tends to bunch up toward the low HEPs .... it is appropriate, for PRA purposes, to select some non-symmetric distribution...'

(ii) '...the lognormal distribution appears to provide an adequate fit to human performance data and is computationally tractable. For example, the parameters of a lognormal distribution can be readily determined by specifying two of its percentiles.'

The lognormal distribution can be specified by two parameters. Usually these two parameters are the median and the error factor (the ratio of the 95% and 5% percentiles; note that many authors define the error factor as the ratio of the 95% and 50% quantiles). Figure 11.5 gives an example in which the error factor is 4. The median is chosen as a 'representative' value of the distribution of HEPs (the median is frequently chosen in preference to the mean as representative value for the distribution of a skewed distribution). The figures quoted in the tables as HEPs are considered to be the median of a lognormal distribution (with standard deviation 0.42).

It is worth noting that in the *Reactor Safety Study* [NRC, 1975] it was

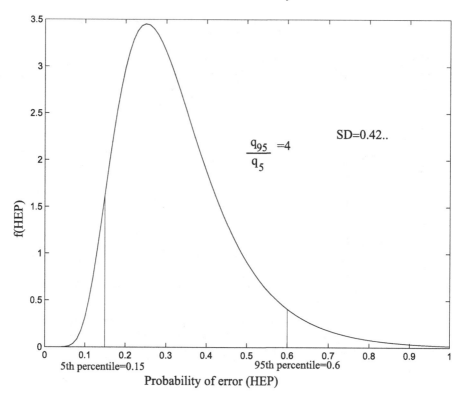

Fig. 11.5. Hypothetical lognormal density of HEPs

decided to replace the 4:1 ratio used above by 10:1 for typical tasks, and by 100:1 for tasks whose nature could not be well defined and those judged to be highly stressful.

11.4 The Success Likelihood Index Methodology

One technique that is widely used for making quantitative estimations of human reliability is the Success Likelihood Index Methodology (SLIM) [Embrey, 1984]. In this methodology the relative effects of different PSFs are estimated in carrying out various tasks. The numbers obtained are used to estimate a Success Likelihood Index for each task, which are in turn used to estimate the **HEPs** of each task.

Example 11.1 *Consider the following five tasks.*

(i) *An operator tightens some bolts to connect a feed line hose.*

11.4 The Success Likelihood Index Methodology

(ii) *An operator presses a button to vent a hose. Two red buttons are next to each other on the panel.*

(iii) *An operator types in the weight of the tanker on a keyboard.*

(iv) *An operator turns a three position valve to the horizontal right position once at the end of every week. The valve is blocked when in the vertical-down position and in the wrong position when in the horizontal-left position.*

(v) *Same as (iv) only now carried out by an operator from another plant.*

The performance shaping factors are

(i) *equipment design,*
(ii) *training level of operator,*
(iii) *task complexity,*
(iv) *distraction proneness.*

For each PSF a numerical PSF scale ranging from 1 to 9 is defined containing an ideal value (for example an ideal level of task complexity). Each task in now rated on each PSF scale (these numbers are supposed to be relative). For each PSF the relative importance weight w is also chosen (either directly or by a paired comparison method).

The rescaled rating R_{ij} of the jth PSF for the ith task is the distance to the ideal point normalized by the greatest distance (over all tasks) to the ideal point. The *Success Likelihood Index* of task i is then defined as

$$\text{SLI}_i = \sum_j w_j R_{ij}.$$

An assumption here is that the different PSFs are sufficiently 'disjoint'. In order to translate the **SLI**s into **HEP**s the following assumption is made:

$$\log_{10}(1 - \text{HEP}) = a \times \text{SLI} + b \tag{11.1}$$

where the real numbers a and b are unknown parameters. These parameters are determined by including two tasks in the study for which **HEP**s are known. This gives two linear equations in a and b from which they can be derived. A practical objection to the use of this method is that the calibration data is required. Users of the SLIM methodology therefore frequently use another method (mostly THERP) to obtain the calibration data.

Alternatives to Equation 11.1 have been suggested: Kirwan [Kirwan, 1994] mentions a failure likelihood index (**FLI**) related to the **HEP** by $\log(\text{HEP}) = a\text{FLI} + b$, and gives an example in which he uses the equation $\log(\text{HEP}) = a\text{SLI} + b$.

This method of subjectively assessing PSFs in the context of particular tasks was motivated by concepts from multi-attribute utility theory. The combined score of SLIM is analogous to a disutility, and the affine uniqueness of utility functions is invoked to justify the linear form of Equation 11.1 (although it does not justify the logarithmic transformation also used in 11.1).

In fact, there is a closer relationship with multiattribute value theory, and in particular with the weighting factors model (see Chapter 13). Kirwan [Kirwan, 1994] suggests that the close connection with decision theory gives SLIM a good theoretical basis. However, the criticisms made of the weighting factor model in Chapter 13 are applicable to SLIM.

It is worth noting that there are many models available from multi-attribute value theory, and that the additive model used in SLIM is only one of the choices that could have been made. The validity of the additive model is therefore one that should be checked in a real application. The most important theoretical point is that of *constant trade-offs*. This means that an increase of $1/w_j$ on the score for PSF j is compensated (that is, the HEP would not be changed) by a decrease of $1/w_i$ on the score for PSF i. A practical consideration is that the PSFs should not 'overlap', so that double counting is avoided: if we use the same PSF twice then its contribution to the overall score will be doubled. The pairwise comparison method for determining the weights is not sensitive to this problem. Some work has been done [Kamminga, 1988] on the use of principal component analysis and linear transformations of the PSFs to eradicate this problem.

11.5 Time reliability correlations

A major difference between human failure and hardware failure is that humans often have and take the opportunity to recover. From a human reliability point of view, it is therefore important to model the time taken to succeed. A probability distribution model for the time required for success is called a *time reliability correlation, TRC*.

THERP contains a nominal diagnosis model in the form of a TRC. This was extended in two TRC 'systems' at approximately the same time. One is due to Fragola and his coworkers [Dougherty and Fragola, 1988]. The other, the human cognitive reliability method (HCR) is described by [Hannaman and Worledge, 1988]. We discuss Fragola's method, and then briefly discuss the differences in HCR.

A fundamental classification is made into failure modes called *slips* and *mistakes*. 'Mistakes are failings of cognition – inadequacies in planning, decision making, and diagnosis; whereas slips (or lapses) are failings in cognitive control – failures in the implementation and monitoring of actions.' ([Dougherty and Fragola, 1988], p. 16). The motivation for choosing this classification is pragmatic, since it is expected that the two categories will exhibit different reliability characteristics.

Examples of slips are spatial reversal (the wrong control of two adjacent controls is used) and time reversal (carrying out a sequence of operations in the wrong order, when the order is critical), where the operator in both cases really knew what was supposed to be done.

Slips are considered to be errors for which there is no time dependence, and their probability is modeled by a simplification of THERP. Mistakes are considered to have time distributions since there is always a possibility of recovery, and their probability is modeled by using TRCs whose parameters are determined by using SLIM.

We shall concentrate here on the time dependent modeling of mistakes. Within the TRC system, a TRC models the response time of the operator in successful diagnosis or decision making (D & D). If the right diagnosis is not made within the critical period of time then this corresponds to a failure. This time is always modeled using a log-normal variable (the lognormal distribution is justified on the basis of some experimental evidence). A lognormal distribution is parameterized by the median and the error factor (defined here as the ratio of the 95 and 50 percentiles). The percentiles of the lognormal distribution can be written in terms of the percentiles of a standard normal distribution as follows. Let X be lognormally distributed, and let $Y = \log(X)$. If f is the error factor of X then it is easy to show that the standard deviation of Y, σ_Y is related to the error factor by

$$\log(f) = \Phi(0.95)\sigma_Y \approx 1.645 \times \sigma_Y.$$

The pth quantile of the lognormal distribution with median m and error factor f is

$$mf^{(\Phi(p)/1.645)}.$$

In the TRC system the time T to successfully respond to the situation is taken to be the product of two log-normally distributed variables $T = \tau_R \times \tau_U$, where the second term has median 1 and error factor f_U and accounts for the model uncertainty. The distribution of the first term is the TRC, and represents the response process. The first term is written as

$$\tau_R = \tau k_C k_{PSF} \times \ell_R$$

where τ is the median response time, k_C is a factor lying between $1/2$ and

2 to take into account 'taxonomic considerations', k_{PSF} is a factor lying between 1/2 and 2 to take account of performance shaping factors, and ℓ_R is a lognormal random variable with median 1 error factor f_R.

The base TRC giving the variable $\tau\ell_R$ is derived from the normal diagnosis curve of [Swain and Guttmann, 1983] by taking the 10 minute and 60 minute values, and solving for the two lognormal parameters. This gives $\tau = 4$ and $f_R = 3.2$. However if 'hesitancy' is present (arising from conflict, burden, uncertainty) then f_R is arbitrarily doubled to 6.4. If the operators are guided by a rule then the factor k_C is set to 0.5. The PSF factor k_{PSF} is defined as $2^{(1-2x)}$ where x is a success likelihood index calculated in SLIM with minimum value 0 and maximum value 1. The uncertainty factor τ_U has been determined on the basis of some pragmatic arguments, and has error factor $f_U = 1.68$.

The representation of uncertainty by multiplication τ_R by the independent lognormal variable τ_U allows a straightforward calculation of the uncertainty in the estimate of

$$P(\text{success after time } t)$$

for each t. The probability of success after time t, given the value of the uncertainty variable τ_U, is

$$P(\text{success after time } t | \tau_U = u) = P(\tau_R \times \tau_U > t | \tau_U = u)$$
$$= P(\tau_R > t/u),$$

which is lognormally distributed. The $q\%$ value of the uncertainty distribution is given by taking the $q\%$ quantile of τ_U, in the last formula. This method of modeling uncertainty is pragmatic. Since the uncertainty is not associated to any observable quantity it is not possible to reduce uncertainty by performing observations and then applying Bayes' Theorem. The mean value of this probability is obtained by averaging over u,

$$P(\text{success after time } t) = P(\tau_R \times \tau_U > t)$$
$$= 1 - \Phi\left(\frac{-\log(t/m)}{\sqrt{(\sigma_R^2 + \sigma_U^2)}}\right),$$

since $\log(\tau_R \times \tau_u)$ is normally distributed with mean $\log(m)$ and variance $\sigma_R^2 + \sigma_U^2$.

Some typical values of the distribution are given in Table 11.6 taken from [Dougherty and Fragola, 1988]. These are for rule based actions without hesitancy, and are based on a mean TRC which is lognormal with median 2 and error factor 3.2.

Table 11.6. *Typical values of the distribution*

Time (mins)	Success Likelihood Index				
	0.1	0.3	0.5	0.7	0.9
5	3×10^{-1}	2×10^{-1}	1×10^{-1}	6×10^{-2}	3×10^{-2}
10	9×10^{-2}	4×10^{-2}	2×10^{-2}	8×10^{-3}	3×10^{-3}
20	1×10^{-2}	5×10^{-3}	2×10^{-3}	5×10^{-4}	1×10^{-4}
30	3×10^{-3}	9×10^{-4}	3×10^{-4}	6×10^{-5}	1×10^{-5}
60	1×10^{-4}	3×10^{-5}	6×10^{-6}	1×10^{-6}	2×10^{-7}

11.6 Absolute Probability Judgement

The idea behind APJ is that expert opinion is utilized in some form to provide HEPs. The particular technique used for expert opinion combination is not fixed. The *Human Reliability Assessor's Guide* [Kirwan et al., 1988] lists five 'approaches', and recommends them in the following order (based on the maximization of information sharing):

   (i) consensus,
  (ii) Nominal Group Technique,
 (iii) Delphi,
 (iv) Aggregated Individuals,
  (v) single expert/engineering judgment.

The Nominal Group Technique mentioned here is similar to Delphi, but allows experts to have a discussion on clarification issues; the description of expert aggregation in [Kirwan et al., 1988] only specifies expert combination using the geometric averaging method (see Chapter 10).

Note that even if the above ordering does reflect the degree of information sharing, it certainly does not reflect the degree of calibration, and therefore does not give any indication of information quality.

The classical model for expert judgement described in Chapter 10 was applied successfully to a human reliability problem in [van Elst et al., 1998] to elicit distributions for completion times of various activities involved in closing movable water barriers on time. The big advantage of the classical model is that it combines the ability of every APJ method of modeling the specific problem situation with the 'internal benchmarking' process of expert calibration. In the study described in [van Elst et al., 1998], this was an important factor in the acceptability of the results.

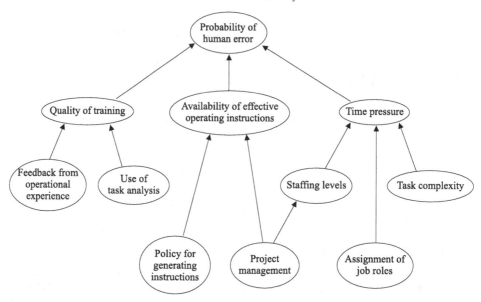

Fig. 11.6. An ID for human error probabilities

11.7 Influence diagrams

An interesting approach to the estimation of human probabilities is the application of influence diagrams (see Chapter 14 for more information about ID modeling).

Within an influence diagram framework it is possible to model the organizational and management factors that affect human failure probabilities, as well as more obvious causal factors. Figure 11.6 shows an influence diagram taken from [Embrey, 1992]. The probability of human error is expressed in terms of other factors such as 'use of task analysis', and 'task complexity'. Expert judgement is used to determine the conditional probability of each state of a node given the states of its parent nodes, and of the unconditional probabilities of the states of each node without parents. The *conditional independence property* of the ID is of extreme importance in correctly modeling the situation. For example, the HEP is assumed here to be independent of the effectiveness of project management if the staffing levels and the availability of effective operating instructions are specified (that is, your uncertainty about the HEP would not be changed if I additionally specified whether project management is effective or ineffective).

The ID approach does require many conditional probabilities to be estimated. Furthermore, it is not really clear how much of the ID structure will be generic. For two distinct but related problems, the conditional and

11.7 Influence diagrams

Table 11.7. *Marginal and conditional probabilities for the ID in Figure 11.6*

1	What is the weight of evidence for use of feedback from operational experience in developing training?			
	Good	Poor		
	0.2	0.8		

2	What is the weight of evidence for use of task analysis in developing training?			
	Used	Not used		
	0.2	0.8		

3	Given feedback and task analysis, what is the weight of evidence for quality of training?			
	Feedback	Task analysis	Quality high	Quality low
	Ggood	Used	0.95	0.05
	Good	Not used	0.80	0.20
	Poor	Used	0.15	0.85
	Poor	Not used	0.10	0.90

4	What is the weight of evidence on the effectiveness of policy for generating instructions?	
	Effective	Ineffective
	0.3	0.7

5	What is the weight of evidence on the effectiveness of project management?	
	Effective	Ineffective
	0.1	0.9

6	Given effectiveness of policy for generating instructions and project management, what is the weight of evidence for availability of good operating instructions?			
	Policy	Project management	Available	Not available
	Effective	Effective	0.90	0.10
	Effective	Ineffective	0.60	0.40
	Ineffective	Effective	0.50	0.50
	Ineffective	Ineffective	0.05	0.95

7	What is the weight of evidence for the assignment of job roles?	
	Good	Poor
	0.5	0.5

8	What is the weight of evidence for task complexity?	
	High	Low
	0.6	0.4

9	Given effectiveness of project management, what is the weight of evidence for adequacy of staffing levels?		
	Project management	Adequate	Inadequate
	Effective	0.6	0.4
	Ineffective	0.2	0.8

Table 11.7. Continued

10	Given staffing levels, assignment of job roles and task complexity, what is the weight of evidence for the degree of time pressure?				
	Staffing	Job roles	Complexity	Low	High
	Adequate	Good	Low	0.95	0.05
	Adequate	Good	High	0.30	0.70
	Adequate	Poor	Low	0.90	0.10
	Adequate	Poor	High	0.25	0.75
	Inadequate	Good	Low	0.50	0.50
	Inadequate	Good	High	0.20	0.80
	Inadequate	Poor	Low	0.40	0.60
	Inadequate	Poor	High	0.01	0.99

11	Given training quality, availability of effective operating instructions and time pressure, what is the probability of success?				
	Training quality	Instructions	Time pressure	Success	Failure
	High	Available	Low	0.9999	0.0001
	High	Available	High	0.9995	0.0005
	High	Not available	Low	0.9992	0.0008
	High	Not available	High	0.999	0.001
	Low	Available	Low	0.999	0.001
	Low	Available	High	0.993	0.007
	Low	Not available	Low	0.991	0.009
	Low	Not available	High	0.990	0.01

marginal probabilities might be different, and, if one is unlucky, the topology of the diagram might change too. The ID will however be useful in indicating what kinds of policy changes might lead to a substantial change in HEP.

The marginal and unconditional probabilities used to quantify the ID shown in Figure 11.6 are given in Table 11.7, which is also taken from [Embrey, 1992]. Confusingly, Embrey calls the numbers 'weight of evidence' instead of 'probability'. The unconditional probability of human failure is obtained by multiplying the marginal and conditional probabilities in the influence diagram using the law of total probability 4.1. For example, the unconditional probability that quality of training is high is

$$0.95 \times 0.2 \times 0.2 + 0.80 \times 0.2 \times 0.8 + 0.15 \times 0.8 \times 0.2 + 0.10 \times 0.8 \times 0.8.$$

Working through the entire diagram gives an unconditional probability of human success as 0.994.

11.8 Conclusions

Despite many years of research, there are still no very satisfactory models for human reliability. The shift towards greater automation in complex systems has moved the emphasis for human reliability problems more towards the cognitive level and away from the sub-cognitive level. The modeling difficulties have been increased considerably by this shift.

11.9 Exercise

11.1 A college professor devises an exam, types it, grades it, and makes a list of student results. The list is then handed in to university administration where the results are typed into the computer. Analyze the process from the point at which the professor starts work on the exam to the moment that a student checks for his/her grade.

12

Software reliability*

In this chapter we discuss some of the problems associated with judging the quality and reliability of software. In particular, we look at a few of the statistical models that have been used to quantify software reliability. Good sources on software reliability are the book of Musa [Musa, 1998] which is quite practical but concentrates on just a couple of software reliability models, and the survey of Littlewood in [Bittanti, 1988] which concentrates on the statistical models.

Although many of the statistical models for reliability prediction were developed in the 1970s and 1980s, [Musa, 1998] makes a strong case for the continued need for reliability prediction, in particular due to the introduction of new programming techniques such as object oriented programming.

12.1 Qualitative assessment – ways to find errors

Few programs run perfectly the first time they are executed, and programmers have to spend a long time on debugging a new program until all errors are eliminated. Many modern development environments contain error detection features so that certain logical, syntax or other errors are displayed to the programmer. In the end the software must be checked to confirm whether the specification is met. We now mention a few of the qualitative methods currently available.

12.1.1 FMECAs of software-based systems

In [ESA, 1993] an outline is given of a procedure to perform a software Failure Modes, Effects, and Criticality Analysis (FMECA). Not all authors agree, however. According to [O'Connor, 1994] (p. 243) it is not practical to perform FMECA on software, since software 'components' do not fail.

*This chapter has been co-authored by Jan G. Norstrøm

12.1.2 Formal design and analysis methods

Several so-called 'formal' software design methods have been developed. The objective is to set up a disciplined framework for specification and programming that will reduce chances of errors being created. There are also formal methods that automatically check programs. This is called 'static analysis'. These methods are still being developed and some are found in [Smith and Wood, 1989].

12.1.3 Software sneak analysis

In a software controlled system it is difficult to error trace a failure due to the interaction between the hardware and software. Hardware can fail due to degradation and the software may interpret the hardware wrongly. The hardware requirements may be wrongly specified, and when the system fails to operate as expected the failure is often only seen in the software.

Sneak analysis (SA) is briefly described in [O'Connor, 1994] as a method or as a set of methods to trace design errors. The sneak analysis or sneak path analysis approaches that exist in [ESA, 1994] can successfully be used to find problems in software that are not found via testing, see [Dore and Norstrøm, 1996]. The analysis starts by identifying a target from where to start backtracking. The software source however is not easy to verify, see [Dore and Norstrøm, 1996]. Other interesting references for sneak analysis are [US Navy, 1987], [Whetton, 1993] and [Buratti, 1981]. These references consider sneak analysis applications to other areas ranging from the chemical process industry to electric switching systems.

12.1.4 Software testing

Testing the program over a range of input conditions is an essential part of the development phase. It is an iterative process in which the designer tests the code while it is being developed. This makes it possible to correct errors while the designer has the coding fresh in memory, and code corrections are less expensive. At least the following tests should be made: operating at extreme conditions (timing, input parameter values and rate of change); ranges of input sequences; fault tolerance (error recovery). Of importance is to test the most critical input conditions. Here historical data and criticality of functions may be used to draw attention to critical combinations. Historical data may indicate what has created problems in similar programs before. Testing and test strategies are briefly treated in [Ramamoorthy and Bastani, 1982], and other references can be found there.

12.1.5 Error reporting

Reporting of software errors is important and should be done by users, developers etc. in a systematic manner. There are various systems for error reporting and various companies have different approaches: [Thayer *et al.*, 1978] contains a chapter on data collection.

12.2 Software quality assurance

The purpose of quality management is to ensure that a pre-selected software quality level has been achieved, on schedule and in a cost effective manner.

12.2.1 Software safety life-cycles

To develop a quality management system one needs a critical life-cycle reference. The IEC 61508 system is applicable to programmable safety related systems. Typical life-cycle phases are:

(i) *System concept, definition and scope of hazard and risk analysis.* An understanding of the equipment under control and its environment is developed and the control system defined.

(ii) *Risk analysis and safety requirements allocation*: Hazard and risk analysis is used to identify hazardous event scenarios of the equipment under control. Overall safety requirements are established and safety functions contained in overall requirements are allocated to the safety related subsystems.

(iii) *Design and development*: The software safety design is based on the concepts and safety requirements for the subsystems. Subsystems architectures are defined. The internal logic of each structure is determined to perform the functions specified in such a way that each module can be tested independently.

(iv) *Coding process*: The software source code is generated and reviewed to check that it conforms to the logic and requirements of the previous steps. It is debugged to that the program becomes executable.

(v) *Software module testing and integration testing*: The modules are tested and integrated to verify the system design and software architecture.

(vi) *Software operation and modification procedures*: These are procedures to ensure the functional safety of the system during operation and modification.

(vii) *Software safety validation and verification*: Here it is verified that the integrated system conforms to the functional specifications for performance, safety and reliability.

(viii) *Overall installation, validation, operation and maintenance*: The software is integrated into a system and is required to operate safely with respect to other subsystems.

The quality of the programming product can be controlled in the first phase life-cycle's in order to achieve the expected performance of the final product. Once phase (iv) is entered the quality is generally fixed.

12.2.2 Development phases and reliability techniques

Software reliability models are often referenced without trying to relate their applicability to the life-cycle phases outlined above. Different approaches should also be taken for different kinds and sizes of software, see [Ramamoorthy and Bastani, 1982]. Figure 12.1 (coming from [Ramamoorthy and Bastani, 1982]) shows a classification of reliability models into various categories related to the phase of a development project.

The classification refers to the following models or modeling concepts.

The development and debugging phase The deterministic error-counting models are the Jelinski–Moranda model and the Poisson models described later. The Bayesian error-counting models are principally due to Littlewood [Littlewood, 1981, Littlewood and Sofer, 1987].

The validation phase The Nelson model and other data domain models apply in this phase. The essential idea is to randomly select input data. The proportion of inputs causing execution errors is used as an estimate of the failure probability of the software. These models are described in [Ramamoorthy and Bastani, 1982].

Operational phase models The data domain models that are described in [Ramamoorthy and Bastani, 1982] can be applied. In many cases, the inputs to a program will not be statistically independent. There may for example be physical constraints forcing sensor inputs to be correlated. In this situation it may be reasonable to take a Markov chain approach to modeling inputs. Inputs may come first from one distribution, then after some random time from another, etc. Littlewood [Littlewood, 1975] and Cheung [Cheung, 1978] have modeled the input distribution selection mechanism as a Markov process.

244 12 *Software reliability*

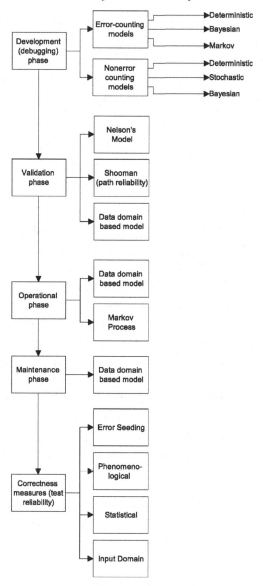

Fig. 12.1. Classification of software reliability models

Maintenance phase models For the maintenance phase only data domain models are suggested, as described in [Ramamoorthy and Bastani, 1982].

Correctness measures (test reliability) models Software for critical applications has its own status. The confidence in the estimate is very important and there exist a few techniques one may apply to measure the correctness.

Error seeding techniques as described below can be used to verify various test strategies. Phenomenological models are also sometimes called axiomatic approaches. Statistical approaches aim to give the level of confidence in the reliability estimate and are outlined in [TRW, 1976].

12.2.3 Software quality

Software quality is defined in the MIL-HDBK-217E as 'the achievement of a pre-selected software quality level within the costs, schedule and productivity boundaries established by management'. Of course, this definition may be of limited value as the quality emphasis may change with respect to the product application environment. It is agreed that managers should monitor the project/product during the life-cycle phases to meet the pre-set goals. The process of assessing the quality of a software product begins with selection of specific characteristics, quality metrics and performance criteria with respect to the system mission and system environment.

12.2.4 Software quality characteristics

This is a field which is still active and no general agreement seems to be near. Several characteristics were suggested in [Lalli and Malec, 1992], but general software characteristics are: maintainability, portability, reliability, testability, understandability, usability, freedom of error. Management will often define software quality in terms of these quality characteristics.

12.2.5 Software quality metrics

The characteristics above can be used to evaluate the quality levels of the software. At the testing stage the characteristics *freedom from error* and *reliability* may be assessed.

Two textbooks from IEEE Computer Society's working group on metrics are [Basili, 1980] and [Perlis *et al.*, 1981]. See also [Redmil, 1989].

12.3 Software reliability prediction

The use of statistical tools for software reliability prediction has been somewhat controversial. Partly this is because software as a set of computer instructions cannot fail in the same way as a physical component. However, when we take the subjective interpretation of the probability we see that there is no reason why software reliability should not be measured in terms of probabilities. Probability represents here the uncertainty of the observer that a particular piece of software will work correctly in a given system (of hardware and system software) given the input data available at that moment. Although the software will work in a deterministic way, the uncertainty of the observer about the conditions of system software, system hardware and input data leads to uncertainty about the successful operation of the software, which may be measured in terms of probability.

Another reason for controversy surrounding the use of reliability prediction is the popularity of management techniques such as total quality management in which the programmer is urged to get it 'right first time', or in which quality is measured simply by the number of defects in a program. From these points of view, reliability is not a relevant quantity.

It seems clear that the qualitative techniques within software reliability should play a very important role. Data is difficult to collect, and the data classification (in which data from 'sufficiently similar' circumstances is pooled) required to produce statistics seems a hopeless task. However, sometimes a quantitative approach is necessary, for example because a regulator requires a quantitative risk analysis of a system containing the software. Aircraft flight control systems, air traffic control systems, flood defense control systems and nuclear plant software are amongst the systems for which quantitative reliability targets have been set.

Musa [Musa, 1998] makes a distinction between software testing for reliability prediction and testing for certification. This is because the system under test is (in principle) stable during certification testing, but is being continually updated during reliability prediction in the debugging phase. In the latter case, the objective is to track reliability growth and aid decision making about the release time of the software. Musa also distinguishes three sorts of test: feature, load and regression tests. In feature tests the operations are executed separately and the interaction of the code with the field operating environment is minimal. In load tests the operations are executed simultaneously under field operating conditions. Regression tests are constituted of a number of feature tests performed after the system has been changed significantly. While certification is performed using load tests, reliability growth can be measured in any of the three sorts of test. Musa [Musa, 1998] discusses the use of the statistical technique called 'sequential testing' in order to reduce the amount of testing required during certification.

We now describe the main models for software reliability prediction. The standard approach to prediction is to use a parameterized statistical model describing the time until next failure. The parameters of the model are estimated from data, usually using the maximum likelihood principle, and are 'plugged in' to the model in order to make a prediction.

12.3 Software reliability prediction

Most models are time-based. The definition of an appropriate notion of time for software reliability prediction is critical if we want to have any reasonable level of predictability. The major contribution of [Musa, 1975] was to point to the need to use execution time instead of calendar time in the prediction models. If necessary, a separate model can be used to recalibrate execution time into calendar time, taking into account information about software use.

12.3.1 Error seeding

A simple approach to testing the quality of debugging is that of error seeding. N errors are added to the code, and if n of these are removed by the debugging team then the maximum likelihood estimate of the proportion of bugs removed is n/N.

This approach relies on the assumption that the errors added are typical of the other errors present in the code. It has the disadvantage that in order to make a good estimate it is necessary to add a large number of bugs, increasing the workload of the debugging team. It is also not clear that the number of bugs in a piece of software is a good way to measure reliability. Most other quantitative techniques try to measure the time until the next crash of the code.

12.3.2 The Jelinski–Moranda model

The Jelinski–Moranda model [Jelinski and Moranda, 1973] is a basic model on which many authors have made variations. Musa's basic model [Musa, 1975] is based on this and stresses the need to measure time using software execution time.

The main assumptions of the model are:

(i) Bugs occur at times T_1, $T_1 + T_2$, ..., and the interfailure times T_1, T_2,... are statistically independent of each other.
(ii) Bugs are corrected immediately after occurrence, without the introduction of new bugs.
(iii) The time to failure due to bug i is exponential with failure rate φ for each i.

These assumptions imply that if there are N bugs in total then the failure time to the first bug (which is the minimum of N independent exponential variables) is exponential with rate $\lambda_1 = N\varphi$. Similarly, the time between the $(i-1)$th and the ith failure is exponential with rate $\lambda_i = (N - i + 1)\varphi$ (see Exercise 1).

The model can be applied to estimate the maximum number of bugs N, the underlying rate φ, and the mean time to next failure. The likelihood of observing inter-failure times $t_1, t_2, \ldots, t_n$ is

$$\prod_{i=1}^{n}(N-(i-1))\varphi \exp(-(N-(i-1))\varphi t_i).$$

By taking logs and setting the derivatives with respect to N and φ equal to zero we obtain the maximum likelihood equations for N and φ:

$$\hat{\varphi} = \frac{n}{\hat{N}T - \sum_{i=1}^{n}(i-1)t_i},$$

$$\sum_{i=1}^{n}\frac{1}{\hat{N}-(i-1)} = \frac{n}{\hat{N}-(1/T)\sum_{i=1}^{n}(i-1)t_i}$$

where $T = \sum_{i=1}^{n} t_i$. The maximum likelihood estimate of N may take any value between n and ∞, including both extremes. The assumption of identical failure rates due to different bugs is often not realistic.

Note also that since the number of bugs is assumed to be finite, we cannot assume *a priori* that asymptotic results such as the asymptotic normality of the maximum likelihood estimate are applicable. This means that we cannot apply the usual procedure of obtaining confidence bounds around the ML estimators using asymptotic normality.

The estimated distribution of the time to *next* failure is exponential with rate $(\hat{N}-n)\hat{\varphi}$.

12.3.3 Littlewood's model

This model [Littlewood, 1981] can be seen as a generalization of the Jelinski–Moranda model in which the number of faults is finite (as in JM) but in which these faults have different associated failure rates. Hence the 'worst' faults manifest themselves first and are removed (it is assumed that all bug fixes are effective).

We suppose that there are N bugs at time 0. Each bug has a time to failing the software X_i that is exponential given its failure rate Φ_i, that is

$$p(x_i|\Phi_i = \phi_i) = \phi_i \exp(-\phi_i x_i).$$

The Φ_i are assumed to be independent identically distributed gamma (α, β) variables. Now, the various bugs, whose unconditional failure times (obtained by integrating out ϕ; a similar calculation was made in subsection 4.2.2) are

Pareto distributed,

$$p(x_i) = \alpha \beta^\alpha / (x_i + \beta)^{1+\alpha}, \qquad (12.1)$$

compete to fail the software. Hence the first failure time, $X_{(1)}$, observed is the minimum of N such Pareto variables, etc. The second, $X_{(2)}$, is the minimum of $N - 1$ Paretos conditioned on being bigger than $X_{(1)}$, etc.

One can show that the inter-failure times $T_i = X_{(i)} - X_{(i-1)}$ are conditionally exponential:

$$p(t_i | \Lambda_i = \lambda_i) = \lambda_i \exp(-\lambda_i t_i),$$

where the Λ_i are a decreasing sequence of rates

$$\Lambda_i = \Phi_{(1)} + \Phi_{(2)} + \cdots + \Phi_{(N-i+1)},$$

and $\Phi_{(j)}$ is the (random) failure rate associated with the jth bug.

By using Bayes' Theorem it is possible to show that rates Φ corresponding to those bugs remaining after time τ are independent gamma $(\alpha, \beta + \tau)$ variables. This means that they are stochastically lower than the original set of Φs, and corresponds to the intuitive idea that the early failures tend to be those with the higher rates.

This model suggests that the time needed to achieve a fault-free program may be very large, but that programs with many faults might still be very reliable. The distribution of the time to next failure after the nth bug has been removed at time τ has density

$$p(x) = \left(\frac{\beta + \tau}{\beta + \tau + x} \right)^{\alpha(N-n)}.$$

To build a predictor, the usual method is to estimate the unknown parameters α, β and N by maximum likelihood.

12.3.4 The Littlewood–Verral model

Littlewood and Verral [Littlewood and Verral, 1973] have suggested a model in which the inter-failure times are treated as random variables with *varying* failure rates Λ_i. The conditional density of T_i is

$$p(t_i | \Lambda_i = \lambda_i) = \lambda_i \exp(-\lambda_i t_i).$$

The sequence Λ_i is usually modeled as a *stochastically decreasing* set of variables (that is, $P(\Lambda_i \leq r) \leq P(\Lambda_{i+1} \leq r)$ for all r and for each i) in order to capture the idea that failure rates should tend to decrease but should not have to (for example when a debugging operation introduces a new bug). The distribution of the Λ_is, is assumed to be gamma $(\alpha, \psi(i))$ and independent of the other Λ_js, that is

$$p(\lambda_i) = \psi(i)^\alpha \lambda_i^{\alpha-1} \exp(-\psi(i)\lambda_i)/\Gamma(\alpha),$$

where $\psi(i)$ is some increasing function of i (in order to make the Λ_is stochastically decreasing). The usual choice is $\psi(i) = \beta_1 + \beta_2 i$ (where $\beta_2 > 0$), so that three parameters have to be estimated, α, β_1, and β_2. The function ψ is also often taken to be a quadratic function of i.

The density of t_i given α and $\psi(i)$ is Pareto,

$$p(t_i|\alpha, \psi(i)) = \alpha[\psi(i)]^\alpha/(t_i + \psi(i))^{\alpha+1}.$$

Estimation can be performed by maximum likelihood. The likelihood equations for α, β_1, and β_2 are obtained by taking the log of the above expression, summing over i, and then setting the derivatives (with respect to α, β_1, and β_2) equal to zero.

The reliability prediction after observing failures at times $t_1, \ldots, t_n$ is

$$R_{n+1}(t) = [\hat\psi(n+1)]^{\hat\alpha}/(t+\hat\psi(n+1))^{\hat\alpha+1}$$

with $\hat\psi(n+1) = \hat\beta_1 + \hat\beta_2(n+1)$, where $\hat\alpha$ $\hat\beta_1$ and $\hat\beta_2$ are the MLE estimates.

12.3.5 The Goel–Okumoto model

Suppose that $N(t)$ is the number of failures observed up to time t. The Goel–Okumoto model [Goel and Okumoto, 1979] assumes that $N(t)$ is a random variable following a non-homogeneous Poisson process (but with a finite number of failure events). The intensity function (or ROCOF) is of the form

$$\lambda(t) = \mu\phi \exp(-\phi t),$$

which can be motivated as follows. The mean number of errors up to time t, $M(t) = E(N(t))$, determines the form of the ROCOF since $\lambda(t) = M'(t)$. It seems reasonable that m should satisfy the boundary conditions $M(0) = 0$ and $\lim_{t\to\infty} M(t) = \mu$ where μ is the total number of software errors. If furthermore the expected number of software errors in a small interval $(t, t + \Delta t)$ is proportional to the number of errors remaining, then

$$M(t + \Delta t) = M(t) + \phi(\mu - M(t))\Delta t + o(\Delta t),$$

so that we obtain a differential equation

$$M'(t) = \mu\phi - \phi M(t).$$

Solving the differential equation under the boundary conditions gives $M(t) = \mu(1 - \exp(-\phi t))$ and $\lambda(t)$ as specified above.

The maximum likelihood equations for μ and ϕ based on n failures with inter-failure times $t_1, t_2, \ldots, t_n$, are

$$n/\mu = 1 - \exp\left(-\phi \sum_{i=1}^{n} t_i\right),$$

$$n/\phi = \sum_{i=1}^{n} i t_i + \mu \left(\sum_{i=1}^{n} t_i\right) \exp\left(-\phi \sum_{i=1}^{n} t_i\right).$$

The Goel and Okumoto model can be seen as a generalization of the JM model in which the number of bugs N is considered uncertain and follows a Poisson distribution with mean μ.

12.4 Calibration and weighting

More recently, there has been a movement towards similar techniques as are used in expert judgement problems. The object of these techniques is to improve the predictive quality of the models which have been described above.

12.4.1 Calibration

When it seems that a predictive tool *consistently* gives an incorrect assessment, it may be possible to recalibrate the tool. This idea can be applied for the models described above. Suppose that we have made observations of failure times $t_1, t_2, \ldots, t_{n-1}$, and wish to estimate $P(T_n \leq t)$. With the models given above we can calculate a predictor $\hat{F}_i$. Now if the estimate equals the true distribution of T_i then $\hat{F}_i(T_i)$ will be uniformly distributed, so the eventual observation $T_i = t_i$ gives an observation of a uniform variable $u_i = \hat{F}_i(t_i)$ for each i. The predictive power of the model may therefore be measured by the extent to which the sequence u_i ($i \geq 2$) approximates independent samples from a uniform distribution. (The reason for working with the u_i instead of the t_i is that the T_is are not assumed to be identically distributed. The quantile transform $u_i = \hat{F}_i(t_i)$ ensures however that the u_i *are* identically distributed if the model is correct.)

A *u-plot* is simply the empirical cumulative distribution plot of a sequence $u_1, u_2, \ldots$ as defined above. If the u_i are uniformly distributed then the cumulative distribution function should approximate the diagonal. A standard goodness-of-fit test which is appropriate is the Kolmogorov-Smirnov test.

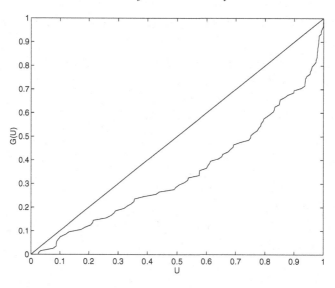

Fig. 12.2. A u-plot

The KS statistic for an empirical distribution function based on n samples is $\sqrt{n}$ times the maximum vertical distance between the graph of the empirical distribution function and the diagonal. Littlewood (in [Bittanti, 1988]) uses the u-plot to show in one example that the JM model gives consistently optimistic results, i.e. the u-plot is always above the diagonal.

A simple graphical technique which will indicate a possible trend in the u_is is to plot u_i against i. If, for large i, there are, for example, more large u_is than there are for small is then there may be a trend in the data.

If the u_is do not appear to have a trend, but are *not* uniformly distributed, then it is possible to *recalibrate* the data. Suppose that the U is distributed on [0, 1] with a continuous distribution function G that is not equal to the identity function $F(x) = x$ (the uniform distribution function). Then $G(U)$ is uniformly distributed. As an example consider what happens when 100 samples from an exponential distribution with parameter $\lambda = 2$ are taken and the cumulative distribution function of the exponential with $\lambda = 1$ is applied. Figure 12.2 shows the distribution function of the resulting sample. This is an estimate of the function G. Hence, if we estimate the distribution function G of the u_is then $G(u_i) = G[\hat{F}_i(t_i)]$ is approximately uniformly distributed, and we can use the distribution function $G\hat{F}_{n+1}$ instead of $\hat{F}_{n+1}$ to predict T_{n+1}.

The use of a recalibration function (where permissible, that is, where the u_i are independent and identically distributed) is a simple way of extending the scope of the parameterized models treated above. Note that we cannot directly estimate the distribution function of T_{n+1} based on the data $t_1, t_2, \ldots,$ as the T_is are not assumed to be identically distributed.

12.4.2 Weighted mixtures of predictors

The basic principle here is to make a new predictor by taking weighted mixtures of old predictors. Suppose we use a number of models such as those described above to obtain predictors $F_i^{(1)}, F_i^{(2)}, \ldots, F_i^{(k)}$ for T_i. We search for non-negative weights $w_1, w_2, \ldots, w_k$ which sum to 1 so that the new predictor

$$F_i^{new} = w_1 F_i^{(1)} + \ldots + F_i^{(k)}$$

is better than the old predictors.

12.5 Integration errors

An important cause of failures is integration error. Here individual components (hard- and software) are designed to individual specifications, but the individual specifications do not produce a properly functioning system. In principle the risk analysis methodology of scenario identification is capable of picking up such problems. In practice however these types of scenarios were traditionally ignored when random component failures were driving the unreliability of the system. With increased reliability (in particular for electronic components) the importance of integration error has grown substantially.

A good example of integration error, in the context of software reliability, was the failure of the maiden flight of the European Space Agency launcher Ariane 5. The failure was caused by the complete loss of guidance and attitude information 37 seconds after the start of the main engine ignition sequence. The loss of information was due to specification and design errors in the software of the inertial reference system. The development program for Ariane 5 did not include checks of all the software systems (which might have detected the faults), partly because parts of the software (including those causing the failure) had been successfully used on Ariane 4.

A board of inquiry [ESA, 1997] set up by ESA after the accident reported a series of errors in the software, and in the management of the software design. The accident was directly caused by a rapid change in direction of the rocket leading to the break-up of the launcher. The change of direction was ordered by the on-board computer on the basis of information sent from

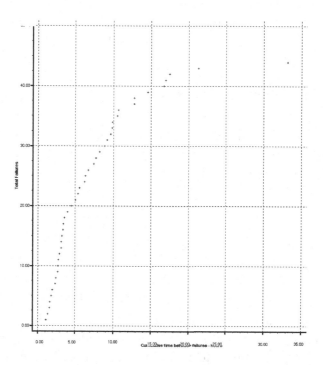

Fig. 12.3. Cumulative times until failure

the active Inertial Reference System (a redundant Initial Reference System malfunctioned at the same time for the same reason). This system was transmitting diagnostic information about an integer overflow, which was incorrectly interpreted by the on-board computer as flight data. The integer overflow was caused because the horizontal acceleration is considerably higher for Ariane 5 than for the Ariane 4 for which the software was originally designed.

The integer overflow problem in itself should not have caused the failure of the spacecraft. However, it had been decided not to protect the code from causing an operand error (as had been done for several other variables). It is not clear why this was done, but it seems fair to assume that the software designers assumed that there was a large margin of safety in the number of bits reserved to specify the integer when the design was made for Ariane 4.

After the loss of the launcher an official inquiry looked into the causes of the accident and made a number of recommendations. Two particularly important recommendations were:

R2 Prepare a test facility including as much real equipment as technically feasible, inject realistic input data, and perform complete, closed-loop, system

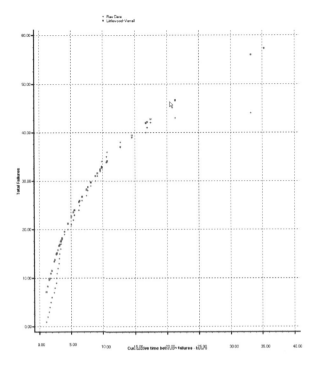

Fig. 12.4. Expected number of failures as function of time using LV model

testing. Complete simulations must take place before any mission. A high test coverage has to be obtained.

R5 Review all flight software (including embedded software), and in particular: identify all implicit assumptions made by the code and its justification documents on the values of quantities provided by the equipment; check these assumptions against the restrictions on use of the equipment; verify the range of values taken by any internal or communication variables in the software; solutions to potential problems in the on-board computer software, paying particular attention to on-board computer switch-over, shall be proposed by the project team and reviewed by a group of external experts, who shall report to the on-board computer Qualification Board.

12.6 Example

We illustrate the use of the Littlewood–Verral model on the data shown in Figure 12.3, which shows the graph of the number of failures up to time t. This has been implemented using the program CASRE (available with [Lyu, 1996]), which allows various models to be fitted. The Littlewood–

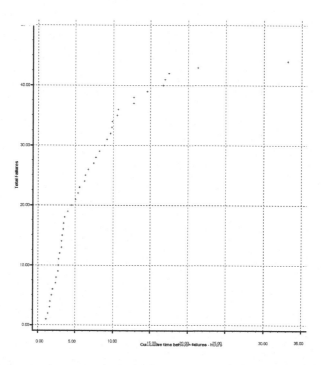

Fig. 12.5. The u-plot

Verral model was fitted using a quadratic ψ function. Figure 12.4 shows the expected number of failures up to time t according to the model. The performance looks reasonable, except at the end of the period, largely due to an unexpectedly long period between failures near the end. Based on the model estimations, the last 23 times were predicted (using the previous failure data), and the u-plot for these times is given in Figure 12.5. The Kolmogorov-Smirnov test does not indicate a significant deviation from the diagonal, suggesting that the model is not unreasonable.

12.7 Exercises

12.1 Show that, according to the JM model, the time between the $(i-1)$th and the ith failure is exponential with rate $\lambda_i = (N-i+1)\varphi$.

12.2 Prove Equation 12.1 by using the formula

$$p(x_i) = \int P(x_i|\varphi_i) f_{(\alpha,\beta)}(\varphi_i) \, d\varphi_i$$

where $f_{(\alpha,\beta)}$ is the gamma density with parameters (α, β).

Part IV

Uncertainty modeling and risk measurement

13

Decision theory

In this chapter we shall give a brief introduction to some of the important ideas in decision theory. Decision analysis is an important area of application for quantitative risk analysis. Although many risk analyses are performed simply to show that an installation conforms to the requirements of a regulator, quantitative risk analyses are increasingly being used as input to a decision making process. Logic seems to dictate that a quantitative risk analysis be followed by a quantitative decision analysis.

The field of decision analysis is far too rich to describe in a single chapter. We shall, then, deal quickly with the basic notions required to give simple examples of multi-attribute decision making under uncertainty for a single decision-maker. See [French, 1988] and [Keeney and Raiffa, 1993] for deeper discussions of all the issues involved.

Decision theory has been long split by heated discussions between protagonists of *normative* theories and those of *descriptive* theories. A descriptive theory tries to represent beliefs and preferences of individuals as they are. A normative theory tries to model how an individual's beliefs *should* be structured *if* they were to follow certain elementary consistency rules which might be expected of a rational individual.†

More recently, these two groups moved towards each other. In particular, the adherents of the normative approach see their methods as being an *aid* to the decision-maker in achieving consistent (rational) preferences. The process of decision analysis is, in this view, one in which the decision-maker's

---

† Arrow's theorem says that axiomatized rational decision making is not possible for groups unless all the members of the group agree that the group should act like an individual, for example by following the wishes of a dictator. The normative theory does not support group decision making directly. In our opinion, where there is no strong consensus within a group, the use of multicriteria decision analysis (MCDA) in itself will not generate such a consensus. An MCDA, possibly combined with a sensitivity analysis, *can* show that even with different standpoints the same conclusion would be reached. Or it could help to zoom in on the reasons for different conclusions, thus aiding the process of bargaining required to achieve agreement.

understanding of his problem increases by an iterative process of discussion with the decision analyst and interaction with the decision model. There is no 'right' answer, simply an increased awareness of the issues involved. The model outcome is seen as a guide to the decision-maker, the DM, instead of dictating to the decision-maker.

This does not mean however that the differences between the different schools of decision analysis have been resolved. Indeed, as there is no objective standard that can be used to measure the success of any of the models, there is no way to resolve these differences.

The advantage of a formal normative decision process, however, is that it makes the whole procedure *documented* and *traceable*. This is always better than an undocumented decision process, as the very process of documentation forces the decision-maker to be clear about his assumptions. Furthermore, the assumptions of rational decision making themselves are hard to disagree with on principle, so that the normative framework seems to be a good framework in which the decision-maker can build up the preferences needed to come to a conclusion.

Our sympathies are on the side of the normative camp. There seems little point in spending time and effort to make a mathematical model that imitates the 'imperfections' in human decision making when the humans are perfectly capable of making imperfect decisions in the first place. Furthermore, the builders of descriptive models do not show that their models deviate from the normative models in the same way that humans do (see [Salo and Hämäläinen, 1997] and the discussion that follows).

As we suggested above, decision analysis seems to be the logical extension of risk analysis. Increasingly risk analysts are applying decision theory techniques. The care with which most analysts build up a risk model is however in sharp contrast to the ease with which a 'standard' (but often inapplicable) decision model is used. It goes without saying that care has to be taken throughout the whole modeling process in order to avoid generating invalid conclusions.

If we agree that normative models should be used to aid the decision-maker to form a rational decision model, then the class of normative models used should be flexible enough to allow the decision-maker to try out different kinds of ideas. The mathematical assumptions used within such a class should be sufficiently weak that the preferences the decision-maker would like to model can be modeled. The most popular classes of normative models used at the moment use rather strong assumptions that are often unreasonable. One of the aims of this chapter is to indicate some of the more common pitfalls, and show what kinds of models are available.

13.1 Preferences over actions

The theory of decision making that we shall expound is based on the von Neumann–Morgenstern [von Neumann and Morgenstern, 1947] and Savage [Savage, 1972] approaches to utility. It is a theory for the decision making of rational individuals, and not of groups.

Central to the theory is the idea of preferences. We write

$$a \geq b$$

when the DM prefers object a to object b. Intuitively, it seems reasonable that when a is preferred to b, and b to c, then also a is preferred to c, that is

$$\{a \geq b \text{ and } b \geq c\} \Rightarrow a \geq c.$$

This property is called *transitivity*. When $a \geq b$ and $b \geq a$ then we write $a \sim b$ and say that the DM is *indifferent* between a and b. We write $a > b$ when $b \not\geq a$ and say that a is *strictly preferable* to b.

When all objects are *comparable*, meaning that for all a and b either $a \geq b$ or $b \geq a$, then a *value* function can be found representing the preference structure (recall that these same two requirements were also used in the first axiom of Savage in Chapter 2). This is a real-valued function v with the property that

$$v(a) \geq v(b) \Leftrightarrow a \geq b.$$

Value functions are not unique. In fact if v represents a preference structure then for any strict monotone function $\phi : \mathbb{R} \to \mathbb{R}$, the composition $\phi \circ v$ is a value function which represents the same preference structure. This is reasonable since only the order is important.

Since most actions of the decision-maker will lead to consequences that are uncertain, the decision-maker must actually determine his preferences between actions with uncertain consequences. To make the model mathematically tractable, Savage artificially extends the space of possible actions, to include all actions with certain consequences (recall that the notion of a constant act was introduced before Axiom 2 of Savage in Chapter 2). His main mathematical theorem (see Theorem 2.3) states that rational preferences over such a space of actions can be represented by a probability on the space of outcomes and a real-valued function (called a utility function) on the space of actions, such that the expected utility (calculated with respect to this probability) orders the actions in the same way as the preferences. The utility function is unique up to a positive affine transformation, that is, if u is a utility function then for any constants $a > 0$ and b, $au + b$ is also a utility function representing the same preferences.

In applications of the expected utility theory to decision making, the logic of Savage's theorem is reversed. One assesses probabilities and a utility function on a set of outcomes, and then uses the expected utility to *define* the preferences between actions with uncertain outcomes on the assumption that the preferences of the decision-maker are rational.

The important practical point is that rational decision theory allows us to split the decision making problem into two parts. The first part is the probability assessment of uncertainties, which can be done by experts. The second part is the specification of preferences (a utility function) which should be done by the decision-maker. If the decision-maker is not completely certain about his preferences, or if there are several stakeholders who want to judge the decision themselves, then a sensitivity analysis can be done *just* on the utility function, leaving the probability assessment intact (assuming that there is agreement on the probabilities – this is the subject of Chapter 10).

In real applications, the rule of maximizing expected utility is often replaced by the rule of maximizing expected monetary value. This avoids the problem of representing the risk attitude of the decision-maker with a utility function, but has limited value as a decision tool as we shall see. In the next section we give a simple example of the use of decision trees to calculate the action with maximum expected monetary value.

Use of an expectation might seem reasonable from a point of view of maximizing utility in the long run, but this is *not* a correct justification as the decision will only be taken once, and so there is no 'long run'.

13.2 Decision tree example

Decision trees provide a particularly convenient way to organize the calculations in a decision problem when the number of actions and states of the world is finite. The theory is best illustrated by an example.

A manager in the company Eurospatiale wishes to award a research contract to a consulting firm. For the purposes of this example we assume that the manager's utility of money is linear so that expected monetary value is used as the decision criterion.

There are two contenders for the contract: firm A and firm B. The bids that the companies make are 10 KAU† and 8 KAU respectively. It is however not clear whether the consultants are capable of completing the research. If the research project fails then the firm still gets paid. If it is successful then Eurospatiale achieves a one-time saving of 15 KAU.

† KAU=kilo accounting units

13.2 Decision tree example

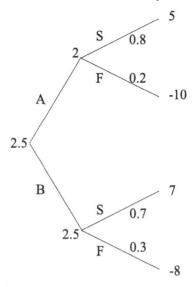

Fig. 13.1. The decision tree for the research project

Based on previous experience, the manager assesses the likelihood of success by the two companies as 80% and 70% respectively. Which consultant should be chosen?

The possible *actions* in this example are A and B, meaning a choice for consultant A or B respectively. The possible *outcomes* are S and F, meaning that the research project is either successful or a failure.

To solve this problem we first use the information above to make a table showing the net savings attained under different actions and outcomes.

	S	F
A	5	−10
B	7	−8

We can solve the above decision problem using a decision tree. This is done in Figure 13.1.

The first branching of the tree represents the two possible decisions open to the decision-maker, namely the choices for consultants A and B. Given a particular choice, the tree then branches again to show the possible 'states of nature', S and F. Along these branches we write the probability of the outcome, and at the end of the branch we write the net saving. We now calculate the expected net saving for each alternative A and B, and write the expected value by the branch of the tree associated to that alternative, for example

$$2 = 0.8 \times 5 + 0.2 \times (-10).$$

These expected values are written on the edges corresponding to the choice of alternative. Finally we choose the alternative giving the largest expected savings, which in this case is consultant B.

It might seem that the formal method involved in decision trees is not applicable for many decisions in which costs and probabilities have not yet been determined. However, the decision tree method can always be applied in a 'back of the envelope' calculation using rough estimates of costs and subjective probabilities. Such a calculation is valuable because it structures the reasoning of the decision-maker.

13.3 The value of information

Typically it is possible for a decision-maker to acquire more information about his decision problem, at some cost (an extra report can always be commissioned). The question arises: is it worth it? This question is very easily answered. In the example above, suppose that the manager wants to pay for the consultants to make extended research proposals. At a cost, the manager can gain more insight, and better estimate the probabilities of success.

The manager can invite the two consultants to make an extended research proposal costing 1 KAU per consultant. He will classify the proposals into *good* and *poor*. He estimates that the probability of a good proposal if the project can finish successfully is 0.7, and 0.1 if the project would fail, for both consultants. The consultants make their proposals independently.

On the basis of just this information we can advise the manager to go ahead with the extended research proposal as it almost doubles his expected savings with respect to no extended research proposal.

The calculation is carried out in a decision tree in Figure 13.2. The calculations required here are repetitive but straightforward. They involve repeated application of two well-known theorems in probability: Bayes and total probability.

Recall that Bayes' Theorem (Theorem 4.1) states that

$$P(A|B) = \frac{P(B|A)P(A)}{\sum_{i=1}^{n} P(B|A_i)P(A_i)},$$

while the theorem of total probability (Equation 4.1) states that whenever $A_1, \ldots, A_n$ are disjoint events (i.e. $A_i \cap A_j = \emptyset$ for $i \neq j$) and $\sum_i P(A_i) = 1$,

13.3 The value of information

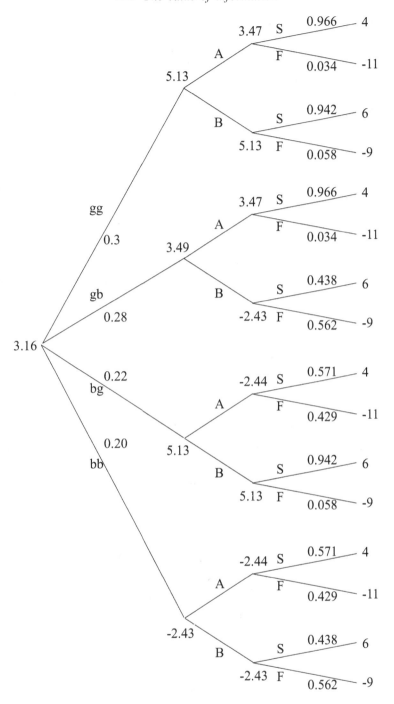

Fig. 13.2. The decision tree for the extended research proposal

then

$$P(B) = \sum_{i=1}^{n} P(B|A_i)P(A_i).$$

The decision tree branches into four subtrees corresponding to the outcomes of the research proposals, where gg means that 'research proposal of A is good' and 'research proposal of B is good', gb means that 'research proposal of A is good' and 'research proposal of B is not good', etc.

Each of the four subtrees is identical in structure to that of the tree in Figure 13.1. However, the net savings have been adjusted to take account of the extra costs of the research proposal, and the probabilities have been adjusted to take account of the information coming from the research proposal. For example, the probability of success of consultant A given that he makes a good research proposal is calculated by

$$P(S \text{ with } A|\text{proposal } A \text{ good}) = \frac{P(\text{proposal } A \text{ good}|S \text{ with } A)P(S \text{ with } A)}{P(\text{proposal } A \text{ good})}$$

using Bayes, which can be further expanded to

$$\frac{P(\text{proposal } A \text{ good}|S \text{ with } A)P(S \text{ with } A)}{P(\text{proposal } A \text{ good}|S \text{ with } A)P(S \text{ with } A) + P(\text{proposal } A \text{ good}|F \text{ with } A)P(F \text{ with } A)}$$

using the theorem of total probability. This now gives

$$P(S \text{ with } A|\text{proposal } A \text{ good}) = \frac{0.7 \times 0.8}{0.7 \times 0.8 + 0.1 \times 0.2} = \frac{56}{58} = 0.966.$$

All the probabilities given in the four subtrees are calculated in this way. We can now determine the optimal decision and net savings in each of the subtrees. This is a 'what if' analysis telling us what we should do if we knew what the outcomes of the research proposals were.

Since we wish to decide whether or not to ask for research proposals, we have to finally determine the expected savings conditional on getting the extended research reports. To calculate this, it is necessary to calculate first the probabilities of getting research proposals from A and B of quality good/good, good/bad, bad/good, and bad/bad. In fact, we already calculated these probabilities above.

$$P(\text{proposal A good}) = 0.58, \qquad (13.1)$$
$$P(\text{proposal B good}) = 0.52. \qquad (13.2)$$

Hence the probability of getting good proposals from A and B is $0.58 \times 0.52 = 0.30$. The probabilities of the other three top-level branches in the tree are calculated in the same way. Finally we can calculate the expected savings obtained by asking for the extended research proposals, by averaging the expected savings in each case after the research proposals have been studied and classified,

$$0.3 \times 5.13 + 0.28 \times 3.48 + 0.22 \times 5.13 + 0.2 \times (-2.43) = 3.15.$$

The expected net savings are much higher than would have been obtained without the extended research reports, even taking into account the extra costs involved. (Note that Theorem 2.4 says that making 'observations' such as the research reports in this example will always give savings, although these may not exceed the costs of making the observations.)

Similar calculations show incidentally that it would be even better for the decision-maker to ask *just consultant B* for an extended proposal. However, it may be impossible to ask just consultant B because of legal problems.

13.3.1 When do observations help?

Observations are not always interesting. They only have value for the decision problem *at hand* when the result of an observation could lead to making a different decision. In particular, perfect information about the state of the world is usually not necessary to make the optimal decision. However, since we can never do better than to have perfect information, it is possible to give an upper bound estimate on the amount of extra benefit (measured in terms of expected utility) that we can possibly get from making observations. This is called the *value of perfect information* and is defined as the mean value

$$\sum_x p(x)U(\mathbf{F}|x),$$

where the sum is over all states of the world x, $p(x)$ is the prior probability of that state of the world, and $U(\mathbf{F}|x)$ is the utility of the optimal decision given that the state of the world is x.

If the cost of making an observation is *more* than the saving we expect to make by getting perfect information then the observation is not worth making.

We conclude this discussion with a proof promised in Chapter 2. Recall that B is some event about which we are uncertain, and X is a binary observation that we can make to update our prior probability on B. The result says that if B really holds then our belief in B increases (in expectation) as a result of observations.

Proposition 13.1 *With B and X as above, we have*

$$E(P(B|X)|B) \geq P(B), \quad E(P(B|X)|B') \leq P(B).$$

Proof We begin with the first statement. We have

$$
\begin{aligned}
E(P(B|X)|B) &= P(B|X=0)P(X=0|B) + P(B|X=1)P(X=1|B) \\
&= \frac{1}{P(B)}[P(B|X=0)P(B,X=0) + P(B|X=1)P(B,X=1)] \\
&= \frac{1}{P(B)}[P(B|X=0)^2 P(X=0) + P(B|X=1)^2 P(X=1)] \\
&\geq \frac{1}{P(B)}[P(B|X=0)P(X=0) + P(B|X=1)P(X=1)]^2 \\
&= \frac{1}{P(B)}[P(B)]^2 = P(B),
\end{aligned}
$$

where the inequality step uses the convexity of the function $x \mapsto x^2$. The second statement of the proposition is proved similarly, the inequality arising by application of the concavity of the function $x \mapsto x(1-x)$. □

13.4 Utility

In the last section we saw how we could calculate the expected net savings. There was an explicit choice to measure the outcomes directly in terms of money. However in many situations it is not desirable to do this. Consider the following alternatives for developing a new product.

(i) Development costs are 95 MAU† with certainty.
(ii) Development costs are uncertain. With probability $\frac{1}{2}$ they are 70 MAU, and with probability $\frac{1}{2}$ they are 110 MAU.

Using the expected monetary value criterion we would choose alternative (ii), as the expected costs at 90 MAU are lower than the 95 MAU of alternative (i). Suppose however that your boss has just sent you a letter in which you are told that the total budget is 100 MAU and that you will be fired if you go over budget. In this case it might be reasonable for you to become *risk averse*, and choose alternative (i). Clearly, your preferences are not being represented well by expected monetary value.

As we indicated earlier, decision making under uncertainty requires use of a utility function to represent the decision-maker's attitude to uncertainty. We briefly indicate how a utility function can be constructed, assuming that there are best and worst alternatives, b and w. Since a utility function is defined only up to a positive affine transformation, we can fix the utility function u by the arbitrary choice $u(w) = 0$ and $u(b) = 1$. Now, for an

† MAU=million accounting units

outcome x which satisfies $w \geq x \geq b$, we can ask the decision-maker to choose between the following two alternatives:

(i) outcome x with certainty;
(ii) the lottery consisting of outcome b with probability p and outcome w with probability $1 - p$.

If p takes a value very close to (i) then the decision-maker will presumably choose alternative (ii). If on the other hand p is very small, he will choose alternative (i). For some p strictly between 0 and 1 he will be indifferent between the two alternatives. Hence the utility of the two alternatives is identical, and

$$u(x) = u(\text{alt.(i)}) = u(\text{alt.(ii)}) = p \cdot u(b) + (1 - p)u \cdot (w) = p.$$

This method can be used to determine the decision-maker's utility of any certain outcome. Knowing the utility of each certain outcome, the utility of any uncertain outcome is obtained simply by averaging the utility over the possible outcomes.

13.5 Multi-attribute decision theory and value models

In many decision problems, more than one factor influences our preferences over the possible outcomes. In the example treated above another factor of importance might be the speed with which the research project is concluded.

Any outcome of the research project can be measured in terms of cost and duration. If the manager decides that only the factors cost and duration are relevant in determining his preferences then he can measure the outcomes by the vector [cost, duration]. These are the *attributes* of the problem.

In utility theory there is *in principle* no problem in dealing with outcomes measured on a two-dimensional scale (or even more dimensions). The decision-maker simply determines his degree of preference for the various multidimensional outcomes using lotteries.

In practice it seems to be more difficult to assess a utility function over several attributes than over one attribute. The area of multi-attribute decision theory gives a number of ways of making an explicit trade-off between several attributes. We begin by considering the situation in which there are no uncertainties, that is, where the consequences of each decision outcome are known precisely. In this case we seek a value function representing the decision-maker's preferences. The single extra step that then needs to be made to support decision making under uncertainty, the recalibration of the value function, will be made in Subsection 13.5.6.

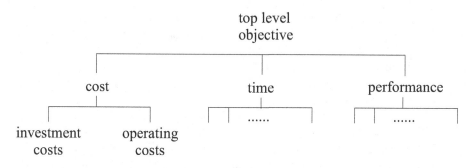

Fig. 13.3. Simple attribute hierarchy

The problem of multi-attribute value theory (MAVT) is to find a simple expression for the decision-maker's value function. If we assume that the decision-maker has determined a set of n attributes each of which can be measured on a real-number scale, then a value function is just a mathematical function $v : \mathbb{R}^n \to \mathbb{R}$. As we remarked earlier, a value function is determined only up to a monotone increasing transformation. This is because the information coded into a value function v consists of

- a collection of indifference sets

$$I_\alpha = \{x \in \mathbb{R}^n | v(x) = \alpha\} \subset \mathbb{R}^n$$

(the decision-maker is indifferent to any outcomes in the same indifference set), and
- a preference ordering of the indifference sets (that is, for each α and β we know whether an outcome in I_α is strictly preferable or strictly less preferable than an outcome in I_β).

The value function is simply a convenient way of labeling the indifference sets. Changing the value function by a monotone increasing transformation changes the labels but not the ordering determined by the labels.

13.5.1 Attribute hierarchies

A common method for structuring the set of attributes is the use of an attribute hierarchy. This is simply a tree in which the top-level objective is split into finer and finer disjoint aspects. Figure 13.3 shows an example attribute hierarchy in which the top-level decision objective is split into cost, time and performance. Cost is then split into investment and operating costs, while performance is split into various operational characteristics.

It is important that the criteria at the lowest level be *measurable*. Many authors take the view that even rather vague criteria ('intangibles') can be measured using subjectively defined scales. For some criteria it might be necessary to use a so-called *proxy* variable, that is, a variable that is (relatively) easy to measure and which, for the purposes of the study, may be taken as sufficiently representative of the criterion under study.

In most multi-attribute methods, the full attribute hierarchy does not play a role in the calculations (although it is sometimes used as an aid to determining weights in a weighting factors model – see below). Its role is restricted to generating a list of distinct criteria.

13.5.2 The weighting factors model

A common model is the *weighting factors* model. Here a number of criteria are defined, and weights w_i chosen for each criterion scale i. Each decision alternative is scored on all criteria, and the overall score of each decision alternative is defined as the weighted sum of the individual criteria scores, $\sum_i w_i x_i$.

The interpretation of the weights is however a critical problem. Many users want to use the weights to reflect the 'relative importance' of the attributes. Unfortunately, the weights have units because the underlying attribute scales have units. A change of $-w_i^{-1}$ units on scale i is always compensated by a change of $+w_j^{-1}$ units on scale j. Changing the units of an attribute *must* therefore lead to a change in the weights (see the exercises). This is a point to which most users are not sensitive, suggesting that they do not understand the model and that the model will not correctly reflect their preferences.

The indifference sets are easily determined: the indifference set corresponding to value c is the hyperplane satisfying the linear equation

$$\sum_i w_i x_i = c.$$

13.5.3 Mutual preferential independence

A slightly more general model is the *additive value model*, valid under the assumption of mutual preferential independence described below.

The additive value model writes a value function in the form

$$v(x_1, \ldots, x_n) = \sum_{i=1}^{n} w_i v_i(x_i)$$

where the w_i are weights and the v_i are so-called *marginal value functions*.

A marginal value function is a value function for one of the attributes in isolation. Unfortunately not every marginal value function can be plugged into the above expression.

In order to explain what mutual preferential independence is we have to make a couple of small definitions. Writing the vector of attributes as $\underline{x} = (x_1, \ldots, x_n)$, we will sometimes want to decompose the vector into two sub-vectors as follows: let I be a proper subset of $\{1, \ldots, n\}$, then write $\underline{a}$ for the vector obtained by removing those elements x_i from $(x_1, \ldots, x_n)$ whose index i does *not* belong to I, and $\underline{b}$ for the vector obtained by removing those elements x_i from $(x_1, \ldots, x_n)$ whose index i does belong to I. The conjoined vector $(\underline{a}, \underline{b})$ contains all the attributes, but possibly in a new order. We call $(\underline{a}, \underline{b})$ a *decomposition* or $\underline{x}$. We say that the attributes indexed by I are *preferentially independent* of the others if for all vectors $\underline{a}$ and $\underline{a}'$, we have that

$$(\underline{a}, \underline{b}) \geq (\underline{a}', \underline{b}) \quad \text{for some } \underline{b}$$
$$\text{implies } (\underline{a}, \underline{b}') \geq (\underline{a}', \underline{b}') \quad \text{for every } \underline{b}'.$$

This definition means that one can talk about preferences for the attributes in the vector $\underline{a}$ without reference to the values being taken by the other attributes in $\underline{b}$. The attributes in the vector $\underline{x}$ are *mutually preferentially independent* if $\underline{a}$ is preferentially independent of $\underline{b}$ for every possible decomposition of $\underline{x}$ into a pair $(\underline{a}, \underline{b})$.

Clearly there are an enormous number of conditions ($2^n - 2$) to be checked to demonstrate that mutual preferential independence holds. Fortunately, it turns out that some of these conditions imply others. The minimum number of conditions that need to be checked is $n - 1$. Various collections of preferential independence conditions imply mutual preferential independence. The simplest set of conditions implying mutual preferential independence is the requirement that each pair of attributes (x_i, x_j) should be preferentially independent of the remaining attributes (see [Keeney and Raiffa, 1993] for more details).

13.5.3.1 Eliciting marginal value functions

One way of eliciting a marginal value function is as follows. Suppose that we want to determine a value function for x_1. We write $\underline{y} = (x_2, \ldots, x_n)$.

We can pick two values for the attribute x_1, say $\ell < h$, and arbitrarily assign $v_1(\ell) = 0$ and $v_1(h) = 1$ (assuming that lower values of the attribute are 'worse' than higher values). We now want to interpolate and find a number $m_{0.5}$ between ℓ and h so that $v_1(m_{0.5}) = 0.5$. To do this we pick

13.5 Multi-attribute decision theory and value models

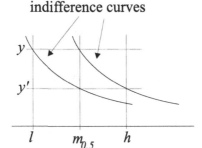

Fig. 13.4. The construction of a marginal value function

a value for the other attributes, $\underline{y}$, and seek a 'worse' value for the other attributes $\underline{y}'$ so that for some $m_{0.5}$ between ℓ and h,

$$(\ell, \underline{y}) \sim (m_{0.5}, \underline{y}'), \quad \text{and}$$
$$(m_{0.5}, \underline{y}) \sim (h, \underline{y}').$$

The situation is illustrated in Figure 13.4. Writing $v_{\underline{y}}$ for the weighted sum of the value functions for the attributes in $\underline{y}$, we then have

$$v_1(\ell) + v_{\underline{y}}(\underline{y}) = v_1(m_{0.5}) + v_{\underline{y}}(\underline{y}'),$$
$$v_1(m_{0.5}) + v_{\underline{y}}(\underline{y}) = v_1(h) + v_{\underline{y}}(\underline{y}'),$$

which together give $v_1(m_{0.5}) = 0.5$. There are some continuity requirements needed to guarantee the existence of $m_{0.5}$ and $\underline{y}'$.

In this (laborious) way we can interpolate the values of the value function in as many points are desired. The laborious method is needed to ensure we obtain a value function that will work in the additive value representation given above. The same procedure is repeated for each attribute.

13.5.3.2 Eliciting weights

Having obtained marginal value functions we need to weight them. This can be done in a fairly straightforward way by using indifferences.

Suppose that x_1 and x_2 are the first two attributes, and $\underline{b}$ is the vector of remaining attributes. Let x_1^*, x_2^* and $\underline{b}^*$ be the attribute values for which the marginal value functions are zero. Then if we can find values $x_1 \neq x_1^*$ and $x_2 \neq x_2^*$ such that

$$(x_1, x_2^*, \underline{b}^*) \sim (x_1^*, x_2, \underline{b}^*)$$

then $w_1 v_1(x_1) = w_2 v_2(x_2)$. Proceeding in this way we can get $n-1$ linear equations relating the weights (without loss of generality we can assume that the weights sum to 1), and solve for the w_i.

13.5.4 Conditional preferential independence

A method that requires weaker mathematical assumptions than the usual mutual preferential independence is the method of *conditional preferential independence*, see [Bedford and Cooke, 1999a]. The trade-off between the attributes is done explicitly by choosing a reference attribute (e.g. cost) and trading the other attributes one after another into the reference attribute.

The heuristic of the method is that one attribute is used to trade off all other attributes into 'standard values'. The trade-off is done in order of 'relative importance' of the attributes. The interpretation of 'relative importance' in the model is that X_2 is more important than X_3 if the way that X_2 is traded off into X_1 is not affected by the value taken by X_3.

Theoretical assumptions *Conditional preferentially independence* means that there is an ordering of the variables $X_1, \ldots, X_n$ and there are values of the variables $x_2^{(0)}, \ldots, x_n^{(0)}$ so that

- X_2 and X_1 are preferentially independent of $X_3, \ldots, X_n$.
- X_3 and X_1 are preferentially independent of $X_4, \ldots, X_n$ given $X_2 = x_2^{(0)}$.
- X_4 and X_1 are preferentially independent of $X_5, \ldots, X_n$ given $X_2 = x_2^{(0)}$, $X_3 = x_3^{(0)}$.
- ....
- X_{n-1} and X_1 are preferentially independent of X_n given $X_2 = x_2^{(0)}, \ldots, X_{n-2} = x_{n-2}^{(0)}$.

The assumptions above imply that we can talk about preferences and indifferences between pairs of (X_1, X_2) independently of the values taken by the other variables $X_3, \ldots, X_n$. Similarly, when $X_2 = x_2^{(0)}$ then the preferences and indifferences between pairs of (X_3, X_1) are the same for all values taken by the other variables $X_4, \ldots, X_n$, and so on. The information about preferences can be summarized in $n-1$ two-dimensional plots or indifference curves of (X_1, X_2), of (X_1, X_3) given $X_2 = x_2^{(0)}, \ldots,$ and of (X_1, X_n) given $X_2 = x_2^{(0)}, \ldots, X_{n-1} = x_{n-1}^{(0)}$. The assumption of conditional preferential independence is clearly slightly weaker than that of mutual preferential independence

Given this information, the preference structure on the attribute space is fully specified. For given any outcome $(x_1, \ldots, x_n)$ we can find y_2 such that the DM is indifferent between $(y_2, x_2^{(0)}, x_3, \ldots, x_n)$ and $(x_1, \ldots, x_n)$. Then we can

13.5 Multi-attribute decision theory and value models

find a y_3 such that the DM is indifferent between $(y_3, x_2^{(0)}, x_3^{(0)}, x_4, \ldots, x_n)$ and $(y_2, x_2^{(0)}, x_3, \ldots, x_n)$. We proceed in the same way using conditional preferential independence until we have found y_n such that the DM is indifferent between $(y_n, x_2^{(0)}, \ldots, x_n^{(0)})$ and

$$(y_{n-1}, x_2^{(0)}, \ldots, x_{n-1}^{(0)}, x_n).$$

In the end we have that the DM is indifferent between

$$(x_1, \ldots, x_n) \text{ and } (y_n, x_2^{(0)}, \ldots, x_n^{(0)}).$$

The preference between two outcomes $(x_1, \ldots, x_n)$ and $(x_1', \ldots, x_n')$ is now determined by calculating the equivalent $(y_n, x_2^{(0)}, \ldots, x_n^{(0)})$ and $(y_n', x_2^{(0)}, \ldots, x_n^{(0)})$ and checking whether $y_n > y_n'$, $y_n < y_n'$ of $y_n = y_n'$.

The above discussion gives a decision model which is valid for decision making under certainty. It can also be easily adapted to make it applicable for decision making under uncertainty. For this, the utility of each outcome must be determined. This is done by eliciting a conditional utility function $u(x_1)$ of x_1 given $X_2 = x_2^{(0)}, \ldots, X_n = x_n^{(0)}$.

The utility of any outcome $(x_1, \ldots, x_n)$ can be calculated in this model by finding an equivalent $(y_n, x_2^{(0)}, \ldots, x_n^{(0)})$ and then taking $u(y_n)$.

The expected utility of each option can be determined by Monte Carlo simulation.

Example 13.1 *We illustrate the above method for modeling preferences by a simple example in which a new product is to be developed. The main aspects of decision making involve the costs of development, possible delay in getting the product onto the market, and the performance of the finished product.*

We assume that the preferences of the decision-maker between different cost–delay outcomes are not affected by the performance of the product. We can then elicit indifference curves from the decision-maker (he is indifferent between outcomes on the same indifference curve).

Indifferences can be elicited by questions of the following type:

Suppose that delay is now at level 10 and cost at level 4. At what cost level c are you indifferent between the current situation and delay 5, cost c?

The indifference curves for cost and delay are given in Figure 13.5. The concave form arises because when delay is already low then the DM is less willing to spend extra to further reduce delay. We then elicit the indifference curves for cost and performance given that delay takes on a reasonable value (say delay=5). These indifference curves are shown in Figure 13.6.

Suppose we want to compare two possible outcomes,

cost = 2, delay = 7.5, performance = 4

276   *13 Decision theory*

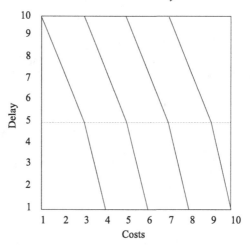

Fig. 13.5. Indifference curves for cost and delay

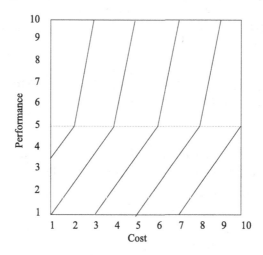

Fig. 13.6. Indifference curves for cost and performance given delay = 5

and

cost = 6, delay = 1, performance = 5.

Using the first set of indifference curves in Figure 13.5 we see that the two possible outcomes are equivalent in preference to

cost = 3, delay = 5, performance = 4

and

cost = 5, delay = 5, performance = 5

respectively. Using the second set of indifference curves in Figure 13.6 we see that the two possible outcomes are equivalent in preference to

$cost = 4$, $delay = 5$, $performance = 5$

and

$cost = 5$, $delay = 5$, $performance = 5$.

Hence the first alternative is preferred.

Implementation The largest task in implementing this model is in determining the indifference curves. Assuming that these curves are smooth, they may be approximated to any degree of approximation by piecewise straight lines. In the weighting factors model, the indifference curves are in fact just straight lines (with slope determined by the ratio of the weighting factors of the two attributes involved).

Generalizing the weighting factors model can therefore be done by assuming that the indifference curves are piecewise straight lines. The DM can be asked to give sets of points corresponding to outcomes to which he/she is indifferent. We then join the points by straight lines.

More mathematical details of the model and its implementation are given in [Bedford and Cooke, 1999a].

If the outcome of a decision alternative is uncertain, then the joint uncertainty distribution of the attribute values should be modelled (for example using the techniques discussed in Chapter 17). Monte Carlo simulation can be used to propagate this uncertainty through the model. An example was given in [Bedford and Cooke, 1999a] based on a case study about a satellite system. A set of 200 Monte Carlo simulations of costs and performance attributes for each of the two decision options are shown in Figure 13.7 together with indifference curves used to trade off performance into cost. The northwest corner of the figure represents a point with lowest costs and highest performance. The southeast corner represents highest costs and lowest performance. The indifference curves may be interpreted as contours of the value function. We would obviously like to choose the design option for which the outcomes (cost and performance) have the most value. In this case, a utility function was elicited over costs given a nominal level of performance. This enabled the expected utility of each decision alternative to be calculated.

Fitting a utility function on the back of a value function is just one way to obtain a utility model. It will be described further below.

13.5.5 Multi-attribute utility theory

Multi-attribute utility theory (MAUT) addresses the problem of decision making under uncertainty when the outcomes are expressed in terms of several attributes.

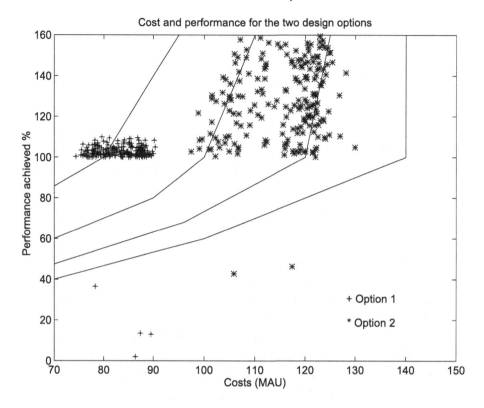

Fig. 13.7. The trade-off between cost and performance

One of the main assumptions used in the literature is that of *utility independence* (which has nothing to do with probabilistic independence). This says that the risk attitude of the decision-maker with respect to any one of the attributes is not affected by the values taken by the other attributes. A consequence of this assumption is that each attribute has its own single-attribute (or marginal) utility function.

It can be shown that, under the assumption of utility independence, the multi-attribute utility function is a multilinear function of the marginal utility functions. Hence the problem of finding a particular function of n variables has been reduced to the simpler problem of finding n functions of one variable and combining them.

Two particular examples of utility independence are *additive independence* and *mutual utility independence* (see [Keeney and Raiffa, 1993] or [French, 1988] for an extended discussion of these concepts). The former ensures that we can write the multi-attribute utility function u as a linear combination of the marginal utility functions $u_1, \ldots, u_n$,

$$u(x_1, \ldots, x_n) = w_1 u_1(x_1) + \cdots + w_n u_n(x_n)$$

($\sum_i w_i = 1$), and holds if for decompositions $(\underline{a}, \underline{b})$ and $(\underline{a}', \underline{b}')$ of $\underline{x}$, the decision-maker is indifferent between the two lotteries

- $(\underline{a}, \underline{b})$ with probability $\frac{1}{2}$,
- $(\underline{a}', \underline{b}')$ with probability $\frac{1}{2}$,

and

- $(\underline{a}, \underline{b}')$ with probability $\frac{1}{2}$,
- $(\underline{a}', \underline{b})$ with probability $\frac{1}{2}$.

An alternative formulation is when the decision-maker is offered lotteries with prizes that are outcomes $x_1, \ldots, x_n$, then the preferences between such lotteries only depend on the marginal distributions of the lotteries, and not on the full joint distributions.

The *mutual utility independence* condition ensures that we can write the multi-attribute utility function u in terms of certain marginal utility functions $u_1, \ldots, u_n$,

$$1 + wu(x_1, \ldots, x_n) = \prod_i (1 + ww_i u_i(x_i)),$$

and holds if for all decompositions $(\underline{a}, \underline{b})$ of $\underline{x} = (x_1, \ldots, x_n)$, $\underline{a}$ is utility independent of $\underline{b}$. The marginal utilities $u_1, \ldots, u_n$ in this expression have to be chosen so that they are zero at the same point as u,

$$u(x_1^0, \ldots, x_n^0) = 0 \text{ implies } u_1(x_1^0) = \cdots = u_n(x_n^0) = 0,$$

otherwise the above expression takes on a more complicated form.

There is an extensive literature on multi-attribute utilities, and various types of utility independence have been identified. See for example the paper of Fishburn and Farquhar [Fishburn and Farquhar, 1982], and the further references therein.

13.5.5.1 Parameter quantification

If utility independence holds in either the additive or the mutual form discussed above, the marginal utility functions appearing in these expressions can be determined by a number of lottery type questions as discussed in Section 13.4. It then remains to determine the values of the other parameters w_i. These can be found quite easily by use of a number of *indifferences* as for additive value functions.

More generally, it is simply a question of obtaining a set of indifferences from the decision-maker in order that a system of simultaneous equations can be solved for the parameters.

13.5.5.2 Problems with utility independence

In practice, for a number of reasons, the usual MAUT is difficult to apply. If the assumption of utility independence is not valid then the nice mathematical theory is not useful. Even in a simple trade-off between cost and performance this assumption does not seem valid outside a small mid-range of values. For example, if performance is known to be very high then the DM might well be more willing to take risks on the cost side than when the performance is known to be low.

In mathematically sound applications of MAUT, a great deal of trouble is taken to ensure that the assumption of utility independence holds. This is done by changing the variables from the 'physically natural' ones in which the problem is defined, to other variables for which the assumptions hold. An example of such a transformation is the one in which two cost variables x and y are replaced by total costs $x + y$ and cost difference $x - y$. The practical problem for many practitioners is that it is sometimes very difficult to see how variables should be adapted.

13.5.6 When do we model the risk attitude?

Any multi-attribute model for n attributes is essentially just a mathematical function from $\mathbb{R}^n$ to $\mathbb{R}$.

From a theoretical point of view, there is no objection to 'converting' a model for decision making under certainty into one for decision making under uncertainty. This can be done by eliciting a utility function u over the output values of the value model v. In effect, this recalibrates the value function to take account of the decision-maker's risk attitude. The model that correctly takes account of the decision-maker's risk attitude is thus

$$u(v(x_1, \ldots, x_n)).$$

Comparing this with the models obtained under the assumption of utility independence we see that the mathematical assumptions are weaker (no requirement that marginal utility functions exist) and that fewer 'lottery' elicitations have to be carried out with the decision-maker (only one utility function has to be determined, rather than n utility functions). On the other hand, the elicitation of marginal value functions requires extensive use of indifferences. Furthermore, the scale on which the value function is defined is not a 'natural' scale, but an artificial one. This makes it difficult for the decision-maker to interpret the lotteries used to determine a utility function over the output values of the value model.

The conditional preferential independence model makes life easier for the decision-maker by trading off everything into a given attribute. For example, we might trade everything into costs. Then the value function scale is costs with units of dollars, euros or whatever else the decision-maker may prefer to use.

13.5.7 Trade-offs through time

Often a decision will impact on a number of different time frames in the future. A comparison of different decision alternatives therefore requires a comparison of benefits or disbenefits gained at different times. The standard way to do this in economics is by discounting future income (or expenditure) by a discount factor, leading to the net present value (NPV) of each decision alternative.

This method, in widespread use by governments, gives odd results when applied over long time scales, since the discounting reduces large future amounts to small present day values. Saving the human race in 1000 years might turn out to be less valuable than a cheap bottle of wine today! For this reason NPV methods have apparently been abandoned in decision making for nuclear waste disposal, where the requirements are for minimal leakage on (for example) a 10 000 year time scale. Atherton and French [Atherton and French, 1998] argue that it is more appropriate to divide the future into different time scales (0–100 years, 100–500 years and 500–10 000 years). Within each of these time scales they then define a multi-attribute hierarchy that describes the most important factors to take into account.

A discussion of time trade-offs of risk is given in [Bedford et al., 1999].

13.6 Other popular models

13.6.1 Cost–benefit analysis

Cost–benefit analysis (CBA) is a very well established method, both within economics and within the regulatory area. CBA is used in the UK nuclear industry [Pape, 1997] to guide decision making in the ALARP region (ALARP is the acronym for 'as low as reasonably practicable').

A list of burdens and benefits is made, and monetary values are assigned per unit (the UK Department of Transport assigns a value to a human life, for example, and updates the value yearly to account for inflation). Then the net cost (burdens minus benefits) of each alternative is determined, and the alternative with the lowest cost is recommended. CBA is, in its more modern form, very close to utility theory. For example, some practitioners allow the use or 'disproportionality factors' to model risk aversion.

The main claimed benefit of CBA is the objectivity of the pricing. After all, the prices are fixed in markets that reflect how society really prioritizes the various attributes. Unfortunately of course markets are not perfect (in the economic sense) so it is not clear that the prices really reflect what is claimed. Another problem is that the *choice* of market in which the prices are determined is highly subjective (think of the huge variation in house prices from one part of the country to another – if we needed the value of a house for a CBA study, from which region would we take the house price?).

For many items essential to applications of CBA no market exists at all. Human lives are the most obvious case. Because of this, the CB analyst is forced to use other methods to infer the price. Wage differentials have been proposed to gauge how much individuals are prepared to pay to avoid risk (lower salary but safer job). However, this method fails because the labor market is highly imperfect. Another nice method is that of *contingent valuation*, that is, 'just ask what the value is'. Applications of this method have given hugely varying values for a human life.

One of the main problems with contingent valuation is that there is a huge difference between 'willingness to pay' (WTP) to obtain a benefit, and 'willingness to accept' (WTA) the loss of that benefit. Whether WTP or WTA values should be used depends on the way the regulatory framework is structured. Does the public have a right to a risk free life, or does industry have a right to cause a certain amount of risk? In the former case the public should be compensated (using WTA values) by a company wanting to generate risk. In the latter case the public should compensate (using WTP values) a company keeping its risk level below the maximum limit.

Most contingent valuation studies concentrate on WTP values, as many WTA studies have been 'spoilt' by the alarming numbers of respondents refusing to accept at any price.

These issues and more are dealt with in great depth by Adams [Adams, 1995], who discusses the role of CBA in the decision to introduce mandatory seat belts in cars.

Upon examination, the 'objective' basis of CBA is extremely weak. It is only capable of creating the basis for a consensus decision when all participants agree with the prices used, and when those benefiting agree to compensate those losing. This rarely happens in practice.

13.6.2 The analytic hierarchy process

The analytic hierarchy process (AHP) is a popular decision tool supported by a large group of practitioners, particularly in the US. Its mathematical foundations are highly disputed however, and the model does not claim to be normative.

The underlying model used in AHP is additive, and is very closely related to the additive value model described above. The determination of weights in AHP is done with the aid of a matrix whose elements represent ratios of 'relative importance'.

The requirement that preference ratios are given is superficially attractive but relies on the assumption that these ratios are well defined. It seems reasonable to say that a cost of 100 units is twice that of a cost of 50 units. But what ratios do we use when one of the alternatives has a cost of -50 (all ratios are supposed to be non-negative)? The use of a ratio scale implies that an 'absolute zero' must exist. Taking a 'zero' in terms of the lowest scoring alternative does not really solve the problem because it makes ratios dependent on the alternatives available.

The interested reader is referred to the discussion of AHP and its relation with multi-attribute value theory in [Salo and Hämäläinen, 1997], and to the discussions that follow, especially that of Saaty, the inventor of AHP.

13.7 Conclusions

Decision theory continues to be an important but frustrating area. There is much disagreement between the theoreticians, and there seems to be much blind use of standard (but not necessarily appropriate) models.

Most decision-makers are prepared to accept the notions of 'rational decision making' as a basis for forming judgements. However, the standard normative models are difficult to use properly.

The majority of decision-makers probably do not know (antecedently) what kinds of trade-offs they believe in. The decision model should help them form and rationalize their beliefs. This requires a degree of understanding in the model that most decision-makers do not have (just adjusting the weights in an additive value model does not give understanding). The challenge is therefore to find models that, with interactive use, do give decision-makers such understanding.

In the area of risk analysis this is of prime importance. Trade-offs between high consequence/ low probability events and costs require a sophisticated approach. We believe that the same level of sophistication should be used

13.8 Exercises

13.1 Using the data from the problem studied in Section 13.3, perform a value of information calculation to determine whether it would be optimal to ask for an extended research proposal just from consultant B.

13.2 Consider a decision problem with three attributes, the first being mass (measured in kilograms). A weighting factors model is applied using weights w_1, w_2, and w_3. If the units of mass are changed to grams, how should the weights be changed in order to give the same outcomes?

13.3 A turbine in a generator can cause a great deal of damage if it fails critically. The manager of the plant can choose either to replace the old turbine with a new one, or to leave it in place.

The state of a turbine is classed as either 'good', 'acceptable', of 'poor'. The probability of critical failure per quarter depends on the state of the turbine:

$$P(\text{failure}|\text{good}) = 0.0001,$$
$$P(\text{failure}|\text{acceptable}) = 0.001,$$
$$P(\text{failure}|\text{poor}) = 0.01.$$

The technical department has made a model of the degradation of the turbine. In the model it is assumed that, if the state is 'good' or 'acceptable', then at the end of the quarter it stays the same with probability 0.95 or degrades with probability 0.05 (that is, good becomes 'acceptable', and 'acceptable' becomes 'poor'). If the state is 'poor' then it stays 'poor'.

(a) Determine the probability that a turbine that is 'good', becomes 'good', 'acceptable', and 'poor' in the next three quarters, and does not critically fail.

The cost of critical failure is 1.5 MAU (including replacement costs). The costs of a new turbine are 5 KAU.

(b) Consider the two alternatives: $a1$: Install new turbine.
$a2$: Continue to use old turbine.
The old turbine is 'acceptable'. Determine the optimal decision using expected monetary value. What would the decision

be if the old turbine were 'good' with 10% probability and 'acceptable' with 90% probability?

It is possible to carry out two sorts of inspection on the old turbine. A visual inspection costs 200 AU, but does not always give the right diagnosis. The probability of a particular outcome of the inspection given the actual state of the turbine is given in the table below:

Outcome	Actual state		
	Good	Acceptable	Poor
Good	0.9	0.1	0
Acceptable	0.1	0.8	0.1
Poor	0	0.1	0.9

The second sort of inspection uses an X-ray, and determines the state of the turbine exactly. The cost of the X-ray inspection is 800 AU.

(c) Determine the optimal choice between no inspection, a visual inspection, and an X-ray inspection.

14

Influence diagrams and belief nets

In this chapter we look at two closely related techniques for representing high-dimensional distributions and decision problems. We assume throughout that all random variables are discrete. The material in this chapter is largely drawn from [Schachter, 1986], [Smith, 1995], [Smith, 1989a], [Smith, 1989b], [Jensen, 1996], [Pearl et al., 1990] and [Götz, 1996]. A reader containing a number of important foundational papers is [Shafer and Pearl, 1990].

14.1 Belief networks

Belief networks, also called Bayesian belief networks, are graphical tools used to represent a high-dimensional probability distribution. They are convenient tools for making inferences about uncertain states when limited information is available. Belief nets are frequently used for making diagnoses, with applications to both medical science and various engineering disciplines, in particular to emergency planning.

A belief net is a directed acyclic graph whose nodes represent random variables (or groups of random variables). An arrow from one node to another represents probabilistic influence. Figure 14.1 shows an extremely simple belief net in which one variable influences another. The probabilistic specification in this belief net is of the marginal distribution of X, and the

Fig. 14.1. A simple belief net

14.1 Belief networks

conditional distribution of Y given X, for every value of X. This specifies the full joint distribution as

$$p(x, y) = p(x)p(y|x).$$

More generally, for n variables $X_1, \ldots, X_n$, one can always decompose the joint distribution by

$$p(x_1, \ldots, x_n) = p(x_1)p(x_2|x_1)p(x_3|x_1, x_2) \ldots p(x_n|x_1, \ldots, x_{n-1}). \quad (14.1)$$

This expression is not usually a compact way to write down a joint distribution. Under certain assumptions however, the expression can be simplified. If the X_i form a Markov chain for example, then for each i, X_i is independent of $X_1, \ldots, X_{i-2}$ given X_{i-1}, so that we can write

$$p(x_i|x_1, \ldots, x_{i-1}) = p(x_i|x_{i-1}) \quad (14.2)$$

for each i. This helps because the left hand side of Equation 14.2 is a function of i variables, whereas the expression on the right is function of just two variables.

A belief net is a way to graphically represent Markov properties similar to that given in Equation 14.2.

Influence diagrams can be used to provide an attractive graphical representation of a decision problem. They have been used increasingly in the past decade in many different areas. For general introductions to the theory the reader is referred to the articles contained in [Oliver and Smith, 1990]. The use of influence diagrams in the context of PRA's was first discussed in [Kastenberg, 1993]. The main difference between influence diagrams and belief nets is the presence of a decision node in an ID. This represents the various decision alternatives available to the decision-maker (as the decision-maker is supposed to choose rationally which alternative to take, the decision node cannot be a random variable!). An ID also contains a value node which represents the value (or often utility) given the states of the other (random and non-random) variables in the problem.

Clearly, the main emphasis in an ID is on determining the optimal decision. In belief nets, by contrast, the emphasis is on (Bayesian) inference by which posterior probabilities can be calculated of certain variables given observations of other variables. In order to do Bayesian inference it is usually necessary to compute a new expression for the joint distribution, corresponding graphically to reversing the direction of some arrows in the belief net. For example, in the net shown in Figure 14.1, the posterior distribution of Y given $X = x$ can be determined directly from the factorization of the joint distribution. In order to determine the posterior distribution of X given $Y = y$ we have to 'reverse the arrow' and determine $p(x|y)$.

The Markov property used in belief nets and IDs can be expressed using the concept of conditional independence.

14.2 Conditional independence

Suppose that X, Y and Z are vectors of random variables. We say that X and Y are *conditionally independent given* Z, written $I(X, Z, Y)$, if given any value z_0 of Z the joint probability density function of (X, Y, Z) factorizes,

$$f_{(X,Y,Z)}(x, y, z_0) = g(x)h(y),$$

where the functions g and h may depend on z_0.

Intuitively, conditional independence says that once we know the value taken by Z, then no information given about Y would change our uncertainty about X.

Example 14.1 *In a motor vehicle, the main light, the starter motor, and the battery can all be in the states failed and working. However, given the state of the battery, the states of the starter motor and the main light might reasonably be considered to be independent of one another. They are not unconditionally independent, since if you observe that the light does not work, there is a higher probability of a flat battery and thus of a non-working starter.*

Example 14.2 *Three fair coins are tossed. X_1 equals 1 if the first coin shows heads, and otherwise equals 0. We similarly define X_2 for the second coin and X_3 for the third coin. Finally define X_4 equal to 1 if $X_1 + X_2 + X_3$ is odd and otherwise 0.*

Clearly $X_1, \ldots, X_4$ are not independent, as knowing X_1, X_2 and X_3 tells you what X_4 is. It is easy to check that $P(X_4 = 0) = \frac{1}{2} = P(X_4 = 1)$. However, given X_1, X_4 is independent of X_2. To check this, one has to consider the two cases $X_1 = 0$ and $X_1 = 1$ separately. If $X_1 = 0$ then

$$\begin{aligned} P(X_2 = X_4 = 0 | X_1 = 0) &= P(X_2 = 0 | X_1 = 0) P(X_4 = 0 | X_2 = 0, X_1 = 0) \\ &= P(X_2 = 0 | X_1 = 0) P(X_3 = 0 | X_2 = 0, X_1 = 0) \\ &= P(X_2 = 0) P(X_3 = 0) = \frac{1}{4}. \end{aligned}$$

Calculating in the same way, we see that the probability table for X_2 and X_4, given $X_1 = 0$, is

$$
\begin{array}{c|cc}
 & \multicolumn{2}{c}{X_2} \\
 & 0 & 1 \\
\hline
X_4 \quad 0 & \frac{1}{4} & \frac{1}{4} \\
1 & \frac{1}{4} & \frac{1}{4}
\end{array}
$$

which clearly factorizes. The probability table for X_2 and X_4, given $X_1 = 1$, is the same. This shows the conditional independence of X_2 and X_4 given X_1.

14.3 Directed acyclic graphs

Belief nets and influence diagrams are directed acyclic graphs. Nodes represent the variables of interest in the problem, and arrows (directed edges) between these nodes indicate the dependency structure in the problem. The requirement that the graph be acyclic means that there is no directed path through the graph that returns to its own starting point.

There are usually four types of nodes in an ID:

- decision nodes (representing alternatives for the decision-maker);
- chance nodes (representing probabilistic quantities);
- deterministic nodes (representing deterministic quantities); and
- value node (a special kind of deterministic node, of which there can be only one, representing the quantity of interest).

In a belief net there are only chance nodes. There is a convention that decision nodes are drawn with a square, chance nodes with a circle, and deterministic nodes with a square whose corners have been rounded off (sometimes the value node is designated with a diamond). When there is an arrow from node X to node Y then we call X a *parent* of Y, and Y a *child* of X.

The arrows drawn between nodes represent qualitative influences which must be quantified by the model builder. For a chance node this means assigning probabilities to each of the possible values it can take, conditional on the values being taken by its parent nodes. For a deterministic node, it means assigning a deterministic function of the values taken by its parent nodes.

It is easy to see that the above specifications uniquely determine a probability distribution P over the nodes *if* we assume a Markov property: a node is conditionally independent of all its non-descendants given the values taken by its parents. This is explained below.

14.4 Construction of influence diagrams

First, a list is made of the relevant variables by starting with the objectives of the analysis, and then describing the primary factors that might influence these objectives. Next the secondary factors which impact on the primary factors are described. We carry on in this way until it seems that all the relevant factors have been found. At this stage it may well be possible to think of new actions for the decision-maker, by looking at the different influencing factors and determining ways in which the influences could be broken or limited.

Next the different influencing factors have to be described in precise terms. Each factor can be in one of a number of different states. For example, 'fire alarm' might be in one of the states 'alarm works' and 'alarm does not work'. The states should be exclusive (that is, it is not possible for more than one state to hold at once) and exhaustive (that is, all possible states have been specified).

The next stage is the construction of the qualitative influence model using a directed acyclic graph. The influencing factors are first ordered, $X_1, \ldots, X_n$. Frequently the ordering arises naturally, for example as a result of temporal or informational ordering. For each i we look for the smallest collection, $\Pi(i)$, of the variables $X_1, \ldots, X_{i-1}$ such that knowing these variables would make the values taken by the *other* variables in $X_1, \ldots, X_{i-1}$ irrelevant for prediction of X_i:

- Does the value taken by X_1 influence the value of X_2?
- Do the values of X_1 and X_2 influence the value of X_3? If so, would knowing X_1 make the value of X_2 irrelevant for predicting X_3, or would knowing X_2 make the value of X_1 irrelevant for predicting X_3?
- ...

The set $\Pi(i)$ is called the *parent set* for node i. Node i is a *child* node for each of the parents. The directed acyclic graph is now built by drawing a node for each variable, and then drawing an arrow from each parent node to the corresponding child node. Note that it is convenient to choose an ordering of the variables so that the parent sets are small. If (some) parent sets are large then it might be worth reordering the variables.

Finally, the influence diagram is quantified. This means that the conditional probability distribution is specified for each variable i given each possible combination of states of the variables in the parent set $\Pi(i)$. Typically this will be done using some form of expert judgement.

By construction, given its parent set $\Pi(i) = \{x_{j_1}, \ldots, x_{j_k}\}$, X_i is independent of the vector of variables X_j where $j < i$ and $j \notin \Pi(i)$,

$$p(x_i|x_1,\ldots,x_{i-1}) = p(x_i|x_{j_1},\ldots,x_{j_k}).$$

With a mild abuse of notation we shall write this as $p(x_i|\Pi(i))$. Hence we can simplify the joint probability density decomposition given in Equation 14.1 to get

$$p(x_1,\ldots,x_n) = \prod_1^n p(x_i|\Pi(i)). \tag{14.3}$$

Example 14.3 *A piece of machinery contains an alarm system to indicate the occurrence of a fault. When the fault is signaled, a technician has to disassemble the machinery. Sometimes the alarm incorrectly indicates a fault, costing money. The alarm never fails to indicate a fault. The possibility of using a second alarm system is to be investigated. If the second alarm is used then the repair technician will only be called if both alarms are activated.*

The various nodes in this problem are:

- *A* decision node *representing the two decision alternatives* second alarm *and* no second alarm.
- Chance nodes fault *taking the states 'yes' and 'no';* primary alarm *taking the states 'alarm rings' and 'alarm does not ring';* secondary alarm *taking the states 'alarm rings', 'alarm does not ring', and 'no second alarm';* repair *taking the states 'repair' and 'no repair'.*
- *A* value node costs *representing the total costs.*

Suppose we order the variables just as above.

(1) *The decision node.*
(2) *The fault node. Its state is not influenced by the choice to implement a secondary alarm.*
(3) *The primary alarm node. Its state is only influenced by the fault node.*
(4) *The secondary alarm node. Its state is influenced by the fault and the decision nodes.*
(5) *The repair node. The decision node, the primary and the secondary alarm nodes influence its state. Its state is not influenced by the fault node, given the primary and secondary nodes, because the repair only depends on the state of the alarms.*
(6) *The cost node. The installation of the second alarm costs money and so influences costs. Since the alarms never miss a fault, the only costs are those caused by a (possibly unnecessary) repair. Hence the primary and secondary alarm nodes do not influence the costs, given the repair node.*

The influence diagram for this problem is shown in Figure 14.2.

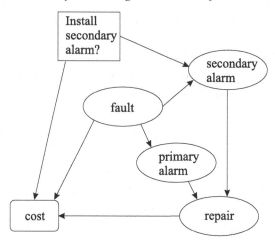

Fig. 14.2. Alarm influence diagram

14.4.1 Model verification

We have indicated above that an influence diagram model is built by first determining a graphical model, and then specifying the required conditional probabilities.

An ideal time for model verification is at the point that the graphical model has been constructed. It is important to check with the domain experts whether the conditional independences that are implicit in the graphical model are valid or not. If the domain expert does not agree with these conditional independences then the graph must be adjusted accordingly.

We now give a method by which such implicit conditional dependences can be found. In general, the set of conditional independences that can be derived from the graph structure with these methods is a subset of all the conditional independences that hold for a probability distribution realizing the graph. This is because influence diagrams do not represent the concept of conditional independence perfectly. There may actually be more conditional independencies that arise from the quantification of the conditional probabilities than are apparent from the graph. As a trivial example, the independent distribution is consistent with *any* influence diagram.

The conditional independencies implicit in a belief net G can be found by building an undirected graph G^m, called the *moralized* or *moral, graph*, from the original belief net. The moralized graph is determined by two operations on the original graph. First, if two nodes sharing the same child are not connected by an edge then an undirected edge is placed between them. Second, all the directed edges are replaced by undirected edges. Given

14.4 Construction of influence diagrams

three disjoint sets of nodes A, B and S, we say that S *separates* A *from* B if all paths in G^m from A to B have to pass through S.

The next result has been adapted from [Lauritzen, 1996].

Theorem 14.1 *For any belief net G, the global Markov propery holds for its moral graph G^m: for all disjoint sets of nodes of the graph A, B and S with the property that S separates A from B in G^m we have $I(A, S, B)$.*

Proof For any set of nodes C denote by G_C^m the subgraph of G^m with node set C and all edges between nodes in C. Write V for the set of all nodes of G.

Let $\tilde{A}$ be the set of nodes of G^m in the connected component of $G_{V \setminus S}^m$ containing A, and define $\tilde{B} = V \setminus (\tilde{A} \cup S)$. Define also

$$C_A = \{i \in V | \{i\} \cup \Pi(i) \subset \tilde{A} \cup S\}.$$

Clearly we have $A \subset C_A$. But we also have $V \setminus C_A \subset \tilde{B} \cup S$. This is because if $i \notin C_A$ then $\{i\} \cup \Pi(i) \not\subset \tilde{A} \cup S$, so there must be an element of $\{i\} \cup \Pi(i)$ in $\tilde{B}$. Hence $i \in \tilde{B} \cup S$.

We can now make a special arrangement of the decomposition of the joint density given in Equation 14.3 as follows:

$$p(x_1, \ldots, x_n) = \prod_{i \in C_A} p(x_i | \Pi(i)) \times \prod_{i \notin C_A} p(x_i | \Pi(i)).$$

The first product term just depends on $i \in \tilde{A} \cup S$. The second product term just depends on $i \in \tilde{B} \cup S$. The only variables in both terms are those in S. When we integrate out the variables not in $A \cup B \cup S$ to get the joint distribution for the variables of $A \cup B \cup S$, the expression obtained also factorizes. Hence $I(A, S, B)$. □

This result does not yet give the optimal way of deriving conditional independence from a belief net. This is because connections may appear in the moral graph due to descendants. If we are interested in checking on possible conditional independence of A and B given S, the best strategy is to work with the moral graph of the *ancestor belief net* for $A \cup B \cup C$. This is the directed subgraph of G just containing nodes i with $i \leq j_0$, where $j_0 = \max\{j \in A \cup B \cup C\}$. The above proof then works with n replaced by j_0.

Figure 14.3 gives an example of a belief net and the corresponding moral graph. You are asked to write down the factorization of the joint distribution in Exercise 2 and check that the above theorem holds.

By looking at the moralized graph we can see directly that B and D are

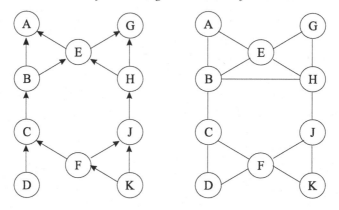

Fig. 14.3. A Bayesian belief net and the corresponding moral graph

conditionally independent given C and H. In fact, B and D are conditionally independent given just C, as one sees by looking at the ancestor belief net of $\{B, D, C\}$.

14.5 Operations on influence diagrams

It is sometimes possible to simplify the structure of an influence diagram by iteratively removing nodes and reversing arrows where necessary.

14.5.1 Arrow reversal

An arrow between two chance nodes X and Y can be reversed as long as the resulting graph remains acyclic [Smith, 1989a, Smith, 1989b]. However, more arrows have to be added in order to keep the model consistent (that is, to imply the same, or a smaller, set of conditional independencies). Arrows have to be added between X's parents and Y, and between Y's parents and X.

The conditional probabilities in the ID or belief net also have to be given for the new arrows: these can be calculated from the old probabilities using Bayes' rule.

A practical reason for using arrow reversal is that the qualitative dependency structure (represented in the directed acyclic graph) may be easy to determine in one form, but difficult to quantify in that same form. Often domain experts find it easier to determine conditional probabilities in one 'direction' than the other.

14.5.2 Chance node removal

A chance node X may be removed, but in order to keep the model consistent more arrows have to be added [Smith, 1989a, Smith, 1989b]:

(i) Arrows are added between all X's parents and X's children.

(ii) If two of X's children are connected, then arrows are added between the parents of the 'parent' child and the 'child' child.

(iii) Arrows are added between all of X's children, in such a way that the diagram remains acyclic.

The conditional probabilities corresponding to the new arrows can be determined by using Bayes' rule.

14.6 Evaluation of influence diagrams

Evaluation of an ID means the determination of the optimal decision. One way to do this is by successively applying the process of chance node removal until all the chance nodes have been removed. Then the only nodes remaining are the decision nodes, deterministic nodes, and the value node. The value corresponding to each combination of decisions can be simply 'read off'.

Another way of evaluating influence diagrams [Smith, 1995] uses the method of junction trees. Junction trees are a convenient representation form for IDs and belief nets that allow the rapid calculation of conditional distributions (and thus also of posterior distributions). We shall not go into the construction of junction trees here.

A third way of evaluating influence diagrams is to construct an equivalent decision tree, and to use the standard dynamic programming techniques to solve the decision tree.

14.7 The relation with decision trees

An influence diagram is simply another way of presenting the information represented in a decision tree. Typical ID software such as DPL [Loll, 1995] is able to switch between the ID representation and the decision tree representation.

To construct a decision tree it is simply necessary to order the variables in such a way that parents are always listed before children (this can be done because the directed graph is acyclic). Then a tree is made with the variables appearing as vertices and appearing in the order just determined. At each vertex the tree branches, with one branch corresponding to each possible value of the corresponding variable. The branches can be assigned probabilities immediately from the conditional probabilities in the ID (the probabilities being 1 in the case of deterministic nodes).

14.8 An example of a Bayesian net application

The example we present is taken from [Guivanessian et al., 1999]. The network is designed to demonstrate the way in which the influence of various factors on the probability of a major fire (fire flashover) for a 25 m^2 office can be modeled.

There are fifteen nodes in the model:

(i) *Fire starts* has the states *yes* and *no*. The probability of *yes* is taken as 0.01.

(ii) *Detection by occupancy* has the states *yes* and *no*. The probability of *yes* given *fire starts=yes* is 0.9. Given *fire starts=no*, the probability of *no* is 1.

(iii) *Actions of occupants* has states *action* and *no action*, and indicates whether the occupants take action to diminish the severity of the fire. Conditional probabilities for this state given its parents are shown together with other conditional probabilities in Table 14.1.

(iv) *Tampering-sd* refers to the possibility of interference (including accidental) with the smoke detection system and has states *yes* and *no*. This node has no parents, and the probability of *yes* is taken as 0.02.

(v) *Smoke detection* takes the states *yes* and *no*.

(vi) *Alarm* refers to the functioning of the acoustic fire alarm, and has the states *yes* and *no*.

(vii) *Tampering-sp* refers to the possibility of interference with the sprinkler system and has states *yes* and *no*. The probability of *yes* is 0.02.

(viii) *Sprinklers* describes the working of the sprinkler system and has states *yes* and *no*.

(ix) *Transmission* has states *yes* and *no*, and indicates the transmission of the alarm to the fire brigade.

(x) *Fire brigade* has states *action* and *no action* and specifies the reaction of the fire brigade to the alarm (within the time required).

(xi) *Flashover* describes the development of the fire as a flashover

(xii) *Decision* is the node describing the decision to install sprinklers or not. If they are not installed then the node *sprinkler* is *no*.

(xiii) The remaining nodes are *cost* nodes representing the installation costs of sprinklers, damage costs caused by the fire brigade when there is no flashover, and the costs of damage given that a flashover occurs. In relative units, the installation costs are set to 1. The costs with no flashover are given in the table. The costs of flashover depend on many factors. For this model we vary the costs of flashover between

14.8 An example of a Bayesian net application

Table 14.1. *Conditional probability tables*

Conditional probability for node *actions of occupants*

Detection by occupancy	yes		no	
Alarm	yes	no	yes	no
Occupants act	0.9	0.5	0.6	0
Occupants do not act	0.1	0.5	0.4	1

Conditional probability for node *smoke detection*

Fire starts	yes		no	
Tampering-sd	yes	no	yes	no
Smoke detected	0.5	0.99	0.5	0
No smoke detected	0.5	0.01	0.5	1

Conditional probability for node *alarm*

Detection by occupancy	yes				no			
Smoke detected	yes		no		yes		no	
Sprinklers	yes	no	yes	no	yes	no	yes	no
Alarm works	0.99	0.98	0.98	0.95	0.98	0.95	0.95	0
Alarm does not work	0.01	0.02	0.02	0.05	0.02	0.05	0.05	1

Conditional probability for node *sprinklers*

Fire starts	yes		no	
Tampering-sp	yes	no	yes	no
Sprinklers act	0.5	0.99	0.5	0
Sprinklers do not act	0.5	0.01	0.5	1

Conditional probability for node *transmission*

Detection by occupancy	yes				no			
Smoke detected	yes		no		yes		no	
Sprinklers	yes	no	yes	no	yes	no	yes	no
Transmission	0.99	0.98	0.98	0.9	0.98	0.95	0.95	0
No transmission	0.01	0.02	0.02	0.1	0.02	0.05	0.05	1

Conditional probability for node *flashover*

Fire starts	yes							
Actions of occupants	yes				no			
Fire brigade	action		no action		action		no action	
Sprinklers	yes	no	yes	no	yes	no	yes	no
Flashover	0.01	0.1	0.02	0.2	0.02	0.1	0.1	0.99
No flashover	0.99	0.9	0.98	0.98	0.98	0.9	0.9	0.01

Costs due to fire brigade and sprinklers use given flashover=no

Fire brigade	action		no action	
Sprinklers	yes	no	yes	no
Costs	20	10	10	0

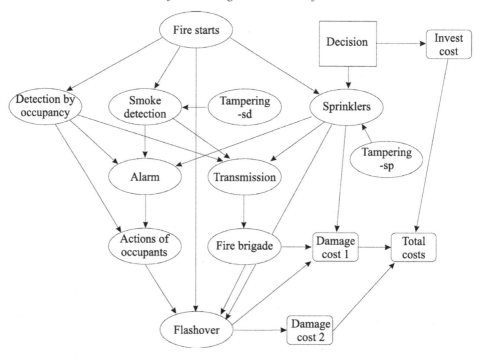

Fig. 14.4. An influence diagram for fire risk

10 and 10^6 to explore the sensitivity of the decision for sprinkler installation.

The conditional probabilities required for the influence diagram are given in Table 14.1. The influence diagram for this example is shown in Figure 14.4.

14.9 Exercises

14.1 Draw a belief net for Example 14.1. Write down the corresponding factorization of the joint distribution for the states of the three components.

14.2 Write down the factorization of the joint distribution for the belief net in Figure 14.3. Show that the conditional independency $I(C, \{E, F\}, H)$ holds by applying Theorem 14.1. Show also that $I(C, F, H)$ holds by considering the ancestor belief net *and* by directly factorizing the joint distribution. Which other conditional independencies can you find?

15

Project risk management*

Project risk management is a rapidly growing area with applications in all engineering areas. We shall particularly concentrate on applications within the construction industry, but the techniques discussed are more widely applicable. Construction risks have been the subject of study for many years. In particular, [Thompson and Perry, 1992] gives a good overall guide to the subject with a large number of references. Several different models and examples of risk analyses for large projects are given in [Cooper and Chapman, 1987]. A description of many projects (in particular high-technology projects) from the twentieth century, the problems encountered during management, and the generic lessons learnt are given in [Morris, 1994].

Large scale infrastructure projects typically have long lead times, suffer from high political and financial uncertainties, and the use of innovative but uncertain technologies. Because of the high risks and costs involved it has become common to apply risk management techniques with the aim of gaining insight into the principal sources of uncertainty in costs and/or time.

A project risk analysis performed by a candidate contractor before it bids for work is valuable because it can give the management quantitative insight into the sources of uncertainty in a project. This gives management a guide to the risks that need to be dealt with in the contract, or in financing arrangements. This is particularly interesting in public/private projects, where the risk attitude of the public authorities may be quite different to that of the contractor, giving room for the parties to divide the risks in a mutually advantageous way. Furthermore a quantitative view of the risks involved (including uncertainties) is important for the company in deciding the bid: Any bid is a gamble, but when a risk assessment has

---

*This chapter is co-authored by Lonneke Holierhoek

been performed the managers know what the odds are. It is clearly better to gamble while knowing the odds than to gamble blindly.

As well as being performed prior to the start of a project, a risk management study can be repeated throughout the lifetime of a project. Such a 'living risk analysis' gives an up-to-date risk profile which allows the incorporation of experience in the project, possible new opinions of experts and new technical developments.

15.1 Risk management methods

A typical example of the way in which risk management is applied is given by the RISMAN method [Rijke et al., 1997] developed for large-scale infrastructure projects in the Netherlands such as flood protection schemes, railway and road projects. Other businesses use similar techniques. Our discussion draws on RISMAN and on the methods used by the construction and dredging company HAM. A number of different steps are possible:

- identification of uncertainties and countermeasures,
- quantification of uncertainties and countermeasures,
- calculation of project risk,
- calculation of the effect of countermeasures,
- decision making and risk handling.

The 'identification' step is purely qualitative, and can be done without the other steps. Counter-measures may be identified and quantified after the calculation of project risk when it becomes clear that countermeasures are necessary.

15.1.1 Identification of uncertainties

The uncertainies involved in any project are caused by many different aspects: insufficient information, unfamiliarity with techniques or locations, lack of experience, unforeseen changes, etc. Recall that not all types of uncertainty can be quantified. In particular, a decision-maker cannot quantify his own uncertainty about his own capacity to react to problems in the course of the project. Probabilities about external events can in principle be quantified.

A tool used to identify the sources of uncertainty is a *Project Uncertainty Matrix*. See Table 15.1. The PUM is divided into sections corresponding to the various phases of the project. The columns in the PUM are labeled *aspect*, *cause*, *consequence*, and *countermeasures*. The *aspect* indicates the general

Table 15.1. *An example Project Uncertainty Matrix*

\multicolumn{4}{	c	}{Project Uncertainties Matrix}
Project :			
Subject : Preparation of Site & Towing of platform		Round : 2nd, evaluation	
pagenr. : B		Date :	
Aspect	Cause	Consequence	Countermeasures
Towing Vessel problems	When problems arise, upfront or during the towing activity, such as: ■ late arrival ■ vessel failure ■ less performance ■ towing vessel not licensed with this equipment the works may be [seriously] delayed	Delay [for instance from towing with 1 vessel instead of 2], Possible LD's	• Put a penalty on late arrival? • For failures etc: make sure that all critical equipment has spares, perhaps a third towing vessel? • Monitor towing force during trip • Review the vessel in detail, check its condition [fit for purpose?]. If needed: Insurance • Penalise in case of no license
Crane barge problems	Problems may arise, such as: ■ late arrival ■ failure of crane ■ failure of ROV	Delay, possible LD's	• Put a penalty on late arrival? • For failures etc: make sure that all critical equipment has spares • Review equipment in detail, test the ROV

area in which the uncertainty is found: political, organizational, technical, legal, social, financial, and geographical. Under *cause* we indicate what is uncertain and why it occurs, what is the usual situation, and what deviations from the usual situation are possible (and potentially undesirable). The *consequence* indicates what effects of the deviation can occur, in particular in terms of project duration and costs. Finally the *countermeasures* show what possible measures can be taken to mitigate the consequences or reduce the probability that the unwanted causes will occur. Countermeasures include not just technical measures (such as the use of different machinery or technology), but also organizational methods (such as the use of another parallel team of workers), financial methods (such as insurance) and legal instruments (the specification of risk-sharing in the contract).

The uncertainties are often classified into three types of uncertainty:

Normal uncertainties These are not (necessarily) normally distributed uncertainties, but those variables in a planning which have a natural variability.

Special events Unplanned low probability/high impact events.

Plan uncertainties These are uncertainties due to the fact that the project will be affected by decisions to be taken in the future.

15.1.2 Quantification of uncertainties

Given that most projects are unique it is unusual to find relevant statistical data with which probabilities or probability distributions can be quantified. Most assessments are therefore based on expert judgement (see below). For the normal uncertainties, a continuous distribution is determined, using expert judgement when no appropriate data is available. Frequently the triangular distribution is used as an approximation.

Special events are assumed to have fixed consequences if they occur. The distribution is therefore characterized by the consequences and the probability of occurrence (for example, 0.8 probability of no extra costs, 0.2 probability of extra costs of $100 000).

15.1.3 Calculation of project risk

This is typically calculated by a combination of CPM and Monte Carlo simulation, and will be described below.

15.2 The Critical Path Method (CPM)

Completion times (and costs) for a project are determined largely by the so-called *critical path*. The various parts of a project are interdependent, the start of one part depending on the prior completion of others. A way to represent this interdependence is with a *network* diagram. A network diagram is shown in Figure 15.1. This diagram is a small part of a real project planning illustrated in Figure 15.4.

Corresponding to each edge of the network is a number representing the time required to complete the corresponding activity. For the moment we assume that this number is known and not uncertain.

The critical path is the path through the network with the longest total completion time. This corresponds to the set of activities determining the completion time of the whole project. The critical path is easily determined by a roll-forward–roll-back algorithm. We illustrate this with the simple example from Figure 15.1 in Figure 15.2. We write c_n for the time duration required for completion of activity n, f_n for the *first time* at which activity n can be completed after all activities before it in the network have been completed, and l_n for the *latest time* at which activity n can be completed

15.2 The Critical Path Method (CPM)

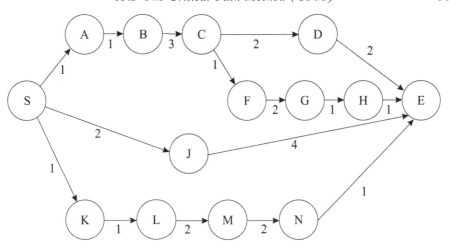

Fig. 15.1. A simple network

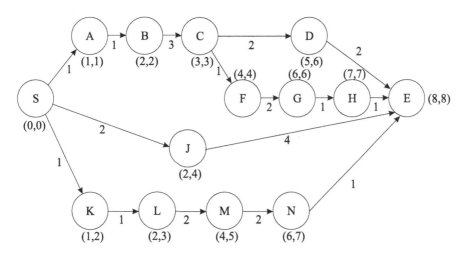

Fig. 15.2. Determining the critical path

without delaying the activities after it in the network. Clearly $f_n \leq l_n$ for all n. We also write $m \to n$ if activity m directly precedes n in the network. The algorithm works in two steps.

Roll forward At each node we write f_n (in the figure written as the first number in the brackets at each node). This is always 0 for the first node, and is calculated in general by taking the maximum of the 'first completion times' for the nodes before it in the net and then adding the completion time of the current activity,

$$f_n = c_n + \max\{f_m | m \to n\}.$$

The numbers are calculated iteratively starting at the first node of the network, hence the name 'roll forward'.

Roll back At each node we write l_n (in the figure written as the second number in the brackets at each node). This is always equal to f_n for the last node n, and is calculated in general by taking the minimum of the 'latest completion times' for the nodes after it in the net after having first subtracted the completion times of those activities,

$$l_n = \min\{l_m - c_m | n \to m\}.$$

The numbers are calculated iteratively starting at the last node of the network, hence the name 'roll back'.

Inspection of the figure shows that there is (at least) one path through the network such that $f_n = l_n$ at each node n on the path. Such a path is called a *critical path*. The critical path is the one determining the total completion time of the project. When the completion times of the activities are unknown they can be modeled as random variables. Then the critical path will also be random. A useful risk management metric that can be used to rank the activities is then the probability that an activity will be on the critical path. This is called the criticality index and is illustrated in the case study discussed below.

15.3 Expert judgement for quantifying uncertainties

As indicated above, it is the exception rather than the rule to have access to reliable and applicable historical data. For this reason most studies make use of some form of expert judgement.

Most studies use some form of linear pool: A number of experts are asked to give their uncertainties in terms of 5%, 50%, and 95% quantiles, or frequently 5% and 95% quantiles and a most likely value (mode). These values are used to estimate a distribution function that is taken to be representative of the expert's uncertainty for the quantity under consideration. (Exercise 1 asks you to determine parameters for the triangular distribution given such information.) Finally the experts are each assigned a weight, and the weighted average of the experts' distributions is taken as representative for the decision-maker's uncertainty over each quantity.

The classical model for expert judgement (described in Chapter 10) has been applied in a number of studies. The main feature of this model is that expert weights are chosen on the basis of empirical calibration. The main motivations for performing such calibration are that one may have more confidence in the outcome, and that the method enables one to persuade different stakeholders prae hoc to place their trust in the outcome post hoc. In other words, the method is one that seeks to build a rational consensus around the uncertainties that are present. For problems in project risk management it is not always necessary to build a rational consensus around the quantification of uncertainties: A private company in which the management has confidence in the abilities of the analysts might not be prepared to spend the extra resources needed to calibrate the experts. On the other hand, where public money is being used and there are different viewpoints within a project about the scale of the uncertainties involved, the use of calibration is an effective way of building up consensus.

15.4 Building in correlations

A potentially important part of the uncertainty quantification is the modeling of correlations. In a building project it is clear that bad weather may affect several different parts of a project simultaneously, and therefore be a factor that correlates completion times.

Correlations can make a substantial difference to the probability that a given path will be critical, and hence to the ranking of paths in terms of criticality. Structured methods of assessing correlation are discussed in [Cooke and Kraan, 1996].

The correlations assigned by the experts are used in a Monte Carlo simulation in order to simulate the distribution of completion times and project costs.

15.5 Simulation of completion times

Without making assumptions on the form of the joint distribution of the activity completion times, it is impossible to calculate an analytic formula for the distribution of the project completion time. Some form of Monte Carlo simulation is therefore used to determine this distribution (the precise form of simulation depending on how the correlations have been modeled).

The output of such a simulation is the distribution of the time to completion, but other interesting information can also be determined. Examples are the activities contributing most to high completion times, the activities con-

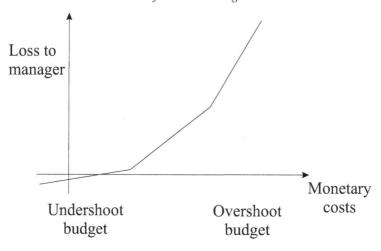

Fig. 15.3. Form of the loss function for overspending

tributing most to the uncertainty in the completion time, and the probability that a given activity will be on the critical path.

Note that in the form presented here, the distribution of completion time does not take into account possible decisions taken during the project lifetime. Sometimes a model is built with decision time points, in which case the network model and distributions pertaining to activities after the decision point should depend on the decision taken and the information available at that point (so before we know which decision has been taken and what information is available we have to evaluate all the possibilities). This effectively makes the network model a decision analysis tool, but obviously requires a lot of input.

15.6 Value of money

The simulation results give a probability distribution over the completion times and the costs. When judging the effect of different management measures on the basis of costs it is necessary to find a single number representing the cost distribution corresponding to each measure. The expected values of the costs are however not always appropriate. A better criterion is expected loss. This criterion allows one to take into account possible budgetary constraints. Often a manager is fairly indifferent to different levels of underspending, but is highly sensitive to an overspend. This suggests a loss function of the form shown in Figure 15.3.

The choice of loss function is a subjective choice of the decision-maker. It enables the decision maker to apply all the tools of decision theory, such as value-of-information techniques (see Chapter 13). The use of such techniques is contingent on acceptance of the individualistic viewpoint (notwithstanding the possible use of sensitivity analysis to find points of consensus between individuals). This leads us to pose the question: 'For whom is the risk analysis carried out?' In many practical applications of project risk analysis the distinct roles of the various stakeholders are ignored. Although there are resource savings to be achieved by sharing the project risk analyses, the different views and interests of the stakeholders inevitably lead to differing assessments of the costs and countermeasures.

15.7 Case study

The case at hand is an offshore project, concerned with the transportation and placing of a production platform and the construction and installation of a pipeline. It was performed using the in-house project risk analysis methods of HAM, an international dredging and marine contractor.

The main objective of the client (an oil company) was to start production before a certain date. Each delay in commencing production meant a considerable loss of income for the client. They had stipulated a scope of work within the tender specifications, according to which the contractor was to perform the activities. However, alternative offers were encouraged, especially if this were to imply acceleration of the program. HAM had decided to enter two offers: one (the standard offer) following the client's specifications and one alternative.

The differences in the two offers were mainly caused by changing one activity (the placing of two types of scour protection was replaced by the placing of mattresses) and omitting another (omission of the *j*-tube extension). In Figures 15.4 and 15.5 the network planning for the two options is presented. The standard offer had a (deterministically) planned duration of 140 days. The alternative was estimated to take 135 days to complete.

A project risk analysis was performed:

  (i) inventory of uncertainties,
  (ii) model building,
  (iii) quantification of uncertainties,
  (iv) model calculations,
  (v) interpretation.

The first step, an inventory of uncertainties, showed that the following issues were of concern in the program:

308   *15 Project risk management*

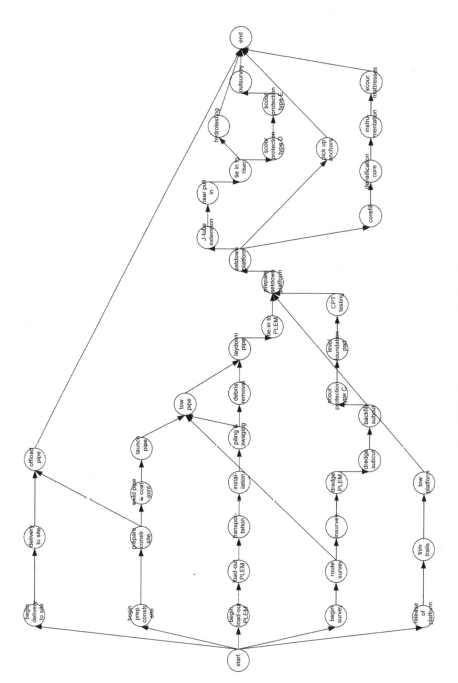

Fig. 15.4. Program standard offer

15.7 Case study

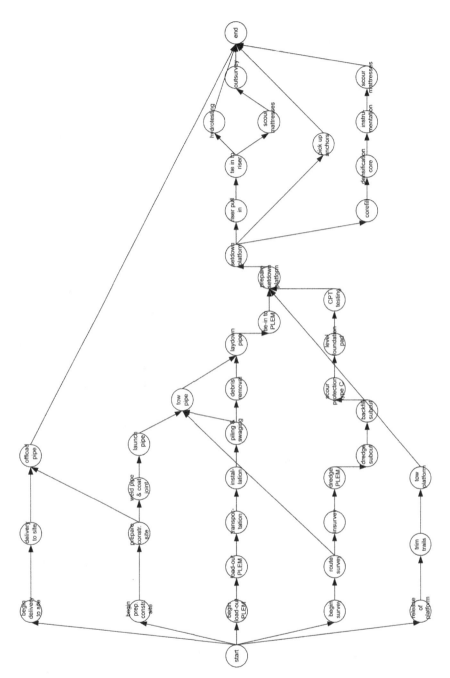

Fig. 15.5. Program alternative offer

- obtaining the necessary licenses and permits in time,
- delay by weather conditions (offshore: ice, wind, current, waves, typhoon, ...),
- uncertainty regarding the specifications and the exact quantities of construction of foundation pad,
- variation in dredging productions due to variation in soil conditions,
- equipment breakdown (dredger, survey equipment, instrumentation, grout pump, mattresses frame, ...),
- availability and/or performance of rented equipment (towing vessels, crane barge, ...),
- late completion of platform before towing (idle time, consequences for weather window, etc.),
- increased draft due to modifications of platform means delay during towing activities,
- delay in filling the platform core, due to suitable fill material at increased distance and/or densification not meeting requirements,
- availability and accessibility of suitable rock material for scour protection,
- delay in construction or installation of pipeline, due to low quality materials and/or facilities, the spoolpiece fabricated to connect pipeline to either platform or PLEM may not fit (The PLEM is a 'Pipeline End Manifold', a construction which anchors the pipes from the platform to the sea bed. Flexible pipes lead from the PLEM to the so-called SBM at which tankers can load with gas, oil, etc.),
- delay in installation of PLEM, due to problems with piling (variation in soil conditions) and/or meeting the specifications,
- technical problems during tie-in of pipeline to platform or of pipeline to PLEM.

The total inventory list of uncertain aspects is much longer, and includes items regarding cost and safety besides project duration. These items are collected using a sort of FMEA (Failure Mode & Effect Analysis), where cause and consequences of the uncertainties are identified. One may also itemize potential countermeasures, which can either remove the uncertainty entirely, or significantly reduce the probability of occurrence or minimize the consequences. At an early stage some decisions will be made regarding countermeasures. The management can decide to base their proposal on certain terms and conditions, some of the risks will be placed with the client. In this case for instance, management decided that the client would be made responsible for obtaining all necessary licenses and permits. Other countermeasures may also be taken, such as

- taking out an insurance,
- setting milestones and placing damages on late completion of activities to be performed by subcontractors,
- choosing an alternative work method,
- bringing sufficient spare parts, etc.

In this case, the management decided to exclude certain risks by submitting a proposal under terms and conditions, and to accept the other uncertainties. The risks and uncertainties which the management found acceptable form the basis for the next step.

A model was built based on the program of the project. The software used was a custom made product, named BIDS. This is a program to perform Monte Carlo simulation, specially designed to deal with a budget and program of a civil engineering project. The programs for the two alternatives were entered, in their form as shown in Figures 15.4 and 15.5.

The third step, quantification of the uncertainties, follows. After decisions regarding the countermeasures, the items to be quantified are

- uncertainty in survey activities, caused by possible breakdown of equipment, weather delay, etc.,
- uncertainty in dredging activities, caused by variation in soil conditions, variation in quantities (specifications) and weather delay,
- uncertainty in activities concerning placing of PLEM, caused by problems with piling (variation in soil conditions), etc.,
- possibility of not meeting the specifications regarding the installation of the PLEM,
- uncertainty in activities concerning construction and installation of pipeline, caused by low quality materials and/or facilities, technical problems during tie-in operation, etc.,
- possibility of an extra spoolpiece needed during tie-in,
- uncertainty in towing activities and placing of platform, caused by problems with availability or performance of towing vessels and crane barge,
- possible failure of the crane barge,
- uncertainty in fill and densification activities caused by breakdown of instrumentation, densification not meeting requirements, breakdown of mattresses frame or grout pump, etc.,
- uncertainty in scour protection activities, caused by breakdown of grout pump and problems with availability and quality of rock material.

The planning experts were asked to give their expert opinion on 39 items with continuous probability distributions (34 for the standard program and

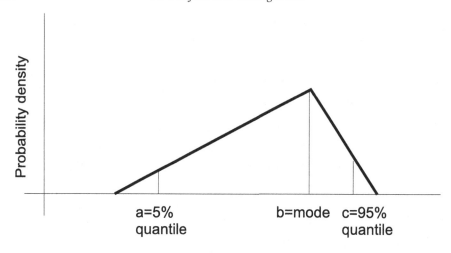

Fig. 15.6. Triangular distribution

an additional 5 for the alternative), and for 3 special events (with discrete probability distributions). For the continuous probability distributions, an expert is asked to give a lower and upper boundary for which he/she is 90% confident that the realization will fall between these values. In other words, he/she gives their 5% quantile and 95% quantile. The value which is entered into the program is assumed to be the most likely value (see previous paragraphs). For this case, at least 2 experts were elicited, and for some items 3. A triangular probability distribution (see Figure 15.6) is fitted to their estimates, and these are combined using equal weights. The 5% and 95% quantiles from the resulting probability distribution are used for fitting a final triangular probability distribution. This distribution serves as input in the model.

For the so-called special events, modeled with a discrete probability distribution, the experts are asked for the probability of occurrence and for an estimate of the consequential delay. Both estimates are combined using equal weights directly. This resulted in the following distribution for the event that there is a delay due to failure of the crane barge: 90% probability nothing happens, 10% probability that 8.75 days delay will occur.

All of these probability distributions are used as input for the model. Some dependencies are modeled as well by means of rank correlation coefficients. For instance the weather will influence all activities going on simultaneously in the same location. In our case two dependency factors are identified: the weather and the soil conditions.

The model calculations are done by means of Monte Carlo simulation.

15.7 Case study

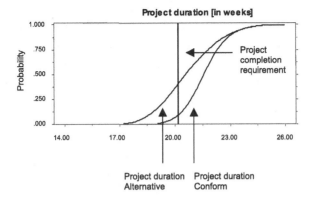

Fig. 15.7. Distribution of project duration

Table 15.2. *Criticality indices for activities*

Activities standard offer	CI	Activities alternative offer	CI
Setdown of platform	100%	Setdown of platform	100%
Prepare setdown	100%	Prepare setdown	100%
Scour protection type D & E	62%	Corefill	100%
Outsurvey	62%	Densification of core	100%
J-tube extension	62%	Instrumentation	100%
Riser pull-in	62%	Scour mattresses	100%
Tie-in to riser	62%	Laydown of pipeline	58%
Laydown of pipeline	58%	Prepare construction site	49%
Tie-in to PLEM	58%	Weld pipe & coat	49%
		Launch pipe	49%
		Tow pipe	49%

For both options 10 000 samples were generated. The resulting probability distributions are shown in Figure 15.7.

In order to compare the importance of the various activities for the overall project, a ranking is made in terms of the so-called criticality index. This is the probability that the activity will be on the critical path. The program BIDS calculates the criticality indices for all activities. These are presented in Table 15.2. Interpreting these results, one notices that for both options the probability of successfully completing the project within the 140 days (or 20 weeks) is not large:

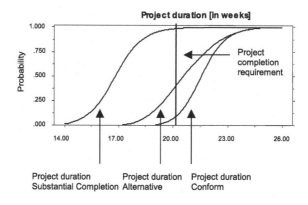

Fig. 15.8. Distribution of project duration, with new option

- for the standard option the probability of success (timely completion) is 7%;
- for the alternative option the probability of success (timely completion) is 36%.

More importantly, the critical activities differ between the two options. For the standard option, the activities concerned with the setdown of the platform are the single most critical group of activities. For the alternative offer, these are joined by activities concerned with filling the platform core. However, the activities of densification of the core and the instrumentation can be completed after the client has commenced production. This would mean a considerable increase of the probability of timely completion. Considerable, because these activities are on the critical path in 100% of the cases for the alternative offer. For the standard offer, this increase is less considerable, as there is a high probability of the activities not being on the critical path. If the activity is not on the critical path, acceleration of this activity will not result in acceleration of the overall program.

New calculations were performed, where the new option, called 'substantial completion', is included. This new option equals the alternative option, omitting the activities of 'densification of core', 'instrumentation' and 'scour mattresses'. In other words, it represents the program up to the point where the client can commence production. The results of these calculations are shown in Figure 15.8. From this graph the probability of success (timely completion) for the option of substantial completion can be read: 98%.

This valuable information was brought to the attention of the client, in order to further promote the alternative. They were very impressed, and easily convinced this was the way to maximize their chances of success. Unfortunately, having been given the idea that this was a much better way to carry out the project, they then awarded the project to another contractor who would carry it out more cheaply. This particular event had not been modeled in the project risk analysis!

15.8 Exercises

15.1 The triangular distribution with parameters (α, β, γ), where $\alpha < \beta < \gamma$, has a piecewise linear density f equal to zero at α and γ, and with a maximum at β. Calculate the maximum value at β (using the fact that the integral of the density must equal 1). Suppose an expert specifies the 5%, 95% and mode of a distribution. Is it always possible to fit a triangular distribution? Find a formula for the parameters (α, β, γ)?

15.2 The critical path determination was illustrated in Figure 15.2 on a subnetwork of that shown in Figure 15.4. Identify the corresponding part of the alternative network in Figure 15.5. Assuming that a new method for scouring the mattresses has been found that enables that activity to be completed in 1 time unit, calculate the critical path for this network.

16

Probabilistic inversion techniques for uncertainty analysis*

Accident consequences are often predicted using physical models. Examples are dispersion and deposition models when modeling the release of dangerous substances in the air, combustion models for modeling the intensity of heat at any given distance from an explosion, etc.

Physical models may be based on a theoretical model of the underlying processes, but are equally often based on empirical fitting of a descriptive model. An example of the latter is the common use of power law functions to predict the lateral plume spread σ_y at a downwind distance x from the source of a release:

$$\sigma_y(x) = A_y x^{B_y}, \qquad (16.1)$$

where the dispersion coefficients A_y and B_y depend on the stability of the atmosphere at the time of the release. This equation is not derived from physical laws, indeed the dimension of A_y must be [meters$^{1-B_y}$]! Similarly, the prediction of the vertical plume spread, σ_z, is given by

$$\sigma_z(x) = A_z x^{B_z}. \qquad (16.2)$$

Almost all physical models contain a wealth of parameters, such as the A_y and B_y above, whose values should be determined in order to use the model. It is usually difficult to fix such values on the basis of data (expert or empirical) as the data is usually insufficient, poor, possibly contradictory and inconsistent with the model.

In any case, since all physical models are idealizations of reality, we should expect that enough real data will eventually prove the model, with any particular choice of parameter values, to be incorrect.

In an uncertainty analysis one works with a joint probability distribution over the parameters instead of with particular parameter values. The model

---
*This chapter is co-authored by Bernd Kraan

no longer makes deterministic predictions because of the uncertainty over the parameter values. Instead, we get a distribution over the output values. (Note that this also applies when the model output was the *distribution* of some stochastic quantity. We now obtain a probability distribution over these distributions. The predictive distribution is then the mixture of all of the output distributions, representing the combination of uncertainties due to 'aleatory uncertainty' and the model parameter uncertainty.)

The distribution in the output values indicates how much the model output is sensitive to the parameters' joint uncertainty distribution.

Although this seems a reasonable approach, the question of how one obtains a 'reasonable' distribution on the parameters is difficult to answer. The aim of this chapter, which draws on [Kraan and Cooke, 1997], is to describe some of the methods for doing this that have been developed in a joint project of the European Union (EU) and the United States Nuclear Regulatory Commission (USNRC). The major objectives of this project [Harper *et al.*, 1994] were

- to formulate generic, state-of-the-art methodology for uncertainty estimation capable of finding broad acceptance,
- to apply the methodology to estimate the uncertainties associated with the predictions of probabilistic accident consequence codes designed for assessing the consequences of accidents at commercial nuclear power plants,
- to better quantify and obtain more valid estimates of the uncertainties associated with probabilistic accident consequence codes, thus enabling more informed and better judgements to be made in the areas of risk comparison and acceptability and therefore to help set priorities for future research,
- to systematically develop credible and traceable uncertainty distributions for the respective code input variables using a formal expert judgement elicitation process.

Expert judgement elicitation was used because of the scarcity of empirical data, and to allow for a diversity of viewpoints. One of the main methodological ground-rules for the study was that experts should only be questioned about *observable quantities* within the area of their expertise. The uncertainty distributions coming from the experts are then translated into distributions on the parameters of the codes, a process called 'probabilistic inversion' (the original name for the procedure was 'post-processing' because it is carried out after the expert judgement study).

There are a number of motivations for asking experts to give uncertainties about observables instead of model parameters (mathematically it would be much more convenient to ask experts directly about their uncertainties over model parameters):

- not all experts may subscribe to the particular model choice that has been made,
- the model parameters might not correspond to physical measurements with which the experts are familiar,
- from a theoretical point of view, it is difficult to define a subjective probability distribution directly over an abstract model parameter (cf. the truth conditions discussed in Chapter 2),
- it is difficult to understand *a priori* what kinds of correlations should be elicited between distributions on abstract parameters to build a joint distribution.

16.1 Elicitation variables and target variables

The unknown model parameters for which we wish to find a (joint) distribution are called *target* variables. *Elicitation variables* are physically meaningful quantities with which the experts are familiar. They may be target variables, or may be related to the target variables through the model. Elicitation variables can in principle be measured by procedures in which experts quantify their uncertainty on the basis of their expertise.

Target variables may be elicitation variables, in which case no probabilistic inversion is necessary. Otherwise the target variable distribution has to be obtained by probabilistic inversion from the distribution on the elicitation variables.

Example 16.1 *(dispersion model) The dispersion model defined in Equation 16.1 above has two parameters A_y and B_y. These parameters are the target variables. Since A_y and B_y are not physically observable they cannot serve as elicitation variables. However, we may take a number of distances downwind of the source $x_1, \ldots, x_n$ and ask experts to assess their uncertainties over the lateral spread $\sigma_y(x_i)$ at each of these distances. The $\sigma_y(x_i)$ are the elicitation variables.*

Example 16.2 *(box model for soil migration) The migration of radioactive material through various depths of soil is modeled using a box model shown in Figure 16.1. The boxes represent layers in the soil, and the transfer or migration rate between any two layers (the proportion of material moved from one layer*

16.2 Mathematical formulation of probabilistic inversion

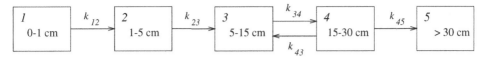

Fig. 16.1. Box model for soil migration

to the other in a short time interval) is assumed constant through time. For this particular case, the model is simply a five-dimensional first order differential equation with six parameters with initial conditions depending on the output of a deposition model.

The transfer coefficients (which are the target variables) do have a physical interpretation but cannot be measured directly, are not observable, and therefore cannot be elicitation variables. A physically measurable quantity is the time T_i at which half an initial mass starting in box 1 has passed beyond box i. These have been used as elicitation variables.

16.2 Mathematical formulation of probabilistic inversion

The elicitation variables $y_1, \ldots, y_n$ can be written as functions of the model parameters $(x_1, \ldots, x_m)$,

$$y_i = G_i(x_1, \ldots, x_m), \quad i = 1, \ldots, n.$$

From an expert judgement study we have obtained uncertainty distributions on the elicitation variables. These can therefore be regarded as random variables $Y_1, \ldots, Y_n$. The full probabilistic inversion problem is to find a distribution on the model parameters $(X_1, \ldots, X_m)$ so that, writing $G_i = G_i(X_1, \ldots, X_m)$, the distribution of $(G_1, \ldots, G_n)$ is the same as that of $(Y_1, \ldots, Y_n)$.

However, the full joint distribution of $Y_1, \ldots, Y_n$ is not usually elicited. A weaker formulation of the probabilistic inversion problem is to find a distribution on the model parameters $(X_1, \ldots, X_m)$ so that the distribution of G_i is the same as that of Y_i for each i. Below we discuss an algorithm solving this weak probabilistic inversion problem that takes the marginals of $Y_1, \ldots, Y_n$ as input. However, the algorithm can be easily extended to take account of dependency information such as rank correlations. Note that the output joint distribution of $\underline{X} = (X_1, \ldots, X_m)$ will normally be dependent even when only the marginal distributions of the Y_i are used as input.

In general the probabilistic inversion problem is either over-determined or under-determined! There may be many solutions, giving us the extra problem of choosing a single distribution fitting the data, or there may be no solution, giving us the problem of finding a distribution which 'best fits' the expert data.

16.3 PREJUDICE

A desperate search for a new acronym finally led to the choice of processing expert judgement into code parameters, PREJUDICE, for the algorithm described in this section.

The first two steps have been discussed above.

(i) Choose physically meaningful elicitation variables $Y_1, \ldots, Y_n$ and obtain via expert judgement marginal distributions for $Y_1, \ldots, Y_n$.
(ii) Define model predictor functions $G_i : \mathbb{R}^m \to \mathbb{R}$ for Y_i.

In the next two steps, choices are made for discretizing the distributions for the $\underline{X}$ and $Y_1, \ldots, Y_n$.

(iii) Choose a finite set $\chi \subset \mathbb{R}^m$.
(iv) Choose values $y_{i,1}, \ldots, y_{i,K}$ in the range of Y_i $(i = 1, \ldots, n)$.

The last two steps are concerned with finding candidate probability distributions, and selecting the most simple one from the family of possible solutions.

(v) Find the class $\mathscr{P}$ of (discrete) probability distributions P on χ with the property that whenever $\underline{X}$ has distribution P,

$$P(G_i(\underline{X}) \in [y_{i,k-1}, y_{i,k}]) = P(Y_i \in [y_{i,k-1}, y_{i,k}]), \quad i = 1, \ldots, n.$$

(vi) Choose the minimally informative probability distribution in $\mathscr{P}$ relative to the uniform distribution on $G(\chi)$.

Some remarks are in order. The numbers $y_{i,k}$ used here are often simply the 5%, 50%, and 95% quantiles of the expert distribution. The choice of the finite set χ is critical for the feasibility of the problem. A heuristic for determining a suitable set will be discussed below. It may turn out that the problem is always infeasible for any choice of χ. One solution to this is discussed in Section 16.4.

The last two steps are performed together by solving a non-linear optimization problem. This is discussed below.

16.3.1 Heuristics

The choice of a 'good' set χ is of great importance in avoiding the problem of infeasibility. Hence the role of the heuristic in steering the algorithm is very important in determining the performance of the method.

16.3 PREJUDICE

The idea is to take a small number of 'pseudo-model outcomes' and find the points in the model parameter space best generating these outcomes. Finally, around each of these points we take a cluster of points to generate the set χ.

To get a collection of 'pseudo-model outcomes', consider the quantiles $y_{i,1}, \ldots, y_{i,K}$ that we have chosen for each of the elicitation variables $1, \ldots, n$. A pseudo-model outcome is a vector whose components are chosen from these quantiles,

$$(y_{1,k_1}, y_{2,k_2}, \ldots, y_{n,k_n}).$$

Some pseudo-model outcomes may be physically impossible, in which case they should be filtered out. Clearly not every pseudo-model outcome is really a model outcome, that is, there may not exist a point $\underline{x}$ in the space of model parameters such that $G_i(\underline{x}) = y_{i,k_i}$ for every i. However, we search for a point $\underline{x}$ whose model outcome reasonably approximates the pseudo-model outcome in the sense that it solves the minimax problem

$$\min \max_{i=1,\ldots,n} (y_{i,k_i} - G_i(\underline{x}))^2.$$

Finally we sample a number of points from a small rectangle around $\underline{x}$: for each model parameter an $\varepsilon_j > 0$ is chosen, and we sample from the rectangle

$$[(1-\varepsilon_1)x_1, (1+\varepsilon_1)x_1] \times [(1-\varepsilon_2)x_2, (1+\varepsilon_2)x_2] \times \cdots \times [(1-\varepsilon_m)x_m, (1+\varepsilon_m)x_m].$$

The set χ is taken to be the union of all the samples around the $\underline{x}$s that we can generate from pseudo-model outcomes.

Note that the parameters generated by this heuristic should be screened for physical feasibility.

16.3.2 Solving for minimum information

The last two steps of the PREJUDICE algorithm are performed together by solving a non-linear optimization problem. We first illustrate how the optimization problem is built using a simple example in which there are just two elicitation variables.

Suppose that the 5%, 50%, and 95% quantiles of the elicitation variables Y_1 and Y_2 are used in step 4. This means that $y_{1,0} = -\infty$ (or the physically possible lower bound), $y_{1,4} = +\infty$ (or the physically possible upper bound), and $y_{1,1} < y_{1,2} < y_{1,3}$ are chosen so that

$$P(Y_i \in [y_{i,0}, y_{i,1}]) = 0.05,$$
$$P(Y_i \in [y_{i,1}, y_{i,2}]) = 0.45,$$

$$P(Y_i \in [y_{i,2}, y_{i,3}]) = 0.45,$$
$$P(Y_i \in [y_{i,3}, y_{i,4}]) = 0.05.$$

The 4 'bins' in the Y_1 direction and 4 'bins' in the Y_2 direction give a total of 16 bins in the product space. For convenience we write

$$U_{ij} = [y_{1,i-1}, y_{1,i}] \times [y_{2,j-1}, y_{2,j}]$$

for $i, j = 1, \ldots, 4$.

We now count how many points of χ are mapped into each U_{ij},

$$n_{ij} = \#\{\underline{x} \in \chi | (G_1(\underline{x}), G_2(\underline{x})) \in U_{ij}\}.$$

Points mapping to the same set U_{ij} will all get the same probability mass, p_{ij} (this follows from the minimum information principle). The total probability mass on U_{ij} is therefore $n_{ij} p_{ij}$.

Recall that the constraints on the probabilities are in terms of the mass in each interval $[y_{i,k-1}, y_{i,k}]$. The constraints on the probabilities can now be expressed in the form

$$n_{k1} p_{k1} + n_{k2} p_{k2} + n_{k3} p_{k3} + n_{k4} p_{k4} = P(Y_1 \in [y_{1,k-1}, y_{1,k}]) \quad (k = 1, \ldots, 4)$$

and

$$n_{1k} p_{1k} + n_{2k} p_{2k} + n_{3k} p_{3k} + n_{4k} p_{4k} = P(Y_2 \in [y_{2,k-1}, y_{2,k}]) \quad (k = 1, \ldots, 4).$$

Given these constraints, and the obvious constraint that $p_{ij} \geq 0$ for every i, j, we minimize the information function

$$\sum_{ij} n_{ij} p_{ij} \log p_{ij}.$$

This optimization problem is convex and can be solved by using optimization software. For more general problems with n elicitation variables and using 5%, 50% and 95% quantiles for each elicitation variable, the number of dimensions in the optimization problem is 4^n. Since this increases rapidly with n it is necessary to use a state-of-the-art optimization technique. We have had good experience with the interior point method of Andersen and Ye [Andersen and Ye, 1998].

16.4 Infeasibility problems and PARFUM

For certain problems it may turn out that no model parameter distribution can be found to generate the required model output distribution. This does not necessarily mean that the model should be abandoned (see the discussion below).

Table 16.1. *Quantile information for elicitation variables at various downwind distances*

Distance	Quantile	σ_Y Experts	σ_Y PREJUDICE	χ_Q Experts	χ_Q PREJUDICE
500 m	5%	42.7	42.7	$8.133E \times 10^{-7}$	$8.12E \times 10^{-7}$
	50%	125.9	126	$1.57E \times 10^{-5}$	$1.57E \times 10^{-5}$
	95%	302	302	$1.213E \times 10^{-4}$	$1.21E \times 10^{-4}$
1 km	5%	83.42	83.4	$2.146E \times 10^{-7}$	$8.12E \times 10^{-7}$
	50%	245.1	245	$4.073E \times 10^{-6}$	$1.57E \times 10^{-5}$
	95%	583.8	584	$3.098E \times 10^{-5}$	$1.21E \times 10^{-4}$
3 km	5%	231.2	231	$2.485E \times 10^{-8}$	$2.49E \times 10^{-8}$
	50%	676.8	676	$4.826E \times 10^{-7}$	$4.83E \times 10^{-7}$
	95%	1615	1610	$3.744E \times 10^{-6}$	$3.74E \times 10^{-6}$
10 km	5%	459.1	459	$1.882E \times 10^{-9}$	$1.88E \times 10^{-9}$
	50%	1691	1690	$8.422E \times 10^{-8}$	$8.42E \times 10^{-8}$
	95%	8546	8540	$6.865E \times 10^{-7}$	$6.87E \times 10^{-7}$

In this case we can try to apply methods from PARFUM, a forerunner of PREJUDICE, described in [Cooke, 1994]. The main idea is to apply PREJUDICE to subsets of the elicitation variables, and then to average the resulting distributions on the model parameters. For example, if there are three elicitation variables and the problem is infeasible, one might get feasibility by taking two elicitation variables at a time. This would give three different distributions on the model parameters, each of which fits the data on two elicitation variables well. The equal weight mixture of these distributions then gives a distribution which best fits the three distributions in the model parameters space in the sense of minimizing information. (The drawback of this mixture is however that it optimizes the fitting in the model parameter space, rather than in the elicitation variable space.)

16.5 Example

We give some of the results of fitting the dispersion coefficients A_y, B_y, A_z and B_z. These coefficients are used in the COSYMA code of the Gaussian plume model. It is assumed that the wind has a constant direction, so that we can define a coordinate system with x as the downwind direction, y as the crosswind direction, and z as the vertical direction. The source of the radioactive pollutant is assumed to be at height H above the ground at the point with coordinates $(0, 0, H)$. The Gaussian plume model then predicts

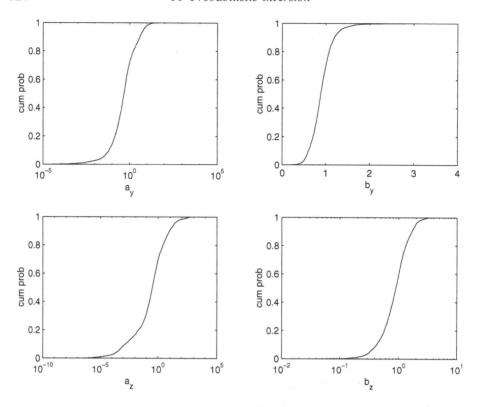

Fig. 16.2. Marginal distributions for the target variables

the time integrated concentration (in Bq × s/m³) at (z, y, z) to be

$$C(x, y, z) = \frac{Q_0}{2\pi \bar{u} \sigma_y \sigma_z} \exp\left(-\frac{y^2}{2\sigma_y^2}\right) \left\{ \exp\left(-\frac{(z - H)^2}{2\sigma_z^2}\right) + \exp\left(-\frac{(z + H)^2}{2\sigma_z^2}\right) \right\}$$

where Q_0 is the initial release quantity, $\bar{u}$ is the mean wind speed, and σ_y and σ_z are the diffusion coefficients. The common model is to take these as functions of x,

$$\sigma_y(x) = A_y x^{B_y}, \qquad \sigma_z(x) = A_z x^{B_z}.$$

The expert data on σ_y and the concentration at various downwind distances (on the centerline at the release height), χ/Q, was gathered during the joint EU/USNRC project, and was discussed in Chapter 10 where a subset of the original expert data was used. The quantiles of the expert distribution have been fitted by applying the PREJUDICE algorithm. A total of 21 150 samples in the target space (A_y, B_y, A_z, B_z) were computed, and propagated

Table 16.2. *Product moment correlation matrix for target variables*

A_y	B_y	A_z	B_z
1	−0.8674	0.4394	−0.3455
−0.8674	1	−0.382	0.2381
0.4394	−0.382	1	−0.8593
−0.3455	0.2381	−0.8593	1

Table 16.3. *Rank correlation matrix for target variables*

A_y	B_y	A_z	B_z
1	−0.5476	0.1617	−0.2856
−0.5476	1	−0.1733	0.1767
0.1617	−0.1733	1	−0.2361
−0.2856	0.1767	−0.2361	1

using the Gaussian plume model to the space of elicitation variables,

$$(\sigma_Y(500 \text{ m}), \sigma_Y(1000 \text{ m}), \sigma_Y(3000 \text{ m}), \sigma_Y(10\,000 \text{ m}),$$
$$\chi/Q(500 \text{ m}), \chi/Q(1000 \text{ m}), \chi/Q(3000 \text{ m}), \chi/Q(10\,000 \text{ m})).$$

The probabilities of the samples were adjusted from the uniform distribution of 1/21 150 for each sample, so that the expert quantiles were matched (as closely as possible). Table 16.1 compares the quantiles of the expert distributions with those produced by PREJUDICE.

The target variables are *not* independent under this distribution, as can be seen from the product moment correlation matrix in Table 16.2 and the rank correlation matrix in Table 16.3.

The reweighted sample of 21 150 points gives a joint distribution for the four target variables with marginal distributions as shown in Figure 16.2.

This example will be continued in Chapter 17, where we explore the dependencies between the target variables and the elicitation variables further.

17

Uncertainty analysis

17.1 Introduction

This chapter gives a brief introduction to the relatively new and expanding field of uncertainty analysis. Fundamental concepts are introduced, but theorems will not be proved here. Since uncertainty analysis is effectively dependent on computer support, the models used in uncertainty analysis are discussed in relation to simulation methods. A good elementary introduction to simulation is found in the book of Ross [Ross, 1990].

Uncertainty analysis was introduced with the Rasmussen Report WASH-1400 [NRC, 1975] which, as we recall, made extensive use of subjective probabilities. It was anticipated that the decision makers would not accept a single number as the probability of catastrophic accident with a nuclear reactor. Instead a distribution over possible values for the probability of a catastrophic accident was computed, using estimates of the uncertainty of the input variables. Since this study uncertainty analyses are rapidly becoming standard for large technical studies aiming at consensus in areas with substantial uncertainty. The techniques of uncertainty analysis are not restricted to fault tree probability calculations, rather they can be applied to any quantitative model. Uncertainty analysis is commonplace for large studies in accident consequence modeling, environmental risk studies and structural reliability.

17.1.1 Mathematical formulation of uncertainty analysis

Mathematically *uncertainty analysis* concerns itself with the following problem. Given some function $M(X_1, \ldots, X_n)$ of uncertain quantities $X_1, \ldots, X_n$, determine the distribution of G on the basis of some information about the joint distribution of $X_1, \ldots, X_n$. Most commonly, information is available about the *marginal* distributions of $X_1, \ldots, X_n$, and rough information about some correlations between the $X_1, \ldots, X_n$ might be available. The distribution of M is determined by computer simulation.

Uncertainty analysis is not to be confused with *sensitivity analysis*. In sensitivity analysis the importance of the variables $X_1, \ldots, X_n$ for the function M is examined individually, for a *fixed* reference value $x_1, \ldots, x_n$. If a distribution for X_i is available then one determines the sensitivity of X_i by computing the distribution of $M(x_1, x_2, \ldots, X_i, x_{i+1}, \ldots, x_n)$. The result depends of course on the reference value $x_1, \ldots, x_n$; and the distribution of $M(x_1, x_2, \ldots, X_i, x_{i+1}, \ldots, x_n)$ is not affected by correlations between X_i and $X_j, j \neq i$. A collection of papers on techniques in both sensitivity and uncertainty analysis can be found in [Scott and Saltelli, 1997].

In performing an uncertainty analysis the following questions must be addressed.

- How should the marginal distributions of $X_1, \ldots, X_n$ be assessed?
- What further information about the joint distribution of $X_1, \ldots, X_n$ is necessary, and how can it be obtained?
- How can an effective sampling procedure be designed to simulate the distribution of M?

17.2 Monte Carlo simulation

17.2.1 Univariate distributions

The basis of all sampling techniques is the following lemma, which can be used to sample from any continuous and invertible univariate distribution.

Lemma 17.1 *Suppose F is a cumulative, continuous and invertible distribution function, and U is a uniformly distributed random variable on $[0, 1]$. Then the random variable $X = F^{-1}(U)$ has cumulative distribution function F.*

Proof By definition, for all $t \in \mathbb{R}$,

$$P(X \leq t) = P(F^{-1}(U) \leq t) = P(U \leq F(t)) = F(t).$$

□

This simple lemma holds also when F is not continuous or invertible, if we define the inverse of F by $F^{-1}(u) = \inf\{t | F(t) \geq u\}$ (Exercise 1).

We can use the lemma to approximate a distribution numerically: If we simulate a sequence $U_1, U_2, \ldots$ of independent uniform variables using a random number generator† then the transformed sequence $X_1 = F^{-1}(U_1)$, $X_2 = F^{-1}(U_2), \ldots$ will be an independent sequence with distribution F. We can then build the sequence of functions

$$\hat{F}_n(t) = \frac{1}{n} \#\{i | X_i \leq t\}$$

which converge to F with probability 1.

17.2.2 Multivariate distributions

Suppose we have a joint distribution $F_{XY}(x, y) = P(X \leq x, Y \leq y)$ for a pair of random variables (X, Y). How might we sample pairs $(X_1, Y_1), \ldots, (X_n, Y_n)$ drawn from F_{XY}?

Since $P(X = x, Y = y) = P(X = x \mid Y = y)P(Y = y)$, we could do this by first drawing a value of Y, say $Y = s$, from the marginal F_Y, and then drawing X from the conditional distribution function

$$F_{X|Y=s}(r) = P(X \leq r \mid Y = s).$$

This requires specifying the distribution function for X conditional on *each* value of Y. This would be rather tedious, and for more than two variables, it becomes extremely tedious, as

$$P(X = x, Y = y, Z = z) = P(X = x \mid Y = y, Z = z)P(Y = y \mid Z = z)P(Z = z).$$

Now we should have to specify the distribution of X conditional on each pair $(Y = y, Z = z)$. There are two strategies for overcoming this problem, both involve replacing a set of random variables $(X_1, \ldots, X_n)$ with another set of variables $(Y_1, \ldots, Y_n)$ having an easily accessible joint distribution, but still looking 'sufficiently' like $(X_1, \ldots, X_n)$.

Each strategy may be implemented using either simple random sampling or some type of stratified sampling. For high-dimensional joint distributions we may require a large number of random samples to represent the joint distribution adequately. Stratified sampling techniques (for example Latin hypercube sampling, [McKay et al., 1979]) attempt to reduce the number of samples needed, but will not be discussed here.

In building a joint distribution for uncertain quantities, we can use the fact that the information available is usually limited. An expert might be persuaded, for example, to give marginal distributions for each uncertain quantile together with the correlation coefficients of each pair of variables.

---

† In practice we are not able to sample variables completely independently because the random number generators implemented in computer codes are imperfect, see [Barry, 1996] for a discussion of the statistical properties of the random number generators frequently used.

17.2 Monte Carlo simulation

The strategies discussed here all use information about marginal distributions together with correlations, and use this information to define a joint distribution that is relatively easy to sample.

17.2.3 Transforms of joint normals

Let X be a random variable with invertible distribution function F_X and let Φ be the cumulative distribution function of a standard normal variable. Since $F_X(X)$ is uniform (0,1), the variable $\Phi^{-1} F_X(X)$ is a standard normal variable, by Lemma 17.1. Similarly, if Y is standard normal, then $F_X^{-1} \Phi(Y)$ has distribution F_X.

There are standard statistical procedures for sampling from a joint normal distribution. Recall also that a joint normal distribution is completely specified by specifying the mean of the marginals and the covariance matrix Σ. If the marginal variances are all 1, then Σ is also the product moment correlation matrix. (NB, a joint distribution with normal marginals is *not* necessarily a joint normal distribution. Let X be a standard normal variable, and let $Y = X$ if $X \in (-0.5, 0.5)$ and $Y = -X$ otherwise; Y is normal, but (X, Y) is not joint normal.)

Suppose we have specified marginal distribution functions $F_1, \ldots, F_n$, and correlation matrix Σ for the functions $\Phi^{-1} F_i$, $i = 1, \ldots, n$. Let $N(\mathbf{0}, \Sigma)$ denote the joint normal distribution with mean $\mathbf{0} = (0, \ldots, 0)$ (n zeros) and covariance matrix Σ (the diagonal elements of Σ are 1). We may then apply the following procedure:

(i) sample $y_1, \ldots, y_n$ from $N(\mathbf{0}, \Sigma)$;
(ii) put $x_i = F_i^{-1} \Phi(y_i), i = 1, \ldots, n$.

This procedure generates n-tuples from a joint distribution having marginals $F_1, \ldots, F_n$ and correlation Σ *after transforming to joint normals*.

This procedure has two advantages and two drawbacks. Advantages are (i) it links up with well understood methods for dealing with joint normals, and (ii) it allows the user to specify the full covariance matrix Σ. The disadvantages are as follows. (i) The user is *required* to specify an entire correlation matrix Σ. This is a difficult task as Σ must be positive definite and the user must specify $n(n-1)/2$ off-diagonal terms. Checking for positive definiteness is a cumbersome business, and a large correlation matrix entered by the user will generally *not* be positive definite. (This problem may be overcome however using a vine representation of the variables, see Subsection 17.2.5.) (ii) The correlation values in Σ are *not* the product moment correlation coefficients of the variables X_i, but rather correlation coefficients of their normal transforms. These two correlations may be quite different, and there is no intuitive relation between them.

To illustrate the last remark, the following simulation can be performed. Take two independent normal variables X, Y with unit variance and mean 5 (effectively excluding negative values). The variable $Z = (X+Y)/\sqrt{2}$ is normal with mean 7.07 and unit variance. The following product moment correlations can be found by simulation.

$\rho(X, Z)$ 0.71
$\rho(X, \exp\{X\})$ 0.74
$\rho X, \exp\{Z\})$ 0.53
$\rho(X^4, Z^4)$ 0.68

17.2.4 Rank correlation trees

A technique has been developed to circumvent the problems noted above. The key notion underlying this technique is the rank correlation tree specification of a joint distribution. We begin by showing how a joint distribution can be built between two variables when only the individual distributions and the rank correlation are known.

17.2.4.1 Copulae

When we transform a continuous random variable X by its own distribution function F_X we get a new random variable $F_X(X)$ which is *always* uniformly distributed on $[0, 1]$. When we have a pair of random variables (X, Y) then we can make another kind of transformation to the unit square by

Definition 17.1 *The copula of two random variables X and Y is the joint distribution of $(F_X(X), F_Y(Y))$.*

Since $F_X(X)$ and $F_Y(Y)$ have uniform distributions, a copula always has uniform marginals. It is easy to see that the joint distribution of (X, Y) is determined by the copula, and the marginals F_X and F_Y. Hence it is easy to model different dependence structures between X and Y while preserving the marginals F_X and F_Y, simply by changing the copula.

A large number of references on copulae can be found in [Hutchinson and Lai, 1990].

Example 17.1 *(the diagonal band distribution)* As a simple example of a copula we mention the diagonal band distribution introduced in *[Cooke and Waij, 1986]* (see Figure 17.1).

17.2 Monte Carlo simulation

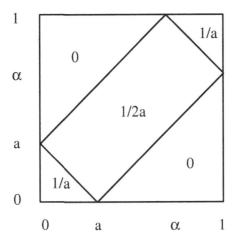

Fig. 17.1. The diagonal band distribution with parameter α

Definition 17.2 *The diagonal band copula with parameter $-1 < \alpha < 1$ has density $db_\alpha(x, y)$ on the unit square $[0, 1]^2$ given as follows.*

(i) *For $0 < \alpha < 1$ let $a = 1 - \alpha$. On the triangle with vertices 0, $(0, a)$ and $(a, 0)$, and also on the triangle with vertices $(1, 1)$, $(1, \alpha)$ and $(\alpha, 1)$, the density $db_\alpha(x, y)$ equals $\frac{1}{a}$. On the rectangle with vertices $(0, a)$, $(a, 0)$, $(1, \alpha)$ and $(\alpha, 1)$, the density $db_\alpha(x, y)$ equals $\frac{1}{2a}$. Elsewhere the density is 0.*

(ii) *For $0 = \alpha$ the density $db_\alpha(x, y)$ equals 1 everywhere.*

(iii) *For $\alpha < 0$ the density is obtained by taking the density for $-\alpha$ and reflecting the unit square in the line $y = \frac{1}{2}$.*

It is a straightforward but long computation to check that the (rank) correlation of the diagonal band distribution with parameter α is given by $\text{sgn}(\alpha)(-\alpha^3 + \alpha^2 + \alpha)$. So all rank correlations between -1 and 1 are possible. The limiting distribution as $\alpha \to 1$ is the Fréchet $+$distribution which has uniform mass on the diagonal (so $P(X = Y) = 1$).

Example 17.2 *(the minimum information copula with given rank correlation) the minimum information copula with given rank correlation is that copula whose information with respect to the independent (uniform) copula is minimal amongst all those with given rank correlation. This copula exists and is unique. There is no closed form solution for the density as a function of the rank correlation, but the density is straightforward to determine numerically. This will be done in the appendix to this chapter.*

Both information and rank correlation are preserved under invertible transformations, so if (X, Y) have the copula with given rank correlation and minimum information with respect to the uniform copula then the joint distribution also has that given rank correlation and has minimum information with respect to the unique joint independent distribution with the same marginals.

17.2.4.2 Tree building

The idea is to use the bivariate distributions to build joint distributions satisfying marginal and rank correlation constraints.

Definition 17.3 *A tree T with n nodes is an acyclic undirected graph on n nodes. The nodes are numbered $1, \ldots, n$, and the set of edges E is the set of pairs of nodes (i, j) connected in the graph. We say further that $(F_1, \ldots, F_n; \tau_{ij}; (i, j) \in E)$ is a rank correlation tree if*

(i) F_i *is an invertible univariate distribution function, $i = 1, \ldots, n$,*
(ii) *for $(i, j) \in E$, $\tau_{ij} \in [-1, 1]$.*

Figure 17.2 gives an example of a tree with nine nodes. A rank correlation tree partially specifies a joint distribution.

Theorem 17.2 (Meeuwissen, 1993, Meeuwissen and Cooke, 1994) *Let $(F_1, \ldots, F_n; \tau_{ij}; (i, j) \in E)$ be a rank correlation tree.*

(i) *The set $\mathcal{R}$ of n-dimensional joint distributions having marginals F_i, $i = 1, \ldots, n$, and rank correlations τ_{ij}, $(i, j) \in T$, is non-empty.*
(ii) *There is a unique distribution $\mathcal{F}$ in $\mathcal{R}$ having minimal information relative to the product distribution $\prod F_i$.*
(iii) *If there is no path of edges in T connecting nodes i and j, then the components i and j are independent under $\mathcal{F}$. If i and j are connected via a (unique) path of edges in T, and if k is between i and j (i.e. there is a path from k to i not containing j and a path from k to j not containing i) then under $\mathcal{F}$, i and j are conditionally independent given k.*
(iv) *If $(i, j) \in T$, then the bivariate marginal F_{ij} of $\mathcal{F}$ is the (unique) distribution having minimal information relative to the product measure F_i, F_j, given the marginals F_i and F_j, and given the rank correlation constraint τ_{ij}.*

When we specify an arbitrary rank correlation tree, we know that there is a unique minimally informative distribution having the specified marginals and specified rank correlations. Moreover, statement (iv) of the above theorem says that we can find this distribution by solving bivariate minimum

17.2 Monte Carlo simulation

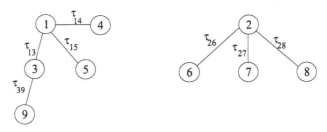

Fig. 17.2. A dependence tree

information problems. Finally, we can simulate $\mathscr{F}$ conveniently using the conditional independence mentioned in (iii). We illustrate these ideas with an example.

Suppose we have uncertain quantities $X_1, \ldots, X_9$ whose joint distribution we wish to simulate. Suppose the marginal distributions F_i for X_i are fixed and are invertible. Suppose also that the rank correlations τ are specified by means of the tree as shown in Figure 17.2. It is easy to design a sampling strategy for this structure. For each pair of variables for which a value of τ is pictured, we construct a minimum information distribution of two uniform variables with rank correlation τ. We then perform a sampling which follows the tree:

(i) sample independently a uniform variable u_1, and compute $x_1 = F_1^{-1}(u_1)$;
(ii) sample u_3 conditional on u_1 from the minimum information distribution with uniform marginals and rank correlation τ_{13}, and compute $x_3 = F_3^{-1}(u_3)$;
(iii) sample u_9 conditional on u_3 from the minimum information distribution with uniform marginals and rank correlation τ_{39}, and compute $x_9 = F_9^{-1}(u_9)$;
(iv) sample u_4 conditional on u_1 from the minimum information distribution with uniform marginals and rank correlation τ_{14} and compute $x_4 = F_4^{-1}(u_4)$;
(v) idem u_5
(vi) sample independently a uniform variable u_2, and compute $x_2 = F_2^{-1}(u_2)$;
(vii) sample u_6 conditional on u_2 from the minimum information distribution with uniform marginals and rank correlation τ_{26} and compute $x_6 = F_6^{-1}(u_6)$;
(viii) idem u_7, and idem u_8.

If we repeat this say 1000 times, the resulting sets of values $(x_1,\ldots,x_9)$ will have approximately the correct marginals with the prescribed rank correlations and will be minimally informative (with respect to the independent distribution) under the pairwise marginal and rank correlation constraints.

Note that X_1 and X_9 are correlated; τ_{19} is not specified as part of the tree structure, but its value is determined by the minimum information distribution. If a value τ_{19} *had* been specified, then the above sampling algorithm could not have been applied. Moreover, in general not all values of τ_{19} in $(-1,1)$ are consistent with τ_{13} and τ_{39}.

17.2.5 Vines

The notion of a vine generalizes that of a rank correlation tree enabling us to model arbitrary correlation matrices. Furthermore, using a vine we can specify a product moment correlation matrix without having to worry about positive definiteness. The general theory of vines was introduced in [Cooke, 1997a] and [Bedford and Cooke, 1999b], and we give only a brief example to show how vines can be used to specify product moment correlations for normally distributed variables.

We first recall the definition and interpretation of partial correlation.

Definition 17.4 *Let $X_1,\ldots,X_n$ be random variables. The* partial correlation *of X_1 and X_2 given $X_3,\ldots,X_n$ is*

$$\rho_{12|3,\ldots,n} = \frac{\rho_{12|4\ldots n} - \rho_{13|4\ldots n}\rho_{23|4\ldots n}}{((1-\rho_{13|4\ldots n}^2)(1-\rho_{23|4\ldots n}^2))^{\frac{1}{2}}}.$$

If $X_1,\ldots,X_n$ follow a joint normal distribution with covariance matrix of full rank, then partial correlations correspond to conditional correlations. In general, all partial correlations can be computed from the correlations by iterating the above equation. Here we shall reverse the process, and use a regular vine to specify partial correlations in order to obtain a correlation matrix for the joint normal distribution.

A regular vine is really just a sequence of trees built one on top of another, in the sense that the set of edges for one form the set of nodes for another. The nodes at the lowest level of the vine correspond to the random variables. These are joined by edges such that each node has at most two edges (much more general constructions are possible, and are detailed in [Bedford and Cooke, 1999b]). At the next level, the edges of the tree just constructed are considered as nodes, and are joined by edges so that two nodes (in the new tree) are joined if and only if they had a common node (as edges in the tree) at the previous level. An example of a standard vine is shown in Figure 17.3. Together with the numbers shown there on each edge, the vine gives a complete normal partial-correlation specification:

Definition 17.5 *A complete normal partial correlation specification is a standard vine together with a number from the interval* $(-1, 1)$ *on each edge.*

We shall interpret the numbers in the specification as partial correlations. To do this, we have to associate a set of variables $A(e)$ with each edge e. Clearly, each edge joins two nodes which are themselves edges for another tree. Moving down from edge to node we eventually reach the original nodes (corresponding to random variables). The set $A(e)$ is simply the set of variables reachable in this way from edge e. It can be shown that whenever i and j are two edges joined at the next level by an edge e,

$$A(e) = A(i) \cup A(j),$$

and furthermore that $A(e) - (A(i) \cap A(j))$ always contains precisely two variables. Hence we can associate the number on an edge e with a partial correlation of $A(e) - (A(i) \cap A(j))$ given $A(i) \cap A(j)$. For example $1, 3|2$.

If $\mathcal{V}$ is a standard vine over n elements, a partial correlation specification stipulates partial correlations for each edge in the vine. There are $\binom{n}{2}$ edges in total, hence the number of partial correlations specified is equal to the number of pairs of variables, and hence to the number of ρ_{ij}. Whereas the ρ_{ij} must generate a positive definite matrix, the partial correlations of a standard vine specification may be chosen arbitrarily from the interval $(-1, 1)$.

The following lemma summarizes some well-known facts about conditional normal distributions (see for example [Muirhead, 1982]).

Lemma 17.3 *Let* $X_1, \ldots, X_n$ *have a joint normal distribution with mean vector* $(\mu_1, \ldots, \mu_n)'$ *and covariance matrix* Σ. *Write* Σ_A *for the principal submatrix built from rows 1 and 2 of* Σ, *etc., so that*

$$\Sigma = \begin{pmatrix} \Sigma_A & \Sigma_{AB} \\ \Sigma_{BA} & \Sigma_B \end{pmatrix}, \quad \mu = \begin{pmatrix} \mu_A \\ \mu_B \end{pmatrix}.$$

Then the conditional distribution of $(X_1, X_2)'$ *given* $(X_3, \ldots, X_n)' = x_B$ *is normal with mean* $\mu_A + \Sigma_{AB} \Sigma_B^{-1} (x_B - \mu_B)$ *and covariance matrix*

$$\Sigma_{12|3\ldots n} = \Sigma_A - \Sigma_{AB} \Sigma_B^{-1} \Sigma_{BA}. \tag{17.1}$$

Writing $\sigma_{ij|3...n}$ for the (i,j)-element of $\Sigma_{12|3...n}$, the partial correlation satisfies

$$\rho_{12|3...n} = \frac{\sigma_{12|3...n}}{\sqrt{\sigma_{11|3...n}\sigma_{22|3...n}}}.$$

Hence, for the joint normal distribution, the partial correlation is equal to the conditional product moment correlation. The partial correlation can be interpreted as the correlation between the orthogonal projections of X_1 and X_2 on the plane orthogonal to the space spanned by $X_3, \ldots, X_n$.

The next lemma will be used to couple normal distributions together. The symbol $\langle v, w \rangle$ denotes the usual Euclidean inner product of two vectors.

Lemma 17.4 *Let $v_1, \ldots, v_{n-1}$ and $u_2, \ldots, u_n$ be two sets of linearly independent vectors of unit length in $\mathbb{R}^{n-1}$. Suppose that*

$$\langle v_i, v_j \rangle = \langle u_i, u_j \rangle \quad \text{for } i, j = 2, \ldots, n-1.$$

Then given $\alpha \in (-1, 1)$ we can find a linearly independent set of vectors of unit length $w_1, \ldots, w_n$ in $\mathbb{R}^n$ such that

(i) $\langle w_i, w_j \rangle = \langle v_i, v_j \rangle$ for $i = 1, \ldots, n-1$,
(ii) $\langle w_i, w_j \rangle = \langle u_i, u_j \rangle$ for $i = 2, \ldots, n$,
(iii) $\langle w'_1, w'_n \rangle = \alpha$, where w'_1 and w'_n denote the normalized orthogonal projections of w_1 and w_n onto the orthogonal complement of the space spanned by $w_2, \ldots, w_{n-1}$.

Proof Choose an orthonormal coordinate system for $\mathbb{R}^n$ so that we can write

$$v_1 = (x_1^{(1)}, x_2^{(1)}, \ldots, x_{n-1}^{(1)}),$$
$$v_i = (0, x_2^{(i)}, \ldots, x_{n-1}^{(i)}) \quad \text{for } i = 2, \ldots, n-1,$$
$$u_i = (x_2^{(i)}, \ldots, x_{n-1}^{(i)}, 0) \quad \text{for } i = 2, \ldots, n-1, \text{ and}$$
$$u_n = (x_2^{(n)}, \ldots, x_n^{(n)}).$$

We can now define

$$w_1 = (x_1^{(1)}, x_2^{(1)}, \ldots, x_{n-1}^{(1)}, 0),$$
$$w_i = (0, x_2^{(i)}, \ldots, x_{n-1}^{(i)}, 0) \quad \text{for } i = 2, \ldots, n-1, \text{ and}$$
$$w_n = (a, x_2^{(n)}, \ldots, x_{n-1}^{(n)}, b),$$

where $\sqrt{a^2 + b^2} = x_n^{(n)}$ and

$$a = \frac{\alpha |x_1^{(1)}||x_n^{(n)}|}{x_1^{(1)}}.$$

Then we have $w'_1 = (x_1^{(1)}, 0)$ and $w'_n = (a, b)$. Clearly, this gives the claimed inner products. □

The corollary to this lemma follows directly using the interpretation of a positive definite matrix as the matrix of inner products of a set of linearly independent vectors.

Corollary 17.5 *Suppose that $(X_1, \ldots, X_{n-1})$ and $(Y_2, \ldots, Y_n)$ are two multivariate normal vectors, and that $(X_2, \ldots, X_{n-1})$ and $(Y_2, \ldots, Y_{n-1})$ have the same distribution. Then for any $-1 < \alpha < 1$, there exists a multivariate normal vector $(Z_1, \ldots, Z_n)$ so that*

(i) *$(Z_1, \ldots, Z_{n-1})$ has the distribution of $(X_1, \ldots, X_{n-1})$,*
(ii) *$(Z_2, \ldots, Z_n)$ has the distribution of $(X_2, \ldots, X_n)$, and*
(iii) *the partial correlation of Z_1 and Z_n given $(Z_2, \ldots, Z_{n-1})$ is α.*

Now, this lemma can be applied to standard vines in the following way. A standard vine over n elements can be cut into two standard vines both with $n-1$ elements (and sharing $n-2$ elements) by removing the top-level edge. If we have normal distributions defined over the two groups of $n-1$ variables, then the above lemma guarantees us the existence of normal distribution over all n variables, *and* additionally satisfying the partial correlation constraint on the top-level vine. This will be illustrated with an example below. This shows

Theorem 17.6 *Given any complete partial correlation vine specification for standard normal random variables $X_1, \ldots, X_n$, there is a unique joint normal distribution for $X_1, \ldots, X_n$ satisfying all the partial correlation specifications.*

Corollary 17.7 *For any regular vine on n elements there is a one-to-one correspondence between the set of $n \times n$ positive definite correlation matrices and the set of partial correlation specifications for the vine.*

We note that unconditional correlations can easily be calculated inductively by using Equation 17.1. This is demonstrated in the following example.

Example 17.3 *Consider the vine in Figure 17.3. We consider the subvine consisting of nodes $1, 2$ and 3. Writing the correlation matrix with the variables ordered as $1, 3, 2$, we wish to find a product moment correlation ρ_{13} such that the correlation matrix*

$$\begin{pmatrix} 1 & \rho_{13} & 0.6 \\ \rho_{13} & 1 & -0.7 \\ 0.6 & -0.7 & 1 \end{pmatrix}$$

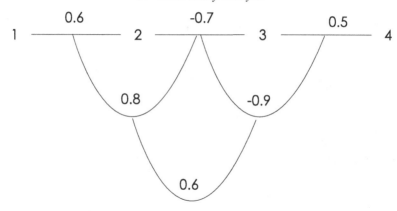

Fig. 17.3. Partial correlation vine

has the required partial correlation. We apply Equation 17.1 with

$$\Sigma_B = (1), \quad \Sigma_A = \begin{pmatrix} 1 & \rho_{13} \\ \rho_{13} & 1 \end{pmatrix}, \quad \Sigma_{AB} = \begin{pmatrix} 0.6 \\ -0.7 \end{pmatrix},$$

$$\Sigma_{13|2} = \begin{pmatrix} \sigma_{1|2}^2 & 0.8\sigma_{1|2}\sigma_{3|2} \\ 0.8\sigma_{1|2}\sigma_{3|2} & \sigma_{3|2}^2 \end{pmatrix}.$$

This gives $\sigma_{1|2} = 0.8$, $\sigma_{3|2} = 0.7141$, and

$$\rho_{13} = 0.8\sigma_{1|2}\sigma_{3|2} - 0.42 = 0.0371.$$

Using the same method for the subvine with nodes 2, 3, and 4, we easily calculate that the unconditional correlation $\rho_{24} = -0.9066$. In the same way we find that $\rho_{14} = -0.5559$. Hence the full (unconditional) product-moment correlation matrix for variables 1, 2, 3, and 4 is

$$\begin{pmatrix} 1 & 0.6 & 0.0371 & -0.5559 \\ 0.6 & 1 & -0.7 & -0.9066 \\ 0.0371 & -0.7 & 1 & 0.5 \\ -0.5559 & -0.9066 & 0.5 & 1 \end{pmatrix}.$$

As this example shows, for the *standard vine* on n elements, the partial correlations can be conveniently written in a symmetric matrix in which the (i,j)th entry $(i < j)$ gives the partial correlation of $ij|i+1,\ldots,j-1$. This matrix, for which all off-diagonal elements of the upper triangle may take arbitrary values between -1 and 1, gives a convenient alternative matrix parameterization of the multivariate normal correlation matrix.

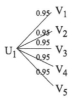

Fig. 17.4. Rank correlation tree for power lines, Case 2

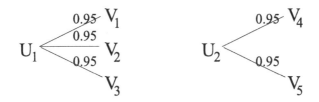

Fig. 17.5. Rank correlation tree for power lines, Case 3

17.3 Examples: uncertainty analysis for system failure

17.3.1 The reactor example

The reactor protection system, for which a BDD was given in Figure 7.8, will be studied in this first example. Assuming that the probability of failure for each power line is 0.01 it was shown that the probability of system failure is 2.95×10^{-4}.

Now suppose that the probability of powerline failure is uncertain (that is, the long term frequency of powerline failure is uncertain), and that this is quantified by a lognormal distribution with mean 0.01 and error factor 10. This information is not enough to completely specify a model for the uncertainty because the joint uncertainty distribution is not determined.

The simplest model (Case 1) is that in which the uncertainties are assumed independent. This does not seem very reasonable if the powerlines are technically identical. If the lines are technically similar then we would expect a high degree of correlation between them. A simple way to model this is to introduce a latent variable. Let V_i be the probability of failure of line L_i, and U be a uniform latent variable. A rank correlation model enables us to build a joint probability model by just specifying a rank correlation tree. We take the tree shown in Figure 17.4. Putting a rank correlation of 0.95 between V_i and U implies a rank correlation of approximately $0.95^2 \approx 0.9$ between each pair V_i, V_j.

If lines 1, 2 and 3 are technically similar, but are different to 4 and 5, then a reasonable model might be as shown in Figure 17.5.

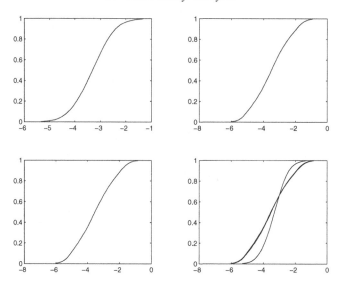

Fig. 17.6. Distributions for powerline system failure probability

Table 17.1. *Moments and quantiles for uncertainty distribution*

Mean	Variance	5% Quantile	50% Quantile	95% Quantile
1.83×10^{-3}	3.59×10^{-5}	3.07×10^{-5}	4.85×10^{-4}	7.39×10^{-3}
4.46×10^{-3}	2.07×10^{-4}	4.98×10^{-6}	3.03×10^{-4}	2.30×10^{-2}
4.84×10^{-3}	2.42×10^{-4}	4.69×10^{-6}	2.85×10^{-4}	2.50×10^{-2}

Monte Carlo simulation can be used to sample from the dependent distributions, and by using the algorithm shown in Figure 7.8 we can calculate the corresponding probability of system failure. The output of the uncertainty analysis is a distribution over the probability of system failure. Figure 17.6 shows the distributions (using a $\log_{10}$ x-axis) for the three cases described here, obtained by generating 10 000 samples. The upper left graph is for Case 1, the upper right for Case 2, the lower left for Case 3, and the lower right graph shows all cases together.

Another way to view the output is to give the means, variances, and important quantiles of the distributions. These are shown in Table 17.1. The results show that the mean probability for system failure is an order of magnitude higher than that suggested by the model in which the uncertainties are ignored. Furthermore, there is a wide spread in the uncertainty about the failure probability, particularly in the second and third models (which do not differ significantly).

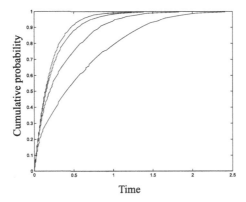

Fig. 17.7. Distribution for series system lifetime

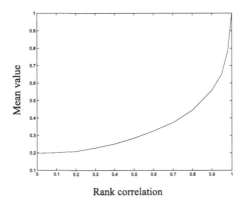

Fig. 17.8. Mean series system lifetime, depending on rank correlation

17.3.2 Series and parallel systems

Correlation between components can have a substantial effect on the distribution of the system. This is illustrated with a simple example of series and parallel systems built from five components with an exponential distributed lifetime and failure rate 1. If the lifetimes are positively correlated to a latent variable then the distribution of the system lifetime depends on the correlation value. Figure 17.7 shows the distribution for the series system for various correlation values. The figure shows four graphs, the upper one corresponding to zero rank correlation, the others (in descending order) corresponding to rank correlations 0.3, 0.6, and 0.9. The system becomes more reliable with increasing rank correlation. This can also be seen by looking at the mean system lifetime as a function of rank correlation, which is shown in Figure 17.8.

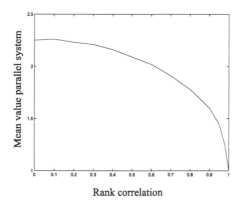

Fig. 17.9. Mean parallel system lifetime, depending on rank correlation

The mean system lifetime for a parallel system is shown in Figure 17.9.

For a series system, the system lifetime increases with increasing rank correlation, for a parallel system it decreases. This is intuitively explained by observing that under positive correlation, the component lifetimes tend to be bunched together. Hence the minimum of these lifetimes (which equals the lifetime of the series system) tends to be higher than the minimum of five independent lifetimes. Similarly, the maximum of these lifetimes (which equals the lifetime of the parallel system) tends to be lower than the maximum of five independent lifetimes.

The effect of correlation on series system lifetimes was noted in [Bedford and Cooke, 1997] in an application to the CLUSTER satellite system (a system with four different satellites), and used to argue that the system was more reliable than the independent model would suggest.

17.3.3 Dispersion model

We now return to the dispersion model discussed in Chapters 10 and 16. This will be used to illustrate the use of *cobweb plots* in understanding and communicating uncertainty. Recall that the Gaussian dispersion model gives a formula for the integrated surface concentration at a position in the neighborhood of a contaminant release. The model uses four parameters A_y, B_y, A_z and B_z. The probabilistic inversion method applied in Chapter 16 yielded a discrete probability distribution p_i on a set of 21 150 points $\underline{k}_i$ ($i = 1, \ldots, 21\,150$) in the four-dimensional parameter space

$$(A_y, B_y, A_z, B_z).$$

17.3 Examples: uncertainty analysis for system failure   343

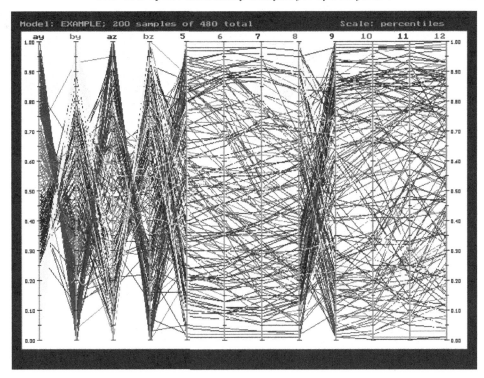

Fig. 17.10. Unconditional cobweb plot

These points do not have equal probability. The cobweb plots however require a random sample drawn with equal probability. We can obtain this from the sample of 21 150 points by going through the whole sample retaining or rejecting each point. A point $\underline{k}_i$ is selected with probability Kp_i (where $K = \max_j\{p_j\}$). This gives a random sample with equal probabilities that can be used in the cobweb plots (if too few points are left over another strategy can be used: model the distribution with a rank correlation tree – see Exercise 3).

A cobweb plot is a picture of a sample from a multivariate distribution. The axes for each variable are drawn vertically. These axes can be drawn using the underlying units of the variables or, more conveniently, using the quantile scale (this ensures that the samples are uniformly distributed over the axis). Each individual multivariate sample is represented by a line joining the coordinates. Figure 17.10 shows the unconditional cobweb plot. The axes represent (left to right) the variables A_y, B_y, A_z, B_z, the crosswind spread σ_Y at downwind distances 0.5 km, 1 km, 3 km, and 10 km, and the centerline concentration χ/Q at downwind distances 0.5 km, 1 km, 3 km, and 10 km.

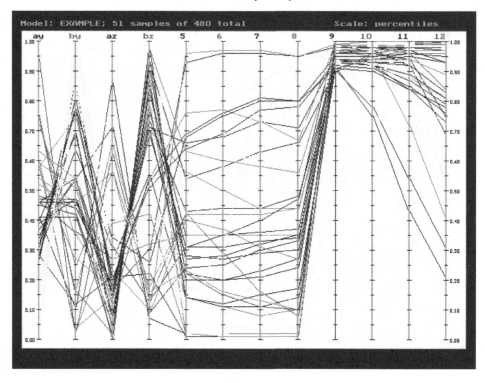

Fig. 17.11. Conditional cobweb plot: high concentration at 0.5 km downwind

The cobweb plot is a useful tool with which one can gain insight into the structure of a multivariate distribution. In particular, it is possible to study the effect of *conditioning* on a given variable or variables. Figures 17.11–17.13 show the effect of conditioning the values taken by the concentration on the centerline at a downwind distance of 0.5 km. We have conditioned on the top 10%, the middle 10% and the bottom 10% values. As here, it is usual that the conditional distributions differ considerably from the 'full' distribution. Typically, variables that are unconditionally (nearly) independent are conditionally dependent. Comparison of the cobweb plots here suggests that the predictive power for centerline concentrations downwind is higher when a high concentration is measured at 0.5 km downwind than when a lower concentration is measured.

The ESRA Technical Committee on Uncertainty Modeling carried out a benchmarking exercise, whose results were published in [Cooke, 1997b]. This compared the use of different computer codes in modeling and propagating uncertainties for the dispersion model used here. One of the main conclusions was that the differences in model output caused by neglecting dependencies

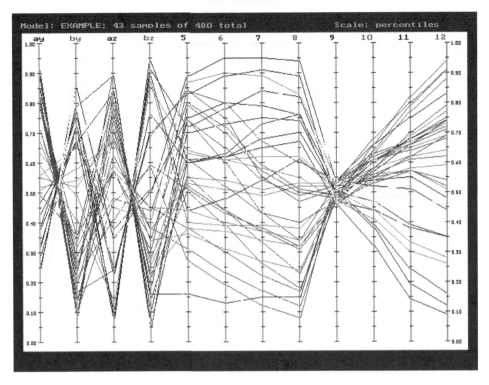

Fig. 17.12. Conditional cobweb plot: mid-range concentration at 0.5 km downwind

are much larger than those caused by different models for the dependence structure.

17.4 Exercises

17.1 Prove that Lemma 17.1 holds also when F is not continuous or invertible, when we define the inverse of F by $F^{-1}(u) = \inf\{t|F(t) \geq u\}$. Draw pictures of the graphs of F which are not continuous or not invertible to see how the inverse is defined.

17.2 Show that the product-moment correlation of the diagonal band distribution with parameter α is equal to $\text{sgn}(\alpha)(-\alpha^3 + \alpha^2 + \alpha)$.

17.3 Find a rank correlation tree that gives a good approximation to the rank correlation matrix of the variables (A_y, B_y, A_z, B_z). Use the approximately true rule that the induced correlation between two variables is equal to the product of the correlations along the branches of the tree joining them.

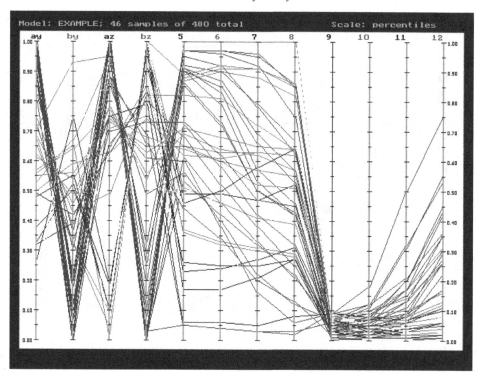

Fig. 17.13. Conditional cobweb plot: low concentration at 0.5 km downwind

17.5 Appendix: bivariate minimally informative distributions

In this appendix we explain briefly how minimally informative bivariate distributions can be generated.

17.5.1 Minimal information distributions

We recall various notions related to entropy and information. Let X be a discrete random variable taking values in $\{x_1, \ldots, x_n\}$, $x_1 < x_2 < \cdots < x_n$, with distribution given by $P(X = x_i) = p_i > 0$ ($i = 1, 2, \ldots, n$).

Definition 17.6 *The entropy $H_n(P)$ of the discrete distribution P is*

$$H_n(P) = -\sum_{i=1}^{n} p_i \log(p_i) .$$

The entropy $H_n(\cdot)$ is non-negative for each P and strictly concave. In the definition we take $0 \log 0 = 0$. It is easy to check that $H_n(P)$ is maximal for the uniform distribution, i.e. $p_i = \frac{1}{n}$. It is minimal if all the mass

17.5 Appendix: bivariate minimally informative distributions

is concentrated on one point. Intuitively one may say that more entropy corresponds to more randomness.

Entropy compares the distribution P to the uniform distribution. The notion of relative information [Kullback, 1959], enables us to compare two discrete distributions.

Let Y be another discrete random variable taking values in $\{x_1, \ldots, x_n\}$ with probabilities $Q(Y = x_i) = q_i > 0$, for $i = 1, 2, \ldots, n$.

Definition 17.7 *The relative information $I_n(Q|P)$ of Q with respect to P is*

$$I_n(Q|P) = \sum_{i=1}^{n} q_i \log\left(\frac{q_i}{p_i}\right).$$

This definition can be suggestively written as

$$\sum_{i=1}^{n} \frac{q_i}{p_i} \log\left(\frac{q_i}{p_i}\right) p_i .$$

Relative information generalizes straightforwardly to a pair of distributions P and Q where Q has a density with respect to P, $\frac{dQ}{dP}$,

$$I(Q|P) = \int_M \frac{dQ}{dP} \log\left(\frac{dQ}{dP}\right) dP . \qquad (17.2)$$

We will typically take P as the uniform distribution on the unit square and Q as a distribution with a density function $f(x, y)$, so that Equation 17.2 becomes

$$\int_0^1 \int_0^1 f(x, y) \log f(x, y) \, dx \, dy.$$

The notion of entropy cannot be directly generalized to continuous densities in a scale independent manner.

The distribution which has minimum information with respect to a given 'background' distribution under given constraints is in a certain sense the smoothest distribution satisfying the constraints which has a density with respect to the background measure.

17.5.1.1 The minimally informative copula with given rank correlation

Recall that the rank correlation between two variables X and Y is equal to the product–moment correlation of $F_X(X)$ and $F_Y(Y)$. This means that the rank correlation is a characteristic number of the copula.

If we have the marginals for two variables and their rank correlation, we can make a model for the joint distribution of the variables by choosing a copula with the specified rank correlation. Assuming that the 'default' choice would be for independence, we look for the copula with minimum information with respect to the uniform (independent) copula. The distribution we seek is the solution to the following convex problem:

$$\begin{aligned} \min & \quad \iint f(x,y)\log f(x,y)\,dxdy \\ \text{such that} & \quad \int f(x,y)\,dx = 1 \quad \forall y, \\ & \quad \int f(x,y)\,dy = 1 \quad \forall x, \\ & \quad \iint f(x,y)xy\,dxdy = \tau/12 + 1/4, \\ & \quad f(x,y) \geq 0. \end{aligned}$$

In [Bedford and Meeuwissen, 1997] we showed how the minimally informative copula with given correlation coefficient could be determined. This copula has density of the form

$$\kappa(y,\theta)\kappa(z,\theta)e^{\theta(x-0.5)(y-0.5)}$$

where $\theta = \theta(\tau)$ is a certain monotone increasing function of the correlation coefficient τ, and the function $\kappa(\cdot,\theta)$ is determined as the solution to an integral equation

$$\kappa(y,\theta) = \left[\int_0^1 \kappa(z,\theta)e^{\theta(x-0.5)(y-0.5)}\,dz\right]^{-1}.$$

Although no closed form solution seems possible for θ, it is easy to determine it numerically (see [Bedford and Meeuwissen, 1997]). When $\theta = 0$ we obtain the independent copula.

A discrete approximation to the minimally informative copula is easily calculated. We seek a discrete distribution of the form

$$P\left(\left(x_i + \frac{1}{2}, x_j + \frac{1}{2}\right)\right) = \frac{1}{n^2}\kappa_i(\theta)\kappa_j(\theta)e^{\theta x_i x_j},$$

where $x_i = \frac{2i-1-n}{n}$ for $i = 1,\ldots,n$. Given θ, we calculate the vector κ by appling a *DAD* algorithm [Nussbaum, 1989]. The classical *DAD* problem is, given a square symmetric non-negative matrix A, to find a diagonal matrix D with positive diagonal entries such that DAD is doubly stochastic. In our case $A = (e^{\theta x_i x_j})$, and $D = \frac{1}{n}\text{diag}(\kappa_1,\ldots,\kappa_n)$. A simple numerical scheme for determining the values of κ_i is iteration of a function which is defined as follows. Let $\mathscr{C}$ be the positive cone in $\mathbb{R}^n$ and define $\Psi : \mathscr{C} \to \mathscr{C}$ by the composition of normalization,

$$y_i \mapsto \frac{y_i n}{\sqrt{\sum_{j,k} y_j y_k e^{\theta x_j x_k}}},$$

17.5 Appendix: bivariate minimally informative distributions

followed by the non-linear map

$$y_i \mapsto \frac{n}{\sum_j y_j e^{\theta x_i x_j}}.$$

This iteration converges geometrically to the fixed point which is the vector κ we required. This enables us to calculate the minimally informative distribution with a given θ, and since we know the distribution we can calculate its τ. This enables us to calculate τ as a function of θ, and as τ is a strict monotone increasing function of θ we can easily numerically invert to solve for the right θ value for any given τ.

18

Risk measurement and regulation

How can we *choose* a probabilistic risk acceptance criterion, a probabilistic safety goal, or specify how changes of risk baseline should influence other design and operational decisions? Basically, we have to compare risks from different, perhaps very different activities. What quantities should be compared? There is an ocean of literature on this subject. The story begins with the first probabilistic risk analysis, WASH-1400, and books which come quickly to mind are [Shrader-Frechette, 1985], [Maclean, 1986], [Lowrance, 1976] and [Fischhoff *et al.*, 1981]. A few positions, and associated pitfalls, are set forth below. For convenience we restrict attention to one undesirable event, namely death.

18.1 Single statistics representing risk

18.1.1 Deaths per million

The most common quantity used to compare risks is 'deaths per million'. Covello *et al.* [Covello *et al.*, 1989] give many examples of the use of this statistic. Similar tables are given by the British Health and Safety Executive [HSE, 1987]. The Dutch government's risk policy statement [MVROM, 1989] gives a variation on this method by tabulating the yearly risk of death as 'one in X'.

Table 18.1 shows a few numbers taken from Table B.1 of [Covello *et al.*, 1989], 'Annual risk of death in the United States'. By each 'cause' the number of deaths per year per million is given. These numbers are obtained by dividing the number of yearly deaths registered per cause by the number of millions of residents in the US.

These numbers may be useful for some purposes, but using them to compare risks from different 'causes' involves a number of serious problems of which the most immediate are:

1. Base rate participation neglected: These risks are not distributed uniformly. For example, tornadoes are restricted to the central US. No one dies of tornadoes in the Northeast or West Coast areas (which are most densely populated). People who do not possess firearms are much less likely to die from firearms than people who do (e.g. hunters and criminals).
2. Overlap: Many falls occur in the home, hence comparing these numbers is misleading at best.
3. No distinction between voluntary and involuntary: Many firearm accidents may be regarded as 'self-inflicted': people who willingly camouflage themselves, and take their hunting rifle into the woods, can expect to get shot at by other hunters. The risks from such activities cannot be compared to risk from exposure to toxic discharges.
4. No account of age distribution: Fatal falls generally occur among the elderly. Other risks affect young and old alike. It may be unreasonable to compare risk from activities which differ in this respect.

18.1.2 Loss of life expectancy

Cohen and Lee [Cohen and Lee, 1979] have developed a measure which is intended to correct the last problem. This is the loss of life expectancy due to various causes. The idea is roughly this: Define the population mortality rate h as a function of age, as follows:

$$h(i) = \frac{\#\text{people who die in their } i+1\text{th year of life}}{\#\text{people of age } i}$$

(this is the discrete time version of the failure rate). The probability $P(N)$ of death at age N is then

$$\begin{aligned} P(N) &= P(\text{death at age } N \mid \text{alive at age } N-1) * P(\text{alive at age } N-1) \\ &= h(N) \cdot \prod_{i=1}^{N-1}(1 - h(i)) \end{aligned} \quad (18.1)$$

and the expected life is

$$E = \sum_{i=1}^{\infty} i P(i). \quad (18.2)$$

The mortality rate due to cause j, h_j, is defined as

$$h_j(i) = \frac{\#\text{people at age } i \text{ who die of } j}{\#\text{people of age } i}.$$

Table 18.1. *Annual risk of death in the United States*

Cause	Deaths per million per year
Motor vehicle accidents	240
Home accidents	110
Falls	62
Drowning	36
Fires	28
Firearms	10
Tornadoes	0.6

Table 18.2. *Loss of life expectancy*

$LLE_{\text{motor vehicle accidents}}$	= 204 days
LLE_{drowning}	= 41 days
LEE_{falls}	= 39 days
$LLE_{\text{air bags in cars}}$	= −50 days

Define $h^{(j)} = h - h_j$, and use this in Formulae 18.1 and 18.2 to compute $E^{(j)}$. Cohen and Lee [Cohen and Lee, 1979] define the *loss of life expectancy due to cause j*, LLE_j, as

$$LLE_j = E^{(j)} - E.$$

It is supposed to represent something like the increase in life expectancy which would be realized if cause j could be completely eliminated. Intuitively, eliminating falls would have relatively less impact than eliminating e.g. motor vehicle accidents. Table 18.2 shows some numbers taken from Table B.4 of [Covello et al., 1989].

By shifting from deaths per million to loss of life expectancy, drowning becomes more important and falls less important. This is explained intuitively by the fact that drowning primarily affects children whereas falls fell people in their last years of life. Note also that the loss of life expectancy can be *negative*; if air bags were mandatory on all cars the life expectancy would *go up*, entailing a negative loss.

Although problems 1, 2 and 3 above also apply to *LLE*, this quantity does seem to address problem 4. However, there are problems here also. Most prominently, $h - h_j$ does not really represent the mortality rate which would exist if cause j could be eliminated, as some of those who die in year i from j would die in year i of something else if j were eliminated. Moreover, the tendency to die of j may be correlated with the tendency to die of some other cause still represented in R; hence $R^{(j)}$ need not represent the population mortality rate after eliminating j. For example, if hunters were prevented from hunting, they might discover some even more effective way of killing each other, say hang-gliding. Eliminating hunting might drive hunters to hang-gliding and increase the mortality rate due to hang-gliding.

It is not clear what LLE_j really represents, and it may not even approximate what Cohen and Lee seem to think it represents.

18.1.3 Delta yearly probability of death

This method of representation computes the intensity at which a given activity is performed (in suitable units) in order to increase the yearly probability of death by 10^{-6}. The corresponding intensities are associated with the same delta probability, and this enables some sort of comparison. For example, according to Table B5 of [Covello et al., 1989], the following activities have the same delta probability:

living 2 days in New York (pollution),
smoking 1.4 cigarettes,
drinking .5 liter wine,
flying 1000 miles by jet.

The problem with this measurement is that it is often difficult or impossible to compute or measure; and one should be suspicious of some of these computations. For example, how does one determine that 1.4 cigarettes add 10^{-6} yearly probability of death? Suppose we had an experiment involving 10^8 people, half of whom smoked 1.4 cigarettes in 1 year. Is the claim that after 1 year, we can expect 50 additional deaths among the smokers? No, the deaths caused by smoking could occur any time after smoking the 1.4 cigarettes. But everyone dies sooner or later. All we could measure would be an increase in mortality rate in the smokers relative to the non-smokers. How do we now measure delta probability: by integrating the difference of mortality rates? The problems with this were discussed above. Moreover, somatic insults usually involve threshold phenomena; below a certain threshold intensity the body can repair the damage from the insult. It is most likely that the mortality rates from the smokers and non-smokers above would be identical, as the insult from such a small dose could easily be repaired.

These problems result from the fact that the delta probability method attempts to compare deaths *caused* by an activity, whereas at best only differential mortality rates could be recovered from data. Mortality rates can

Table 18.3. *Hourly specific mortality rates*

Activity	Hourly mortality rate
Plague in London in 1665	1.5×10^{-4}
Rock climbing (on rock face)	4×10^{-5}
Civilian in London air-raids, 1940	2×10^{-6}
Jet travel	7×10^{-7} (assuming 700 mph)
Automobile travel	10^{-7} (assuming 30 mph)
Bicycling	10^{-6}
Coal mining (accidents)	3×10^{-7}
Background Living (US)	1.03×10^{-6}

be compared only from populations with the same background mortality rate, differing only with regard to exposure to the given activity. Even then, integrating differential mortality rates has no clear empirical meaning. Data suitable for determining differential mortality rates is seldom, actually never, available.

18.1.4 Activity specific hourly mortality rate

This quantity reflects the probability per hour of dying while engaged in a specified activity. For purposes of cross-activity comparison, this rate will be treated as constant and estimated as an average: mortality rate from activity j per hr,

$$\lambda_j = \frac{\text{\#deaths during activity } j}{\text{\#number of hours spent in activity } j}.$$

This measure is only meaningful if the deaths *caused by* j actually occur while doing j. For example, risks from automobile driving or commercial aviation accidents can be represented in this way, but risks from smoking cannot. One could of course replace the numerator by '# deaths *caused* by activity j', but this would involve complications discussed above.

Some activity specific mortality rates from [RS, 1992] p.81, are given in Table 18.3 together with some rates computed from [Covello *et al.*, 1989], Table B.5 and Table B.2.

The mortality of living dominates other activities such as coal mining because only relatively healthy people engage in these activities. This points to the difficulty of defining a 'background mortality rate' for comparing activity specific mortality rates. A background rate should be computed with

Table 18.4. *Deaths per 10^9 km traveled*

Sector	Class	1967–71	1972–76	1986–90
Rail travel	passengers, from train accidents	0.65	0.45	1.1
Air travel	UK airlines, passengers on scheduled services	2.3	1.4	0.23
Road travel	public service vehicles, passenger or driver	1.2	1.2	0.45
	car or taxi driver and passenger	9	7.5	4.4
	pedal cyclist	88	85	50
	two wheeled motor vehicle driver	163	165	104
	two wheeled motor vehicle passenger	375	359	
	pedestrian, based on 8.7 km per week	110	105	70

respect to a population which is comparable to the population engaging in the activity. Note that this remark does not apply for cross-activity comparisons, as the mortality of for example air, automobile or bicycle travel would not be strongly dependent on the background mortality rate of the participants.

18.1.5 Death per unit activity

A variant on the *deaths per year per million* measurement is to replace the time unit by a unit measuring the amount of activity. For example, the number of deaths of bicyclists may be expressed relative to the number of hours spent cycling or alternatively to the number of kilometers traveled. This is illustrated in Table 18.4 from the Royal Society study group report [RS, 1992].

18.2 Frequency *vs* consequence lines

The different representations of risk above have in common that the consequences (number of deaths etc.) have been averaged. In fact our assessment of the risk inherent in a particular activity depends also on the balance between low probability/high casualty incidents on the one hand and high probability/low casualty incidents on the other.

The frequency *vs* consequence line (fC line) is the graph of the function giving the frequency of events with n or more deaths per year plotted against the number of deaths n. The scales are chosen to be logarithmic. The first use of fC lines was made by Farmer in 1967 for releases of iodine 131 from thermal nuclear reactors (although his definition was slightly different to

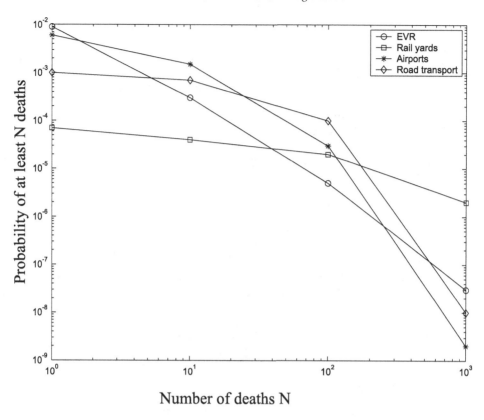

Fig. 18.1. Frequency consequence lines

that given here). These graphs are often called Farmer curves, or FN curves (where the N stands for the number of fatalities). A number of fC lines are shown in Figure 18.1.

18.2.1 Group risk comparisons; ccdf method

The Dutch government has taken the initiative to introduce a group risk criterion in the management of societal risk [Versteeg, 1988]. The criterion is based on the random variable 'number of prompt fatalities per year' associated with a given activity. Set on a logarithmic scale, the complementary cumulative distribution function, ccdf, is confined to the positive quadrant shown in Figure 18.2. The criterion divides this quadrant into three areas. If the activity's ccdf enters the 'unacceptable' region, then the activity is judged to entail an unacceptably large group risk; if the ccdf is confined to the 'acceptable' region, then the group risk of the activity is considered negligible; in other cases risk reduction should be pursued via the 'as low as reasonably achievable' (ALARA) principle.

18.2 Frequency vs consequence lines

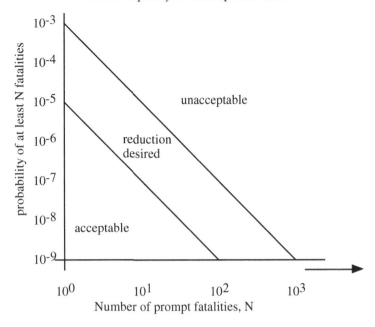

Fig. 18.2. The Dutch group risk criterion

Note that the Dutch group risk criterion uses an fC graph and requires that the whole fC line lie within the *acceptable* region. The shape of the acceptable region implies a policy decision about an acceptable balance between low probability/high casualty incidents and high probability/low casualty incidents.

Similar criteria have been used elsewhere. A report of the UK Health and Safety Executive [HSE, 1991] suggested tolerability bounds which are straight lines with slope -1 (as compared with -2 for the Dutch criteria, which therefore are more averse to large-scale accidents).

The use of a straight line to bound the acceptable region is rather arbitrary and corresponds, because of the logarithmic scales of the axes, to a constraint of the form

$$C(n) \leq \alpha/n^{\beta}, \qquad (18.3)$$

where C is the frequency per year of accidents with at least n fatalities. As noted in [Evans and Verlander, 1997], this kind of constraint suffers from a peculiar inconsistency. They give the following example. An aircraft company operates small and big aircraft. The small aircraft carry 10 people and a crash of one of these aircraft (which then kills all people on board) occurs with a frequency of 0.08 per year. The large aircraft carries 100 people and a

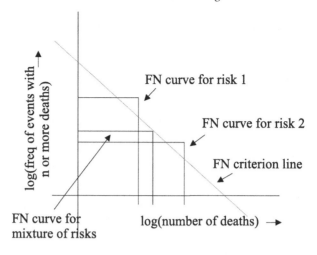

Fig. 18.3. Unacceptable risks

crash of one of these planes occurs with a frequency of 0.008 per year. Under the criterion given in Formula 18.3 with $\alpha = 1$ and $\beta = -1$, the operations of this company would be judged just acceptable. A second company operating with exactly the same aircraft and the same frequencies of crashes would also be judged acceptable. Suppose now that the companies decide to concentrate on small and large aircraft, respectively. One swaps all its small planes for the large planes of the other. We now have two companies, one whose operations kill exactly 10 people with a frequency of 0.16 per year, and one whose operations kill exactly 100 people with a frequency of 0.016 per year. Both of these operations are now judged unacceptable under the risk criterion, even though the numbers of aircraft and the total frequency of casualties have not changed! Similar examples can be made for any values of α and β.

What is really going on in this example is that one can find two fC curves C_1 and C_2 that are not acceptable, while the 'probabilistic mixture' $\frac{1}{2}C_1 + \frac{1}{2}C_2$ (corresponding to the original activities of the company) is acceptable. See Figure 18.3. This example suggests the following type of criterion for comparing risk: writing $C_1 \succeq C_2$ if C_1 is more acceptable than C_2, then whenever $C_1 \succeq C_2$ and C_3 is any third fC curve, we should require

$$\frac{1}{2}C_1 + \frac{1}{2}C_3 \succeq \frac{1}{2}C_2 + \frac{1}{2}C_3.$$

This is really the same as the sure thing principle axiom used in utility theory. We return to this shortly.

18.2.2 Total risk

Vrijling *et al.* [Vrijling *et al.*, 1995], recognizing some of the problems arising in the use of fC curves for risk management, have suggested a different kind of criterion, based on a function of the fC curve. Writing N for the number of fatalities per year, they suggest that the expected value of N would be a good criterion. However, in order to make the criterion risk averse (that is, to take into account the variability in N), they suggest adapting it by adding a term containing the standard deviation. The criterion is thus that

$$E(N) + 3\sigma(N) \leq 100\beta$$

for some β. The number 3 has been chosen so that existing 'acceptable' activities are accepted. The number β is a policy factor depending on the degree of voluntariness of the activity (for example $\beta = 10$ for mountaineering, but is 0.01 for an imposed risk with no direct benefit). Arguments are given in [Vrijling *et al.*, 1995] for this choice of bound.

This criterion suffers, because of the addition of the $3\sigma(N)$ term, from the same problem as the Dutch group risk criterion. We can find acceptable N_1 and N_2 such that the probabilistic mixture

$$N = \begin{cases} N_1 & \text{with probability } \frac{1}{2}, \\ N_2 & \text{with probability } \frac{1}{2} \end{cases}$$

(this means that if C_i is the fC curve for N_i ($i = 1, 2$), then $C = \frac{1}{2}C_1 + \frac{1}{2}C_2$ is the fC curve for N), is *not* acceptable. In fact we always have

$$E(N) + k\sigma(N) \geq \frac{1}{2}(E(N_1) + k\sigma(N_1)) + \frac{1}{2}(E(N_2) + k\sigma(N_2))$$

(by applying Jensen's inequality). This shows that a mixture of acceptable risks may turn out to be unacceptable under the total risk criterion.

Cassidy [Cassidy, 1996] has suggested using the so-called risk integral defined by

$$RI = \int_0^\infty x(1 - F_N(x))\, dx,$$

where $F_N(x)$ is the probability of at most x deaths in a year. Vrijling and van Gelder [Vrijling and van Gelder, 1997] show that this equals the expression $\frac{1}{2}(E(N)^2 + \sigma(N)^2)$. This index suffers the same problem as the Dutch group risk criterion and the total risk criterion.

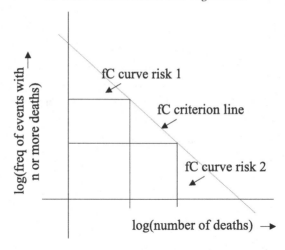

Fig. 18.4. Equally risky activities

18.2.3 Expected disutility

In risk management it is necessary to compare different risky activities. This suggests using preference relations $\succeq$ between different activities. As we noted above, it seems reasonable to require that the usual axioms of rational preference hold for comparing risks. Hence it would seem appropriate to see if we can find a utility function over the outcomes (yearly numbers of deaths) so that expected utility gives a reasonable ranking of risks. It seems more appropriate to talk about disutility of death so we shall consider the disutility, which is simply minus the utility.

In general a risk criterion such as that in Formula 18.3 does not specify a disutility function. However, let us consider just those risky activities which can only kill a fixed number of individuals per year (so $N = k$ with probability p and $N = 0$ with probability $1 - p$). The fC curves of such activities form two sides of a rectangle, see Figure 18.4. Now suppose we consider two such activities to be equally risky when the corners are on the line defined by Formula 18.3. As observed by Evans and Verlander [Evans and Verlander, 1997], the disutility function is now fixed. To see this, suppose that we have two risks, the first of which kills N_1 people per year ($N_1 = k_1$ with probability p_1 and $N_1 = 0$ with probability $1 - p_1$), the second of which kills N_2 people per year ($N_2 = k_2$ with probability p_2 and $N_2 = 0$ with probability $1 - p_2$). Assuming that the fC curves both 'just touch' the criterion line defined by Formula 18.3 we get $p_i = \alpha / k_i^\beta$, for $i = 1, 2$. Hence, taking the disutility of 0 deaths to be 0 (which we can do since any utility –

and therefore any disutility – is only fixed up to affine scale),

$$p_1 D(k_1) = D(N_1) = D(N_2) = p_2 D(k_2).$$

This gives

$$D(k_1) \propto \frac{1}{p_1} \propto k_1^\beta,$$

and since the proportionality constant can be taken arbitrarily (again due to the fact that disutilities are only fixed up to affine scale) we can fix $D(k) = k^\beta$.

Unfortunately, there is a very large difference between risk criteria based on the disutility of the number of deaths being $D(k) = k^\beta$, and those based on bounds on the fC curve as in Formula 18.3. To see this consider the 'worst' fC curve just permissible under Formula 18.3. This is of the form $P(N \geq n) = \alpha/n^\beta$. Let us suppose that $\beta = 1$ (as has been used by the HSE in the UK). In this case expected disutility is simply the mean number of deaths. The expected disutility of the worst case curve is easily calculated, for $P(N = n) = \alpha/n - \alpha/(n+1) = \alpha/n(n+1)$ so that the expected disutility is

$$\sum_n P(N = n)n = \sum_n \frac{\alpha}{(n+1)} = \infty.$$

Hence the fC criterion curve with $\beta = 1$ permits activities with an infinite number of expected deaths per year. In practice this means that real activities with arbitrarily high humbers of expected deaths would be permitted.

Any criterion involving an expectation (expected disutility, total risk, etc.) suffers from the problem of *instability*. We are usually concerned with high consequence/low probability events. It is clear that when evaluating the integral needed to determine the expected disutility, the actual value determined (and hence the relative ranking when compared to other hazards) will be highly sensitive to the estimates of low probabilities *and* to the particular choice of utility function.

The only way to avoid stability problems is to choose a disutility function which does not rapidly increase with large numbers of deaths. In an extreme case, this would mean that society would view all accidents with large numbers of deaths to be equally bad, irrespective of the numbers of deaths.

18.2.4 Uncertainty about the fC curve

Every calculation of an fC curve is just an estimation. Within a Bayesian framework it is possible to explicitly model uncertainty in the fC curve. It seems not unreasonable to take into account this uncertainty in risk management. Of the criteria mentioned above only the disutility is really able to take into account uncertainties.

18.2.5 Benefits

One of the many problems with comparing risks in terms of the negative consequences is that only a small part of the picture is illustrated. The analogy with decision-making made above is not really correct since we are not making a choice between different risks. We can only choose between different risks when they arise between competing technologies in the same sector: nuclear powered electricity generation as opposed to conventional electricity generation; long distance train travel as opposed to intra-continental air travel, etc. In making such choices we have to analyze the benefits as well as the disbenefits. If two technologies have the same benefits but one is seen as riskier than the other then the riskier one will be difficult to justify to the public, even if it satisfies official 'acceptable risk' criteria.

18.3 Risk regulation

We have already described the specification of group risk criteria using fC curves. Risk based regulation is now common in many different sectors. An overview of some of the quantitative risk goals currently set is given in Table 18.5, adapted from [Kafka, 1999].

18.3.1 ALARP

In addition to quantitative goals, there are qualitative principles which may be invoked. We give one example below, and return to the issue of principles in the concluding section.

Within the UK, the basic underlying principle of risk regulation is that of ALARP: as low as reasonably practicable. The typically British concept of reasonableness has been made somewhat more precise in policy statements of the Health and Safety Executive [HSE, 1992, Pape, 1997], where quantitative safety goals are defined. The basic safety objective (BSO) and basic safety limit (BSL) give 'ideal' and upper limits to the probability of death as a result of activity or operation of plant; the ALARP region lies between. Here the trade-off between safety and the cost of implementing improvements is made on the basis of what is 'reasonably practicable'. This trade-off depends on a value for a human life, and will be discussed further below.

Table 18.5. *Risk goals for various technologies*

Technology	Risk goal
Marine structures	Failure probability for different accident classes 10^{-3}–10^{-6}
Aviation, air planes	Catastrophic failure per flight hour: less than 10^{-9}
Nuclear power plants (USA)	Large scale core melt: less than 1 in 10 000 per year of reactor operation; individual risk for NPP less than 0.1% of total cancer risk from all causes.
(Canada)	Early fatality (public): 10^{-5} per site-year. Large early release: 10^{-5} per site-year
Space vehicles	Catastrophic consequence for a Crew Transfer Vehicle (CTV) smaller than 1 in 500 CTV missions
Process industry (Netherlands)	Consequence $> 10^n$ fatalities must be smaller than 10^{-3-2n}. Individual risk less than 10^{-6} per plant year.
IEC 61508 (for all technologies) Electrical/electronic safety systems with embedded software	Average probability of failure per demand: 10^{-1} to 10^{-5} for different safety levels 1–4 and low demand mode of operation.

18.3.2 The value of human life

Government agencies are frequently faced with the problem of prioritizing some life-saving measures above others. Cost–benefit analysis uses the notion of the 'value of human life' to judge which life-saving measures are economic. The UK government for example uses figures from the Ministry of Transport in several policy areas.

It has been noted however that government policy, subject as it is to pressure from the public and lobbies, is frequently inconsistent in the sense that differing amounts may be spent to save a statistical life in different sectors. In many areas even simple measures that could cheaply save lives are not implemented for a variety of reasons. A survey of 587 different potential life-saving measures is given in [Tenga *et al.*, 1995] together with the expected costs per saved life (in the US; [Ramsberg and Sjöberg, 1997] presents a similar study for Sweden). The costs differ dramatically from net savings (for example for the installation of car windshields with adhesive bonding as opposed to rubber gaskets) to costs of nearly 100 billion dollars (for chloroform private well emission standard at 48 paper mills). The median

Table 18.6. *Median costs of life-saving measures per sector*

Sector	Cost per statistical life ($)
Medical	19,000
Injury reduction	48,000
Toxin control	2,800,000
Overall	42,000

costs of life-saving measures in their survey are shown in Table 18.6. Note that the median reflects the costs of those life-saving measures that have been considered in a particular sector, rather than directly what the 'true' costs are of saving a life in that sector. The survey was of potential, and not necessarily implemented, life-saving measures.

If government authorities want to choose an appropriate amount to spend on saving lives, how should this be found? The appropriate value of a human life used in cost–benefit analyses is usually determined on the basis of what economists call 'contingent evaluation', or in other words surveying members of the public. The framing of such questions is generally divided into *willingness to pay* (WTP) and *willingness to accept* (WTA) forms. Typically the amount people are willing to pay to avoid exposure to a risk is considerably less than what they would accept to become exposed. In determining the value of a human life one therefore has to be clear about whether we are talking about WTP or WTA values. As [French et al., 1999] (following [Adams, 1995]) points out, in ALARP it is unclear whether one should use WTP or WTA. Consider the difference between basic safety limits (BSLs) and basic safety objectives (BSOs):

- If the aim is to get as near to the BSO as possible, then a utility should accept a penalty for a (commercial) benefit which will increase risk away from the BSO. The public, on the other hand, will be willing to accept compensation for taking the extra risk.
- If the aim is to simply keep under the BSL, then the public should be willing to accept a penalty for the benefit of a further risk reduction towards the BSO. The utility, on the other hand, will be willing to accept compensation for forgoing the right to use the extra risk margin.

The presentation of the ALARP principle seems most in keeping with the former. Thus one may argue that WTA valuations should be used, contrary to common practice and the advice of the regulator [HM Treasury, 1996].

18.3.3 Limits of risk regulation

As we have seen above, there are many ways to measure risks even for the most simple consequence, death. Attempts to regulate risk based on a single risk metric will often lead to changes of operation in those bodies being regulated so that their systems are optimized under the regulation constraint. This may lead to risks, as measured by other plausible metrics, increasing past a level that might be considered acceptable.

Both organizations and individuals have a multitude of possible objectives, besides risk reduction, that might be optimized. Businesses optimize profitability subject to constraints on risk imposed by society. Individual car drivers usually tend to optimize journey time subject to constraints on safety. This suggests that improvements in the safety of cars or roads will tend to be used to reduce journey time rather than to reduce the actual level of risk. Of course, such decisions are not taken consciously by individuals, and it is difficult to make an objective assessment of risk. These considerations suggest therefore only that measures introduced with the aim of reducing risk could have positive or negative net effects on risk depending on the way in which the individuals involved change their behaviour in reaction to the risk measures.

A discussion of this problem, with particular emphasis on road transport, is given in [Adams, 1995]. Figure 18.5 is reproduced from that book. It shows the decrease in the number of deaths of motorcyclists in the US around the time that the use of crash helmets was made compulsory prior to 1970. Later, many states repealed this legislation and the number of casualties increased. However, as Figure 18.5 shows, the increase appears to be larger in the states that did *not* repeal the legislation than those that did! Note that the statistical significance of this appearance cannot be judged from the graph. In particular, we should need to know the absolute numbers of fatalities in the 'repeal' and 'non-repeal' states.

18.4 Perceiving and accepting risks

In this section we shall very briefly look at some of the problems associated with risk perception. The subject is largely outside the scope of this book, but a large and growing literature is available. A good starting point is the chapter on risk perception in [RS, 1992].

Much of the debate on risk perception is born from the dichotomy of 'objective' and 'subjective' measures of risks. Objective measures of risk, such as those discussed in the first two sections of this chapter, have physical, measurable dimensions. Subjective measures typically involve *re-*

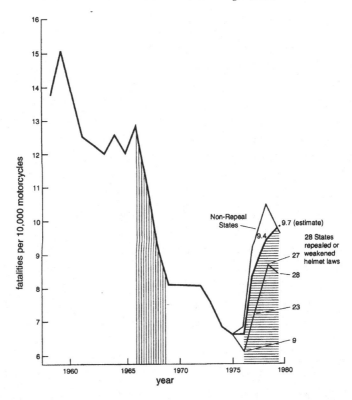

Fig. 18.5. Changes in numbers of deaths of motor cyclists in different US states

vealed or *expressed* preferences. Revealed preferences are based on actual risk taking or risk acceptance behavior. Expressed preferences are based on statements of what one would like to have, as opposed to what one de facto accepts. For an introduction to subjective risk measures, see [Slovic et al., 1990].

There is ample evidence that people's risk taking behavior is not always consistent with 'objective' measures of risk. This has prompted genuflections by risk analysts and policy makers toward risk communication. Communication should help the public to perceive risks objectively. On the other hand, social scientists offer analysts and policy makers various tools for understanding and predicting people's perceptions. These would enable policy makers to anticipate and mold perceptions, thereby enhancing public acceptance. We discuss some recent topics in the perception and acceptance of risk. In the final section we argue that the distinction of objective versus subjective risk has outlived itself.

18.4.1 Risk perception

A number of studies have been made with the aim of identifying the 'dimensions' which could explain our perception of risk. The basic idea is to work with a list of hazards (e.g. nuclear accidents, firearms, bicycle riding), and a list of *qualities*, or *constructs* (e.g. natural/man-made, specific/non-specific location, most/least dangerous). The respondent assesses (on a numerical scale, say 1–10) for each quality the degree to which it is an appropriate descriptor of each given hazard. One can then compute correlations between various qualities, and using a principal component analysis one can calculate the 'dimensionality' of the system.

The best-known model in this area is the Psychometric Model which developed from [Fischhoff *et al.*, 1978]. Here the scores of the various participants on a given scale are averaged to produce a mean score for each risk on each scale. Typically, a large amount of the variance can be 'explained' by a small number of factors. This has been an important reason behind the popularity of the Psychometric model. However, as pointed out in [Sjöberg, 2000], a model for risk perception should try to explain the individual scores, and not work with averaged data. Unfortunately, the unaveraged data shows considerably more noise. According to [Sjöberg, 2000], only about 20% of the variance can be explained for the individual data, in comparison to 60–70% for the averaged data.

A more fundamental critique of such studies would point out that Pearson or product moment correlation is meaningful only for variables measured on an affine scale; that is, a scale, like degrees Fahrenheit, which is determined by a 'freezing' and 'boiling' point. Suppose 'dreadfulness' is a quality scored by respondents on the scale 1 to 10. Suppose in response to an inquiry you scored the dreadfulness of hazards 'home appliances', 'vaccinations', 'pesticides', and 'nuclear power' as 1, 2, 7, 9, respectively. Then it should follow that the difference in dreadfulness between vaccinations and home appliances is half the difference in dreadfulness between nuclear power and pesticides. If your responses merely intended to give the ordering of dreadfulness then we could just as well represent your reponses with the natural logarithm of your scale values – $(0, 0.3, 0.85 0.95)$; but now the difference in dreadfulness between nuclear power and pecticides is smaller than that between home appliances and vaccinations. If such difference statements are meaningless, then so are the results of such studies. When scores are averaged across respondents, the question whether the 'average respondent' uses an affine scale becomes acute, and acutely impossible to answer.

18.4.2 Acceptability of risks

Instead of applying sophisticated multivariate techniques to data whose suitability cannot be verified, we could also just let brainstormers or focus groups generate criteria influencing the acceptability of risk. The UK Department of the Environment and Health and Safety Executive jointly ran a seminar in 1979 which came up with the following list of criteria by which the 'public' judge the acceptability of various risks.

(i) Concentrated, obvious risks (e.g. of motorway pile-ups or major industrial explosions) are regarded as worse than diffuse risks like those from general road accidents or an equal number of deaths scattered around as a result of smaller scale industrial accidents.

(ii) Risk to non-beneficiaries (e.g. public exposed to emissions from nuclear power stations, or people living alongside railways) are regarded as worse than risks to beneficiaries (e.g. recipients of radio-therapy or railway workers).

(iii) Involuntary risks (e.g. of receiving carcinogens in food) are regarded as worse than voluntary risks (e.g. rock climbing)

(iv) Risk imposed for the benefit of others (e.g. whooping cough vaccinations imposed on older children for the benefit of younger age groups) are regarded as less acceptable than risks undertaken for self-protection.

(v) Risks that are isolated, and not compensated for by associated benefits (e.g. exposure to X-rays when fitting shoes) are regarded as less acceptable than risks obtained in a largely beneficial context (e.g. risks from radon emissions in buildings that otherwise provide warmth at a low energy cost).

(vi) Immediate hazards (e.g. of new electrical equipment) are regarded as worse than deferred hazards (e.g. resulting from bad maintenance).

(vii) Unfamiliar, unnatural or 'new' hazards (e.g. from new food additives or radiation from nuclear industry) are regarded as worse than risks from familiar, natural, and established causes (e.g. traditional foods, cosmic radiation or emissions from Aberdeen granite).

(viii) Risks arising from secret activities (e.g. in the defense field) are regarded as worse than those derived from open activities.

(ix) Risks evaluated by groups who are suspected of partiality (e.g. statements by an industry about the safety of its own installations) are regarded as worse than risks evaluated by impartial groups.

(x) Risks that some other person pays to put right are regarded as worse than risks individuals have to pay themselves to remedy!

The Royal Society study group [RS, 1983] added an eleventh criterion: Risks to named persons (e.g. someone trapped on a sinking ship) are given higher priority than risks to a 'statistical' life (e.g. those killed in traffic accidents).

It is important to note that this list is one produced by civil servants in a brainstorming exercise and not a list produced by a study of public perception. However, the above listed criteria are all intuitively reasonable and all show how far public perception of risk is from the 'objective' criteria discussed before.

18.5 Beyond risk regulation: compensation, trading and ethics

The focus of this book has been on the assessment of technological risks. Once risks are measured – and this chapter demonstrates that measurement of risk is no mean task – the issue of dealing with risks poses itself. The measures developed for man-made risk apply just as well for natural hazards and catastrophes. In this concluding section we cluster a number of issues in dealing with risks. This at once summarizes issues raised in this chapter, and directs the reader to wider issues beyond the scope of this book. Ample material and references may be found in [Lowrance, 1976], [Maclean, 1986], [Glickman and Gough, 1990], [Andersson, 1999], [Shrader-Frechette, 1985] and [Shrader-Frechette, 1993]. Natural hazards are extensively treated in [Woo, 1999].

The first cluster of issues concerns the use of 'objective' versus 'subjective' measures of risk. Recent work of [Sjöberg, 1999] provides a critical perspective on efforts to explain risk behavior with simple appeals to psychological, cultural and/or political dimensions. When applied to risks relevant to personal decision making objective measures may involve sparse reference classes, noisy data, problematic assumptions and expert judgement. More commonly, these measures are simply not available. Thus, suppose I am deliberating whether driving my motorcycle is too risky. I could use the data in Table 18.4, but I could also reflect that I am not just a 'two wheeled motor vehicle driver'. Rather, I am a middle-aged well-educated non-drinking church-going risk averse driver of a two wheeled motor vehicle with a 125 cc motor. The information I need for my personal decision is not found in Table 18.4. The objective/subjective dichotomy may well have outlived itself. A better distinction might juxtapose public and personal decision making with regard to risk.

Public decision making with regard to risk naturally involves the attempt to standardize the measurement of risks across diverse domains. Inevitable differences between assessment of risks in different sectors burdens our institutional structures for risk management with problems of risk communication. [Granger Morgan, 1990] distinguishes four types of structures for implementing risk management procedures:

(i) tort law, particularly laws related to negligence, liability, nuisance and trespass,
(ii) insurance,
(iii) voluntary standard-setting, e.g. the Underwriters Laboratory or the National Fire Protection Association, and
(iv) mandatory governmental standards and regulations.

Societies differ greatly in their reliance on these structures. [Thompson, 1986] distinguishes the 'consensus' style of the United Kingdom from the 'adversary' style of the United States. The former is said to involve a top-down consultative approach, whereas the latter is statutory and bottom up. The ALARA and ALARP principles discussed in this chapter are invoked to guide decisions. Other principles commonly encountered in some of the above domains include the following [Lowrance, 1976]:

(i) generally recognized as safe (GRAS);
(ii) best available practice (BAT);
(iii) the Delaney principle – 'no [food] additive shall be deemed safe if it is found...to induce cancer in man or animal';
(iv) the precautionary principle – '...where there are threats of serious or irreversible damage, lack of scientific certainty shall not be used as a reason for postponing cost-effective measures to prevent environmental degradation' [UN, 1992].

In the corridors of scientific conferences the CATNIP principle (cheapest available technology not involving prosecution) is also mentioned. The applicability of such principles is a subject of ongoing debate.

If risks can be measured, then they can be compared, and if they can be compared, then they could in principle be compensated. Workers in hazardous occupations may be offered a 'risk premium' in terms of salary, insurance or other benefits. According to a 1987 amendment to the US Nuclear Waste Policy Act, states, localities or Native American tribes willing to accept a radwaste storage facility may be offered monetary compensation ranging from $5 million to $20 million. Such compensations raise ethical issues. Should all forms of compensation be sanctioned between consenting parties? Is the consent 'free and informed'? and 'How does one compensate future generations?' [Shrader-Frechette, 1985] and [Shrader-Frechette, 1993].

18.5 Beyond risk regulation: compensation, trading and ethics

Compensating risks is one form of trading in risks. The notion of trading in risk has taken new forms in recent years. Risk, by which we do not just mean financial risk, has become a commodity that can be traded just like any other. The development of a market in risks creates a buffer for societal dislocation caused by major disasters by distributing the risks of rebuilding over the global financial community. In this way 'global' society has a mechanism for absorbing the impact of 'local' disasters. Catastrophe bonds are new 'financial instruments' for distributing risk. Such bonds pay an attractive return if a specified catastrophe does not occur, and return nothing if the catastrophe does occur. These may be attractive to investors since the occurrence of catastrophes is uncorrelated with common economic indicators.

Such new financial instruments form a risk in themselves. Insurance of whatever form is a tool for spreading negative consequences rather than removing them. The effect of global insurance is to spread risk globally rather than have it concentrated locally. This ultimately exposes the global community to risk from combinations of disasters occurring over the world. The difficulties experienced by Lloyds of London in the 1990s were at least partially caused by exposure to a sequence of natural disasters. Larger natural disasters occurring in the global economy in the future could easily lead to a global economic recession.

Where does this leave technical risk? The nuclear power industry has never been required to have full commercial insurance for its activities. The consequences of a large scale nuclear accident are considered uninsurable, and governments seeing a strategic need for a nuclear power industry have historically accepted part of the risk burden on behalf of society. Nowadays the strategic requirement has been lessened as the relationship between military and civilian programs has been weakened, and it would seems strange not to require insurance for a future nuclear industry. Such a requirement might well be beneficial for the industry by giving a premium to those future technological developments that would reduce the less insurable risks.

Maybe new financial instruments can play a role in dealing with other problems too. A major difficulty for regulators in finding 'appropriate' levels of risk is that societal preferences (expressed through either politicians, public opinion surveys, whatever) develop and change through time. Major investments in plant or infrastructure usually require a period of many years to give a return, by which time the societal view of the risk imposed may have changed (either because there is a prevailing view that plant should pose a smaller risk or because of changes in town planning). Governments in Sweden and Germany have announced the intention of closing down nuclear power plants before the end of plant lifetime, but have been faced with huge compensation claims from plant owners. Could the financial markets have a role to play in bringing the matching together of the different dynamics of societal risk acceptance and investment cycles?

Bibliography

[Abernathy et al., 1983] Abernathy, R.B., Breneman, J.E., Medlin, C.H., Reinman, G.L., *Weibull Analysis Handbook*, USAF Wright Aeronautical Laboratories, AFWAL-TR-83-2079, 1983.

[Adams, 1995] Adams, J., *Risk*, UCL Press, London, 1995.

[Andersen and Ye, 1998] Andersen, E.D. and Ye, Y., "A computational study of the homogeneous algorithm for large-scale convex optimization", *Computational Optimization and Applications*, **10**, 243–269, 1998.

[Andersson, 1999] Andersson, K. (ed), *Values in Decisions on Risk*, European Commission, DG XI., 343–351, 1999.

[Andrews and Moss, 1993] Andrews, J.D. and Moss, T.R., *Reliability and Risk Assessment*, Longman, Harlow, Essex, 1993.

[Appignani and Fragola, 1991] Appignani, P.L. and Fragola J.R., *Shuttle Integrated Risk Assessment Program: Main Propulsion System Propellant Management System: Reliability Failure Data*, SAIC, New York, 1991.

[APS, 1975] American Physical Society Study Group on Light Water Reactor Safety, "Report to the American Physical Society", *Reviews of Modern Physics*, **47**, suppl. no. 1, 1975.

[Aralia, 1995] Aralia Groupe, Computation of prime implicants of a fault tree within Aralia, in *Proceedings of the European Safety and Reliability Association Conference ESREL'95*, ed. Watson I.A. and Cotton M.P., pp. 160–202, SaRS, Manchester, 1995.

[Ascher and Feingold, 1984] Ascher, H. and Feingold, H., *Repairable Systems Reliability*, Lecture Notes in Statistics 7, Marcel Dekker, New York, 1984.

[Atherton and French, 1998] Atherton, E. and French, S., "Valuing the future: a MADA example involving nuclear waste storage", *Journal of Multi-Criteria Decision Analysis*, **7**(6), 304–321, 1998.

[ATV, 1987] ATV office, *T-Book Reliability Data of Components in Nordic Nuclear Power Plants*, Vattenfall AB, S-162 87, Vallingby, Sweden, 1987.

[Atwood, 1986] Atwood, C.L., "The Binomial failure rate common cause model", *Technometrics*, **28**, 139–148, 1986.

[Aven and Jensen, 1999] Aven, T. and Jensen, U., *Stochastic Models in Reliability*, Springer, New York, 1999.

[Azalarov and Volodin, 1986] Azalarov, T.A. and Volodin, N.A., *Characterization problems associated with the exponential distribution*, Springer, New York, 1986.

[Bain, 1978] Bain, L.J., *Statistical Analysis of Reliability and Life-Testing Models*, Marcel Dekker, New York, 1978.

[Ballard, 1985] Ballard, G.M., "An analysis of dependent failures in the ORNL-Precursor study" in *Proceedings of the ANS/ENS International Topical Meeting PSA Methods and Applications*, San Francisco, Calif. 24–28 February, 1985.

[Barlow and Proschan, 1965] Barlow, R.E. and Proschan, F., *Mathematical Theory of Reliability*, 1965, reissued in series Classics in Applied Mathematics, 17 (Society for Industrial and Applied Mathematics (SIAM), Philadelphia, Pa, 1996).

[Barlow and Proschan, 1986] Barlow, R.E. and Proschan, F., "Inference for the exponential distribution", in *Theory of Reliability: Proceedings of the International School of Physics "Enrico Fermi", course 94*, ed North-Holland, Amsterdam, 1986.

[Barry, 1996] Barry, T.M., "Recommendations on the testing and use of pseudo-random number generators used in Monte Carlo analysis for risk assessment", *Risk Analysis*, **16**, 93–105, 1996.

[Basili, 1980] Basili, V. R., *Tutorial on Models and Metrics for Software Management and Engineering*, IEEE Computer Society Press, Los Almitos, 1980.

[Bedford and Cooke, 1997] Bedford, T. and Cooke, R.M., "Reliability methods as management tools: dependence modelling and partial mission success", *Reliability Engineering and System Safety*, **58**, 173–180, 1997.

[Bedford and Cooke, 1999a] Bedford, T. and Cooke, R.M., "A new generic model for applying MAUT", *European Journal of Operations Research*, **118**, 589–604, 1999.

[Bedford and Cooke, 1999b] Bedford, T. and Cooke, R.M., *Vines – a New Graphical Model for Dependent Random Variables*, Preprint, TU Delft, 1999.

[Bedford and Meeuwissen, 1997] Bedford, T. and Meeuwissen, A.M.H. "Minimally informative distributions with given rank correlation for use in uncertainty analysis", *Journal of Statistics and Computer Simulation*, **57**, 143–174, 1997.

[Bedford and Mcilijson, 1995] Bedford, T. and Meilijson, I., "The marginal distributions of lifetime variables which right censor each other", in *Analysis of Censored Data*, IMS Lecture Notes Monograph Series, 27, eds H.L. Koul and J.V. Deshpande, Hayward, 1995.

[Bedford and Meilijson, 1996] Bedford, T. and Meilijson, I., "A new approach to censored lifetime variables", *Reliability Engineering and System Safety*, **51**, 181–187, 1996.

[Bedford and Meilijson, 1997] Bedford, T. and Meilijson, I., "A characterisation of marginal distributions of (possibly dependent) lifetime variables which right censor each other" *Annals of Statistics*, **25**, 1622–1645, 1997.

[Bedford et al., 1999] Bedford, T., Atherton, E. and French, S., "Time at risk: ALARP trade-offs over time", in *Safety and Reliability Proceedings of ESREL'99*, eds G.I. Schueller and P. Kafka, pp. 1441–1446, Balkema, Rotterdam, 1999.

[Bell and Esch, 1989] Bell, T.E. and Esch, K. "The space shuttle: a case of subjective engineering", *IEEE Spectrum*, 42–46, June 1989.

[Bernardo, 1979] Bernardo, J.M., "Reference posterior distributions for Bayesian inference", with discussion. *Journal of the Royal Statistical Society Series B*, **41**, no. 2, 113–147, 1979.

[Bernstein, 1996] Bernstein, P., *Against the Gods; the Remarkable Story of Risk*, Wiley, New York, 1996.

[Birnbaum, 1962] Birnbaum, A., "On the foundations of statistical inference" (with discussion), *Journal of the American Statistical Association*, **57**, 269–362, 1962.

[Bittanti, 1988] Bittanti, S. (ed), *Software Reliability Modelling and Identification*, Springer Lecture Notes in Computer Science, 341, Springer, Berlin, 1988.

[Borel, 1924] Borel, E., *Valeur pratique en philosophie des probabilités* in Traite du calcul des probabilités, ed E. Borel, Gauthier-Villars, Paris, 1924.

[Borel, 1965] Borel, E., *Elements of the Theory of Probability*, 1965 (translated from Éléments de la théorie des probabilités, 1909.

[Bouissou, 1997] Bouissou, M., Bruyère, F. and Rauzy, A., "BDD based fault-tree processing: a comparison of variable ordering techniques", in *Proceedings of ESREL'97*, ed C.G. Soares, pp. 2045–2051, Pergamon, 1997.

[Box and Tiao, 1992] Box, G.E.P. and Tiao, G.C., *Bayesian Inference in Statistical Analysis*, Wiley, New York, 1992.

[Bryant, 1986] Bryant, R., "Graph-based algorithms for Boolean function manipulation", *IEEE Transactions on Computers*, **C-35**, 677–691, 1986.

[Bryant, 1992] Bryant, R., "Symbolic Boolean manipulation with ordered binary decision diagrams", *ACM Computing Surveys*, **24**, 193–211, 1992.

[Buratti, 1981] Buratti, D.L., Pinkston, W.E. and Simikins, R.O., "Sneak software analysis", 4th Digital Avionics Systems Conference, November 1981.

[Carlsson et al., 1998] Carlsson, L., Johanson, G. and Wolfgang, W., "Recent development of the ICDE Project", in *Probabilistic Safety and Management (PSAM 4)*, Vol. 1, eds A. Mosleh and R.A. Bari, Springer, London, 1998.

[Cassidy, 1996] Cassidy, K. "Risk criteria for the siting of hazardous installations and the development in their vicinity", *Proceedings of ESREL'96/PSAM 3*, Springer, 1996.

[CCPS, 1989] Center for Chemical Process Safety, *Guidelines for Process Equipment Reliability Data* with Data Tables, American Institute for Chemical Engineers, New York, 1989.

[Cheung, 1978] Cheung, R.C., "A user-oriented software reliability model", in *Proceedings of COMPSAC 1978*, Chicago, Il., Nov. 1978, pp. 565–570.

[Cohen and Lee, 1979] Cohen, B.L. and Lee, I., "A catalogue of risks", *Health Physics* **36**, 707–722, 1979.

[Colglazier and Weatherwas, 1986] Colglazier, E.W. and Weatherwas, R.K., "Failure estimates for the space shuttle", *Abstracts for Society for Risk Analysis Annual Meeting 1986, Boston, Mass.*, p. 80, Nov. 9–12, 1986.

[Colt and Priore, 1985] Colt, D.W. and Priore, M.G., "Impact of nonoperating periods on equipment reliability", RADC-TR-85-91, Rome Air Development Center, Griffiss Air Force Base, New York, 1985.

[Cooke, 1986] Cooke, R.M., "Conceptual fallacies in subjective probability", *Topoi*, **5**, 21–27, 1986.

[Cooke, 1991] Cooke R.M., *Experts in Uncertainty*, Oxford University Press, 1991.

[Cooke, 1993] Cooke, R.M., "The total time on test statistic and age-dependent censoring", *Statistics and Probability Letters* **18** no. 4, 1993.

[Cooke, 1994] Cooke, R.M., "Parameter fitting for uncertain models: modelling uncertainty in small models", *Reliability Engineering and System Safety*, **44**, 89–102, 1994.

[Cooke, 1995] Cooke, R.M., *UNICORN: Methods and Code for Uncertainty Analysis*, AEA Technology for ESRA, Risley (UK) 1995.

[Cooke, 1996] Cooke, R.M., "The design of reliability data bases", Part I and II, *Reliability Engineering and System Safety*, **51**, no. 2, 137–146 and 209–223, 1996.

[Cooke, 1997a] Cooke, R.M., "Markov and entropy properties of tree- and vine-dependent variables", *Proceedings of the ASA Section on Bayesian Statistical Science*, 1997.

[Cooke, 1997b] Cooke, R.M., "Technical Committee Uncertainty Modelling: Benchmark Workshop", Delft University Press, 1997.

[Cooke and Bedford, 1995] Cooke, R. and Bedford, T., "Analysis of reliability data using subsurvival functions and censoring models", in *Recent Advances in Life-Testing and Reliability*, ed N. Balakrishnan, pp. 11–43, CRC Press, Boca Raton, Fla, 1995.

[Cooke and Goossens, 1990] Cooke, R.M. and Goossens, L., "The accident sequence precursor methodology for the European post-Seveso era", *Reliability Engineering and System Safety*, **27**, 117–130, 1990.

[Cooke and Kraan, 1996] Cooke, R.M. and Kraan, B., "Dealing with dependencies in uncertainty analysis", in *Probabilistic Safety and Management '96*, eds C. Cacciabue and I.A. Papazoglou, pp. 625–630, Springer, London, 1996.

[Cooke and Paulsen, 1997] Cooke, R.M. and Paulsen, J., "Concepts for measuring maintenance performance and methods for analysing competing failure modes", in *Reliability Engineering and System Safety*, **55**, 135–141, 1997.

[Cooke and Waij, 1986] Cooke, R.M., and Waij, R., "Monte Carlo sampling for generalised knowledge dependence with application to human reliability", *Risk Analysis*, **6** (3), 335–343, 1986.

[Cooke et al., 1993] Cooke, R., Bedford, T., Meilijson I. and Meester, L., "Design of reliability data bases for aerospace applications", Report to the European Space Agency, Department of Mathematics Report 93-110, TU Delft, 1993.

[Cooper and Chapman, 1987] Cooper, D.F. and Chapman, C.B., *Risk Analysis for Large Projects: Models, Methods and Cases*, Wiley, Chichester, 1987.

[Coudert and Madre, 1992] Coudert, O. and Madre, J.C., "A new method to compute prime and essential implicants of Boolean functions", in *Advanced Research in VLSI and Parallel Systems: Proceedings of the 1992 Brown/MIT conference*, eds T. Knight and J. Savage, pp. 113–128, MIT Press, Cambridge, Mass., 1992.

[Covello and Mumpower, 1985] Covello, V.T. and Mumpower, J., "Risk analysis and risk management: an historical perspective", *Risk Analysis*, **5**, 103–120, 1985.

[Covello et al., 1989] Covello, V.T., McCallum, D.B. and Pavlova, M.T. (eds) *Effective Risk Communication*, Plenum Press, New York, 1989.

[COVO, 1982] COVO, "Risk analysis of six potentially hazardous industrial objects in the Rijnmond area: a pilot study", Reidel, Dordrecht, Netherlands, 1982.

[Cox, 1959] Cox, D.R., "The analysis of exponentially distributed life-times with two types of failure", *Journal of the Royal Statistical Society B* **21**, 414–421, 1959.

[Cox and Hinkley, 1974] Cox, D.R. and Hinkley, D.V., *Theoretical Statistics*, Chapman and Hall, London, 1974.

[Cox and Lewis, 1966] Cox, D.R. and Lewis, P., *The Statistical Analysis of Series of Events*, Methuen, London, 1966.

[CPR, 1985] Commission for prevention of disasters from dangerous substances, *Methoden voor het bepalen en verwerken van kansen* (Commissie voor de Preventie van Rampen door Gevaarlijke Stoffen: Methods for determining and calculating probabilities), CPR 12, ISSN 0166-8935, Arbeidsinspectie, The Hague, 1985.

[Crowder, 1991] Crowder, M., "On the identifiability crisis in competing risks analysis" *Scandinavian Journal of Statistics* **18**, 223–233, 1991.

[Crowder, 1994] Crowder, M., "Identifiability crises in competing risks", *International Statistical Review*, **62**, 3, 379–391, 1994.

[Crowder et al., 1991] Crowder, M., Kimber, C., Smith, R. and Sweeting, T., *Statistical Analysis of Reliability Data*, Chapman and Hall, London, 1991.

[David and Moeschberger, 1975] David, H.A. and Moeschberger, M.L., *The Theory of Competing Risks*, Charles Griffin and Co., London, 1975.

[De Finetti, 1964] De Finetti, B., "La prévision; ses lois logiques, ses sources subjectives", *Annales de l'Institut Henri Poincaré*, **7**, 1–68, 1937. English translation in *Studies in Subjective Probability*, eds H. Kyburg and H. Smokler, Wiley, New York, 1964.

[De Finetti, 1974] De Finetti, B., *Theory of Probability*, vols I and II, Wiley, New York, 1974.

[De Groot, 1970] De Groot, M., *Optimal Statistical Decisions*, McGraw-Hill, New York, 1970.

[Dore and Norstrøm, 1996] Dore, B. and Norstrøm, J.G., Pilot application of sneak analysis on computer controlled satellite equipment in *Proceedings of the Conference on Probabilistic Safety Assessment and Management 96 ESREL'96 – PSAM III, 24–28 June*, Springer, London, 1996.

[Dougherty and Fragola, 1988] Dougherty, E.M. and Fragola, J.R., *Human Reliability Analysis: a Systems Engineering Approach with Nuclear Power Plant Applications*, Wiley, New York, 1988.

[Dudley, 1989] Dudley, R.M., *Real Analysis and Probability*, Wadsworth and Brooks/Cole, Belmont, Calif., 1989.

[EEC, 1982] EEC, "Major accident hazards of certain industrial activities", Council Directive, 82/501/EEC, 24 June, 1982.

[Efron, 1978] Efron, B., "Controversies in the foundations of statistics", *American Mathematic Monthly*, **85**, no. 4, 231–246, 1978.

[Efron, 1986] Efron, B., "Why isn't everyone a Bayesian?" (with discussion and a reply by the author). *American Statistician* **40**, no. 1, 1–11, 1986.

[EIREDA, 1991] *EIREDA European Industry Reliability Data Handbook*, C.E.C.-J.R.C./ISEI 21020 ISPRA (Varese) Italy, EDF - DER/SPT 93206 Saint Denis (Paris) France, 1991.

[Embrey, 1984] Embrey, D.E., Humphreys, P.C., Rosa, E.A., Kirwan, B. and Rea, K., *SLIM-MAUD: an Approach to Assessing Human Error Probabilities Using Structured Expert Judgement*, NUREG/CR-3518, US Nuclear Regulatory Commission, Washington, DC, 1984.

[Embrey, 1992] Embrey, D.E., "Incorporating management and organisational factors into probabilistic safety assessment", *Reliability Engineering and System Safety*, **38**, 199–208, 1992.

[EPA, 1976] EPA, Reactor Safety Study, oversight hearings before the Subcommittee on Energy and the Environment of the Committee on Interior and Insular Affairs, House of Representatives, 94th Congress, second session, serial no. 94-61, Washington, DC, 11 June, 1976.

[ESA, 1993] Product Assurance and Safety Department, ESA, *Guide to Software Reliability and Safety Assurance for ESA Space Systems*, Draft 2 of ESA PSS-01-213 Issue 1, ESA, Aug. 1993.

[ESA, 1994] Product Assurance and Safety Department, ESA, *Sneak Analysis Methods and Procedures for ESA Space Programmes*, ESA PSS-01-411, Issue 1, January 1994.

[ESA, 1997] European Space Agency, "Ariane 5 Board of Inquiry report", ESA 1997.

[ESReDA, 1999] ESReDA Working Group Report, *Handbook on Quality of Reliability Data*, Det Norske Veritas, Havik, Norway, 1999.

[Evans and Verlander, 1997] Evans, A.W. and Verlander, N.Q., "What is wrong with criterion FN-lines for judging the tolerability of risk?", *Risk Analysis*, **17**, 157–167, 1997.

[Fine, 1973] Fine, T., *Theories of Probability*, Academic Press, New York, 1973.

[Fischhoff et al., 1978] Fischhoff, B. et al, "How safe is safe enough? A psychometric study of attitudes towards technological risks and benefits", *Policy Sciences*, **9**, 127–152, 1978.

[Fischhoff et al., 1981] Fischhoff, B. et al, *Acceptable Risk: a Critical Guide*, Cambridge University Press, 1981.

[Fishburn and Farquhar, 1982] Fishburn, P.C. and Farquhar, P.H., "Finite degree utility independence", *Mathematics of Operations Research*, **7**, 348–353, 1982.

[Fleming, 1975] Fleming, K.N., "A reliability model for common mode failure in redundant safety systems", *Proceedings of the Sixth Annual Pittsburgh Conference on Modeling and Simulation*, General Atomic Report GA-A13284, 23–25, 1975.

[Fleming et al., 1983] Fleming, K.N., Mosleh, A. and Kelley, A.P. Jr, "On the analysis of dependent failures in risk assessment and reliability evaluation, *Nuclear Safety*, **24**, 637–657, 1983.

[Fleming et al., 1986] Fleming, K.N., Mosleh, A. and Deremer, R.K., "A systematic procedure for the incorporation of common-cause events into risk and reliability models", *Nuclear Engineering and Design*, **93**, 245–275, 1986.

[Fragola, 1987] Fragola, J.R., "Reliability data bases: the current picture", *Hazard Prevention*, Jan/Feb. 1987, 24–29.

[Fragola, 1995] Fragola, J.R., *Probabilistic Risk Assessment of the Space Shuttle*, SAIC doc. no. SAICNY95-02-25, New York, 1995.

[Fragola, 1996] Fragola, J.R., "Reliabilility and risk analysis data base development; an historical perspective", *Reliability Engineering and System Safety*, **51**, no. 2, 125–137, 1996.

[Frank, 1999] Frank, M.V., "Reentry safety: probability of fuel release', in *Safety and Reliability, Proceedings of ESREL '99*, eds G.I. Schuëller and P. Kafka, Balkema, Rotterdam, 1999.

[French, 1988] French, S., *Decision Theory: an Introduction to the Mathematics of Rationality*, Ellis Horwood, Chichester, 1988.

[French et al., 1999] French, S., Atherton, E. and Bedford, T., "A comparison of CBA and MAUT for ALARP decision-making", in *Risk Analysis: Facing the New Millennium* (Proceedings of the 9th Annual Conference of the Society for Risk Analysis Europe), ed L. Goossens, pp. 550–553, Delft University Press, 1999.

[Gail, 1975] Gail, M., "A review and critique of some models used in competing risk analysis", *Biometrics*, **31**, 209–222, 1975.

[Galambos, 1978] Galambos, J., *The Asymptotic Theory of Extreme Order Statistics*, Wiley, New York, 1978.

[Garrick, 1984] Garrick, B.J., "Recent case studies and advancements in probabilistic risk assessments", *Risk Analysis*, **4**, 267–279, 1984.

[Garrick, 1989] Garrick, B.J., "Risk assessment practices in the space industry: the move toward quantification", *Risk Analysis*, **9**, 1–7, 1989.

[Glickman and Gough, 1990] Glickman, T. and Gough, M. (eds) *Readings in Risk*, Resources for the Future, Washington, DC, 1990.

[Goel and Okumoto, 1979] Goel, A.L. and Okumoto, K., "Time-dependent error-detection rate model for software reliability and other performance measures", *IEEE Transactions on Reliability* **28**, 206–211, 1979.

[Good, 1983] Good, I.J., *Good Thinking. The Foundations of Probability and its Applications*, University of Minnesota Press, Minneapolis, 1983.

[Goossens et al., 1998] Goossens, L.H.J., Cooke, R.M. and Kraan, B.C.P., "Evaluation of weighting schemes for expert judgement studies", in *Probabilistic Safety Assessment and Management, (Proceedings of PSAM4)*, eds A. Mosleh and R.A. Bari, pp. 2389–2396, Springer, London, 1998.

[Götz, 1996] Götz, H., *Influence Diagrams and Decision Trees in Severe Accident Management*, Report of Fac. of Technical Mathematics and Informatics, TU Delft, 96-84, ISSN 0922-5641, 1996.

[Götz et al., 1996] Götz, H., Seebregts, A.J. and Bedford, T.J., Influence diagrams and decision trees for severe accident management, in *Proceedings of PSA'96, Utah*, pp. 1996.

[Granger Morgan, 1990] Granger Morgan, M., "Choosing and managing technology-induced risk", in [Glickman and Gough, 1990] pp. 17–28.

[Granger Morgan and Henrion, 1990] Granger Morgan, M. and Henrion, M., *Uncertainty: a Guide to Dealing with Uncertainty in Quantitative Risk and Policy Analysis*, Cambridge University Press, 1990.

[Green and Bourne, 1972] Green A.E. and Bourne, A.J., *Reliability Technology*, Wiley, London, 1972.

[Grimmett and Stirzaker, 1982] Grimmett, G.R. and Stirzaker, D.R., *Probability and Random Processes*, Oxford University Press, 1982.

[Guivanessian et al., 1999] Guivanessian, H., Holicky, M., Cajot, L.G. and Schleich, J.B., "Probabilistic analysis of fire safety using Bayesian causal network", in *Safety and Reliability, Proceedings of ESREL '99*, eds G.I. Schuëller and P. Kafka, pp. 725–730, Balkema, Rotterdam, 1999.

[Hacking, 1975] Hacking, I., *The Emergence of Probability: a Philosophical Study* of Early Ideas about Probability, Induction and Statistical Inference, Cambridge University Press, London, 1975.

[Halmos, 1974] Halmos, P., *Measure Theory*, Van Nostrand, Princeton, NJ, reprinted Springer, New York, 1974.

[Hannaman and Worledge, 1988] Hannaman, G.W. and Worledge, D.H., "Some developments in human reliability analysis approaches and tools", *Reliability Engineering and System Safety*, **22**, 235–256, 1988.

[Hardy et al., 1983] Hardy, G.H., Littlewood, J.E. and Pólya, G., *Inequalities*, Cambridge University Press, 1983.

[Harper et al., 1994] Harper, F.T., Goossens, L.H.J., Cooke, R.M., Helton, J.C., Hora, S.C., Jones, J.A., .Kraan, B.C.P., Lui, C., McKay, M.D., Miller, L.A., Päsler-Sauer, J. and Young, M.L., *Joint USNRC/CEC Consequence Uncertainty Study: Summary of Objectives, Approach, Application, and Results for the Dispersion and Deposition Uncertainty Assessments*, NUREG/CR-6244, EUR 15855, SAND94-1453, Sandia National Laboratories and Delft University of Technology, 1994.

[HFRG, 1995] Human Factors Reliability Group, *Human Reliability Assessor's Guide*, SRD Association, Warrington, Cheshire, 1995.

[HM Treasury, 1996] HM Treasury, *The Setting of Safety Standards*, HMSO, London, 1996.

[Hogarth, 1987] Hogarth, R., *Judgement and Choice*, Wiley, New York. 1987.

[Hokstadt and Jensen, 1998] Hokstadt, P. and Jensen, R., "Predicting the failure rate for components that go through a degradation state" , in *(Proceedings of ESREL '98) Safety and Reliability*, eds S. Lydersen, G.K. Hansen and H.A. Sandtorv, pp. 389–396, Balkema, Rotterdam, 1998.

[Hora, 1996] Hora, S., "Aleatory and epistemic uncertainty in probability elicitation with an example from hazardous waste management", *Reliability Engineering and System Safety*, **54**, 217–223, 1996.

[Hora and Iman, 1989] Hora, S.C. and Iman, R.L., "Expert opinion in risk analysis. The NUREG-1150 methodology", *Nuclear Science and Engineering*, **102**, 323–331, 1989.

[Howard, 1988] Howard, R.A., "Uncertainty about Probability: a decision analysis perspective", *Risk Analysis*, **8**, 91–98, 1988.

[Hoyland and Rausand, 1994] Hoyland, A. and Rausand, M., *System Reliability Theory; Models and Statistical Methods*, Wiley, New York, 1994.

[HSE, 1978] 'Canvey Island (1978): an investigation', Health and Safety Executive, HMSO, 1978.

[HSE, 1981] 'Canvey Island (1981): a second report', Health and Safety Executive, HMSO, 1981.

[HSE, 1987] Health and Safety Executive, *The Tolerability of Risk from Nuclear Power Stations*, HMSO, London, December 1987.

[HSE, 1991] *Major Hazard Aspects of the Transport of Dangerous Goods*, HMSO, London, 1991.

[HSE, 1992] Health and Safety Executive, *Safety Assessment Principles for Nuclear Plants*, HMSO, London, 1992.

[Hutchinson and Lai, 1990] Hutchinson, T.P. and Lai, C.D., *Continuous Bivariate Distributions, Emphasizing Applications*, Rumsby Scientific Publishing, Adelaide, 1990.

[IAEA, 1988] IAEA, "Component reliability data for use in probabilistic safety assessment", IAEA TECHDOC 478, Vienna, 1988.

[IAEA, 1992] IAEA, "Procedures for conducting common cause failure analysis in probabilistic safety assessment", International Atomic Energy Agency, Vienna, 1992.

[Ibrekk and Granger Morgan, 1987] Ibrekk, H. and Granger Morgan, M., "Graphical communication of uncertain quantities to nontechnical people" *Risk Analysis*, **7** no. 4, 1987.

[IEEE, 1977] Nuclear Power Engineering Committee of the IEEE Poser Engineering Society, *IEEE Guide to the Collection and Presentation of Electrical, Electronic and Sensing Component Reliability Data for Nuclear Power Generation Stations*, IEEE STD-500, June 1977.

[IEEE, 1984] IEEE Standard 500-1984, Piscataway, NJ, 1984.

[Jäger, 1983] Jäger, P., "The question of quality of risk analysis for chemical plants: discussing the results for sulphuric acid plant", in *Proceedings of the Fourth International Symposium on Loss Prevention and Safety Promotion in the Process Industries*, Harrowgate September 12–16, 1983 ed Wilkinson, W.L., Chem E. Symposium Series 80. Rugby.

[Jaynes, 1983] Jaynes, E.T., *Papers on Probability, Statistics and Statistical Physics*, ed R.D. Rosenkrantz, D. Reidel, Dordrecht, Netherlands, 1983.

[Jelinski and Moranda, 1973] Jelinski, Z. and Moranda, P.B., "Application of a probability based method to a code reading experiment", *Record 1973: IEEE Symposium on Computer Software Reliability*, pp. 78–82, IEEE, New York, 1973.

[Jensen, 1996] Jensen, F.J., *An Introduction to Bayesian Networks*, UCL Press, London, 1996.

[Johanson and Fragola, 1982] Johanson, G. and Fragola, J., "Synthesis of the data base for the Ringhals 2 PRA using the Swedish ATV data system", presented at ANS/ENS International Meeting on Thermal Reactor Safety, August 1982, Chicago, Ill.

[Johnson, 1964] Johnson, L.G., *Theory and Technique of Variation Research*, Elsevier, New York, 1964.

[Kafka, 1999] Kafka, P., "How safe is safe enough? – An unresolved issue for all technologies", in *Safety and Reliability, Proceedings of ESREL99*, eds G.I. Schueller and P. Kafka, Balkema, Rotterdam, 1999.

[Kahneman *et al.*, 1982] Kahneman, D., Slovic, P. and Tversky, A., (eds) *Judgment under Uncertainty, Heuristics and Biases*, Cambridge University Press, 1982.

[Kamminga, 1988] Kamminga, S., MSc thesis, TU Delft report, 1988.

[Kaplan and Garrick, 1981] Kaplan, S. and Garrick, B.J., "On the quantitative definition of risk", *Risk Analysis*, **1**, 11–27, 1981.

[Kaplan and Meier, 1958] Kaplan, E.L. and Meier, P., "Nonparametric estimation from incomplete observations", *Journal of the American Statistical Association*, **53**, 457–481, 1958.

[Kapur and Lamberson, 1977] Kapur, K.C. and Lamberson, L.R., *Reliability in Engineering Design*, Wiley, New York, 1977.

[Kastenberg, 1993] Kastenberg, W.E. (ed), *A Framework for the Assessment of Severe Accident Management*, NUREG/CR-6056, Washington, DC, 1993.

[Keeney and Raiffa, 1993] Keeney, R.L. and Raiffa, H., *Decisions with Multiple Objectives*, Cambridge University Press, 1993.

[Kelly and Seth, 1993] Kelly, D. and Seth, S., *Data Review Methodology and Implementation Procedure for Probabilistic Safety Assessments*, MITRE corporation, McLean, Va MTR-93W0000153, HSK-AN-2602, 1993.

[Kemeny et al., 1979] Kemeny, J. et al., *Report of the President's Commission on the Accident at Three Mile Island*, Washington, DC, 1979.

[Keynes, 1973] Keynes, J.M., *Treatise on Probability*, Macmillan, London, 1973. (First edition 1921)

[Kingman, 1993] Kingman, J.F.C., *Poisson Processes*, Oxford University Press, 1993.

[Kirkegaard and Kongsø, 1999] Kirkegaard, P. and Kongsø, H.E., "Comparison between the BDD method and conventional reliability analysis techniques", in *Safety and Reliability, Proceedings of ESREL99*, eds G.I. Schuëller and P. Kafka, 1021–1026, Balkema, Rotterdam, 1999.

[Kirwan, 1994] Kirwan, B., *A Guide to Practical Human Reliability Assessment*, Taylor and Francis, London, 1994.

[Kirwan and Ainsworth, 1992] Kirwan, B. and Ainsworth, L.K., *A Guide to Task Analysis*, Taylor and Francis, London, 1992.

[Kirwan et al., 1988] Kirwan, B., Embrey, D.E. and Rea, K., *Human Reliability Assessor's Guide*, Report RTS 88/95Q, NCSR, UKAEA, Culcheth, Cheshire, 1988.

[Kolmogorov, 1933] Kolmogorov, A.N., *Grundbegriffen der Wahrscheinlichkeitsrechnung, Ergebnisse der Mathematik und ihrer Grenzgebiete*, Springer, 1933.

[Kotz et al., 1983] Kotz, S., Johnson, N.L. and Read, C.B. (eds.) *Encyclopedia of Statistical Sciences*, Wiley, New York, 1983.

[Kraan and Cooke, 1997] Kraan, B. and Cooke, R.M., "Post-processing techniques for the joint CEC/USNRC uncertainty analysis of accident consequence codes", *Journal of Statistical Computation and Simulation*, **57**, 243–259, 1997.

[Kullback, 1959] Kullback, S., *Information Theory and Statistics*, Wiley, New York, 1959.

[Kumamoto and Henley, 1996] Kumamoto, H. and Henley, E., *Probabilistic Risk Assessment and Management for Engineers and Scientists*, IEEE Press, Piscataway, NJ, 1996.

[Lalli and Malec, 1992] Lalli, V.R. and Malec, H.A., *Reliability Training*, NASA Reference Publication 1253, 1992.

[Lannoy, 1996] Lannoy, A., *Analyse quantitative et utilité du retour d'expérience pour la maintenance des matériels et la sécurité*, Editions Eyrolles, Paris, 1996.

[Laplace, 1951] Laplace, P.S., (Essai philosophique sur les probabilités, introduction to third edition of *Théorie analytique des probabilités* (1820), vol 7, Laplace's collected works), (English translation, *A Philosophical Essay on Probabilities*), Dover, New York, 1951.

[Lauritzen, 1996] Lauritzen, S.L., *Graphical Models*, Oxford University Press, 1996.

[Lenoir, 1992] Lenoir, P., "Hermes failure rate data base", Aerospatiale Tech. Doc. H-AS-1-07-ASPR ed 2, 1992.

[Lewis et al., 1979] Lewis, H. et al., *Risk Assessment Review Group Report to the U.S. Nuclear Regulatory Commission*, NUREG/CR-04000, 1979.

[Littlewood, 1975] Littlewood, B., "A reliability model for Markov structured software", in *Proceedings of the International Conference Rel. Software*, Los Angeles, Calif., pp. 204–207, 1975.

[Littlewood, 1981] Littlewood, B., "Stochastic reliability growth: a model for fault removal in computer programs and hardware designs", *IEEE Transactions on Reliability*, **30**, 313–320, 1981.

[Littlewood and Sofer, 1987] Littlewood, B. and Sofer. A., "A Bayesian modification to the Jelinski–Moranda software reliability growth model", *IEE/BCS, Software Engineering Journal*, **2**, Pt. 2, March 1987, 30–41.

[Littlewood and Verral, 1973] Littlewood, B. and Verral, J.L., "A Bayesian reliability growth model for computer software", *Journal of the Royal Statistical Society Series C*, **22**, 332–346, 1973.

[Loll, 1995] Loll, S. (ed), *DPL User Guide*, ADA Systems, Duxbury Press, Belmont, Calif., 1995.

[Lowrance, 1976] Lowrance, W., *Of Acceptable Risk*, Los Altos, Calif., 1976.

[Lyu, 1996] Lyu, M.R., *Handbook of Software Reliability Engineering*, IEEE/McGraw-Hill, 1996.

[Maclean, 1986] Maclean, D. (ed), *Values at Risk*, Totowa, NY, 1986.

[Marshall and Olkin, 1967] Marshall, A.W. and Olkin, I., "A multivariate exponential distribution" *Journal of the American Statistical Association*, **62** (317), 30–44, 1967.

[Martin, 1959] Martin, *Procedure and Data for Estimating Reliability and Maintainability*, Engineering Reliability, Martin, M-M-P-59-21; 1959.

[Martin-Löf, 1970] Martin-Löf, P., "On the notion of randomness" in *Intuitionism and Proof Theory*, eds A. Kino and R.E. Vesley, pp. 73–78, North-Holland, Amsterdam, 1970.

[McConway, 1981] McConway, K. J., "Marginalization and linear opinion pools", *Journal of the American Statistical Association,* **76**, 410–414.

[McKay et al., 1979] McKay, M.D., Beckman, R.J. and Conover, W.J., "A comparison of three methods for selecting values of input variables in the analysis of output from a computer code", *Technometrics*, **21**, 239–245, 1979.

[Meeuwissen, 1993] Meeuwissen, A., *Simulating Correlated Distributions*, PhD dissertation, Delft, 1993.

[Meeuwissen and Cooke, 1994] Meeuwissen, A. and Cooke, R., "Tree dependent random variables" Preprint TU, Delft, 1994.

[Meeuwsen, 1998] Meeuwsen, J.J., *Reliability Evaluation of Electric Transmission and Distribution Systems* PhD dissertation, Delft, 1998.

[Minarick and Kukielka, 1982] Minarick, J. and Kukielka, C., *Precursors to Potential Severe Core Damage Accidents 1969–1979*, NUREG/CR-2497, 1982.

[Mood et al., 1987] Mood, A.M., Graybill, F.A. and Boes, D.C., *Introduction to the theory of statistics*, 3rd ed, McGraw-Hill, 1987.

[Morris, 1994] Morris, W.P.G., *The Management of Projects*, Thomas Telford, London, 1994.

[Mosleh and Apostolakis, 1985] Mosleh, A. and Apostolakis, G., "The development of a generic data base for failure rates", in *Proceedings for the ANS/ENS International Topical Meeting on Probabilistic Safety Methods and Applications*, San Francisco, Calif., Feb. 24–March 1, 1985.

[Mosleh and Apostolakis, 1986] Mosleh, A. and Apostolakis, G., "The assessment of probability distributions form expert opinions with an application to seismic fragility curves", *Risk Analysis*, **6**, no. 4, 447–461, 1986.

[Mosleh et al., 1988] Mosleh, A., Fleming, K.N., Parry G.W., Paula H.M., Rasmuson D.M. and Worledge D.H., *Procedures for Treating Common Cause Failures in Safety and Reliability Studies*, NUREG/CR 4780 US Nuclear Regulatory Commission, vol. 1 and 2, 1988 and 1989.

[Muirhead, 1982] Muirhead, R.J., *Aspects of Multivariate Statistical Theory*, Wiley, New York, 1982.

[Musa, 1975] Musa, J.D., "A theory of software reliability and its applications", *IEEE Transactions on Software Engineering*, **1**(3), 312–327, 1975.

[Musa, 1998] Musa, J.D., *Software Reliability Engineering*, McGraw-Hill, New York, 1998.

[Musa and Okumoto, 1984] Musa, J.D. and Okumoto, K., "A logarithmic Poisson execution time model for software reliability measurement", in *Proceedings of the 7th International Conference on Software Engineering, Orlando*, pp. 230–238, 1984.

[MVROM, 1989] Ministerie van Volkshuisvesting, Ruimtelijke Ordening en Milieubeheer, Directoraat-Generaal Milieubeheer, *Omgaan met Risico's*, Leidschendam, March, 1989.

[MVROM, 1990] Ministerie van Volkshuisvesting, Ruimtelijke Ordening en Milieubeheer, Directoraat-Generaal Milieubeheer, *Omgaan met Risico's van Straling*, Leidschendam, 1990.

[Nagel, 1961] Nagel, E., *The Structure of Science*, Routledge and Kegan Paul, London, 1961.

[Natvig, 1985] Natvig, B., "Recent developments in multistate reliability theory", in *Probabilistic Models in the Mechanics of Solids and Structures*, ed S. Eggwertz and N.C. Lind, pp. 385–93, Springer, Berlin, 1985.

[NEI, 1994] Nuclear Energy Institute, *Severe Accident Issue Closure Guidelines*, NEI91-04, Rev 1, December 1994.

[Nelson, 1982] Nelson, W., *Applied Life Data Analysis*, Wiley, New York, 1982.

[Nelson, 1990] Nelson, W., *Accelerated Testing*, Wiley, New York, 1990.

[Norstøm et al., 1999] Norstøm, J.G., Cooke, R.M. and Bedford T., "Value of information based inspection strategy of a fault tree", in *Safety and Reliability, Proceedings of ESREL '99*, eds G.I. Schueller and P. Kafka, pp. 621–626, Balkema, Rotterdam, 1999.

[NRC, 1975] Nuclear Regulatory Commission, *Reactor Safety Study: an Assessment of Accident Risks in US Commercial Nuclear Power Plants*, WASH-1400, NUREG-75/014, 1975.

[NRC, 1979] Nuclear Regulatory Commision, "Nuclear Regulatory Commission issues policy statement on Reactor Safety Study and Review by the Lewis Panel", NRC Press Release, no. 79-19, 19 January, 1979.

[NRC, 1983] Nuclear Regulatory Commision, *PRA Procedures Guide*, US Nuclear Regulatory Commission NUREG/CR-2300, 1983.

[NRC, 1986] Nuclear Regulatory Commision, *Safety Goals for Nuclear Power Plants*, US Nuclear Regulatory Commission NUREG-0880, 1986.

[NRC, 1989] Nuclear Regulatory Commission, *Severe Accident Risks: An Assessment for Five US Nuclear Power Plants*, Nuclear Regulatory Commission, NUREG-1150, Washington, DC, 1989.

[Nussbaum, 1989] Nussbaum, R.D., 'Hilbert's projective metric and iterated non-linear maps, II', *Memoirs of the American Mathematic Society* **79**, Number 401, 1989.

[O'Connor, 1994] O'Connor, P.D.T., *Practical Reliability Engineering*, Wiley, 3rd ed, 1994.

[OECD, 1996] OECD Nuclear Energy Agency, *Chernobyl Ten Years On; Radiological and Health Impact*, OECD, Paris, 1996.

[O'Hagan, 1994] O'Hagan, A., *Kendall's Advanced Theory of Statistics, Volume 2B: Bayesian Inference*, Edward Arnold, London, 1994.

[Oliver and Smith, 1990] Oliver, R.M. and Smith, J.Q., *Influence Diagrams, Belief Nets and Decision Analysis*, Wiley, Chichester, 1990.

[OREDA, 1984] OREDA, *Offshore Reliability Data*, Hovik, Norway, 1984.

[Otway and von Winterfeldt, 1992] Otway, H. and von Winterfeldt, D., "Expert judgement in risk analysis and management: process, context and pitfalls", *Risk Analysis*, **12**, 83–93, 1992.

[Pape, 1997] Pape, R.P., "Developments in the tolerability of risk (TOR) and the application of ALARA", *Nuclear Energy*, 36, 457–463, 1997.

[Paulsen et al., 1996] Paulsen, J., Dorrepaal, J., Hokstadt, P. and Cooke, R., *The Design and Use of Reliability Data Base with Analysis Tool*, Risø-R-896(EN) NKS/RAK-1(96) R6, RisøNational Laboratory, Roskilde, Denmark, 1996.

[Pearl et al., 1990] Pearl, J., Geiger, D. and Verma, T., "The logic of influence diagrams", in *Influence Diagrams, Belief Nets and Decision Analysis*, eds R.M. Oliver and J.Q. Smith, Wiley, Chichester, 1990.

[Perlis et al., 1981] Perlis, A.J., Sayward, F.G. and Shaw, M., *Software Metrics: an Analysis and Evaluation*, MIT Press, Cambridge Mass. 1981.

[Peterson, 1976] Peterson, A.V., "Bounds for a joint distribution function with fixed subdistribution functions: application to competing risks", *Proceedings of the National Academy of Science of the USA*, **73** (1), 11–13, 1976.

[Peterson, 1977] Peterson, A.V., "Expressing the Kaplan–Meier estimator as a function of empirical subsurvival functions", *Journal of the American Statistical Association*, **72**, 854–858, 1977.

[PLG, 1983] PLG Inc (Pickard, Lowe and Garrick), *Seabrook Station Probabilistic Safety Study*, prepared for Public Service Company of New Hampshire and Yankee Atomic Electric Company, PLG-0300, December 1983.

[Pollard, 1984] Pollard, D., *Convergence of Stochastic Processes*, Springer Series in Statistics, New York–Berlin, 1984.

[Popper, 1959] Popper, K., *The Logic of Scientific Discovery*, Hutchinson, London, 1959, translated from *Logik der Forschung*, Vienna 1935.

[Pörn, 1990] Pörn, K., *On Empirical Bayesian Inference Applied to Poisson Probability Models*, Linköping Studies in Science and Technology Dissertation No. 234, Linköping, Sweden, 1990.

[Ramamoorthy and Bastani, 1982] Ramamoorthy, C.V. and Bastani, F.B., "Software reliability – status and perspectives", *IEEE Transactions on Software Engineering*, **SE-8**, no. 4, 354–371, July 1982.

[Ramsberg and Sjöberg, 1997] Ramsberg, J.A.L. and Sjöberg, L., "The cost-effectiveness of lifesaving interventions in Sweden", *Risk Analysis*, **17**, 467–478, 1997.

[Ramsey, 1931] Ramsey, F., "Truth and probability" in *The Foundations of Mathematics*, ed R.B. Braithwaite, pp. 156–198, Kegan Paul, London, 1931.

[Rasmussen *et al.*, 1981] Rasmussen, N., Pedersen, O.M., Carnino, A., Griffon, M., Mancini, C. and Gagnolet, P., *Classification System for Reporting Events Involving Human Malfunctions*, Report Risø-M-2240, Risø National Laboratories, Denmark, 1981.

[Rauzy *et al.*, 1997] Rauzy, A., Dutuit, Y. and Signoret, J.-P., "A new notion of prime implicants and its implementation by means of binary decision diagrams", in *Proceedings of Mathematical Methods on Reliability*, Bucharest, September 1997 RENEL-GSCI-CIDE, Bucharest, 1997.

[Reason, 1990] Reason, J., *Human Error*, Cambridge University Press, 1990.

[Redmil, 1989] Redmil, F.J., *Dependability of Critical Computer Systems 2*, Elsevier, London, 1989.

[Reichenbach, 1932] Reichenbach, H., "Axiomatik der Wahrscheinlichkeitsrechnung", *Mathematische Zeitschrift*, **34**, 568–619, 1932.

[Rijke *et al.*, 1997] Rijke, W. de, *et al.*, "Risman, a method for risk management of large infrastructure projects", *Proceedings of ESREL'97*, ed C. Guedes Soares, pp. 265–272, Pergamon, 1997.

[Robert, 1994] Robert, C.P., *The Bayesian Choice*, Springer Texts in Statistics, New York, 1994.

[Rogovin and Frampton, 1980] Rogovin, M. and Frampton, G.T., *Three Mile Island, a Report to the Commissioners and to the Public Government Printing Office*, January, 1980.

[Ross, 1990] Ross, S.M., *A Course in Simulation*, Macmillan, New York, 1990.

[Ross, 1997] Ross, S.M., *Introduction to Probability Models*, 6th ed, Academic Press, Boston, Mass., 1997.

[RS, 1983] Royal Society Study Group, *Risk Assessment*, London, January 1983.

[RS, 1992] The Royal Society, *Risk: Analysis, Perception and Management*, Royal Society, London, 1992.

[Salo and Hämäläinen, 1997] Salo, A. and Hämäläinen, R., "On the measurement of preferences in the Analytic Hierarchy Process" (with discussion), *Journal of Multi-Criteria Decision Analysis*, 6, 309–343, 1997.

[Savage, 1972] Savage, L.J., *The Foundations of Statistics*, 2nd ed, Dover, New York, 1972. (First published 1954.)

[Sayers, 1988] Sayers, B.A. (ed.), *Human Factors and Decision Making: their Influence on Safety and Reliability*, Elsevier Applied Science, London, 1988.

[Schachter, 1986] Schachter, R.D., "Evaluating influence diagrams", *Operations Research*, **34**, 874–882, 1986.

[Schlick, 1936] Schlick, M., "Meaning and verification", *The Philosophical Review* **45**, 339–369, 1936.

[Schnorr, 1970] Schnorr, C.P., *Zufälligkeit und Wahrscheinlichkeit*, Lecture Notes in Mathematics 218, Springer, Berlin, 1970.

[Scott and Saltelli, 1997] Scott, M. and Saltelli, A., *Journal of Statistical Computation and Simulation*, Special issue on theory and applications of sensitivity analysis of model output in computer simulations, **57**, 1997.

[Seebregts and Jehee, 1991] Seebregts, A.J. and Jehee, J.N.T., *Comparison of Guidelines for Level-2 Probabilistic Safety Assessment*, ECN Energy, ECN-C-91-085, 1991.

[Shafer and Pearl, 1990] Shafer, G. and Pearl, J. (eds), *Readings in Uncertain Reasoning*, Morgan Kaufmann, Palo Alto, 1990.

[Shrader-Frechette, 1985] Shrader-Frechette, K., *Risk Analysis and Scientific Method*, Reidel, Dordrecht, Netherlands, 1985.

[Shrader-Frechette, 1993] Shrader-Frechette, K., *Burying Uncertainty*, University of California Press, Berkeley, 1993.

[Siu and Mosleh, 1989] Siu, N. and Mosleh, A., "Treating data uncertainties in common cause failure analysis", *Nuclear Technology*, **84**, 265–281, 1989.

[Sjöberg, 1999] Sjöberg, L., Risk perception in Western Europe 10 years after the Chernobyl accident, *Values in Decisions on Risk*, ed K. Andersson, pp. 343–351, European Commission, DG XI., 1999.

[Sjöberg, 2000] Sjöberg, L., "Factors in risk perception", *Risk Analysis*, **20**, 1–12, 2000.

[Slovic et al., 1990] Slovic, P., Fischhoff, B. and Lichtenstein, S., "Rating the risks" in [Glickman, Gough], pp. 61–75. 1990.

[Smith, 1989a] Smith, J.Q., "Influence diagrams for statistical modelling", *Annals of Statistics*, **17**, 654–672, 1989.

[Smith, 1989b] Smith, J.Q., "Influence diagrams for Bayesian decision analysis", *European Journal of Operations Research*, **40**, 363–376, 1989.

[Smith, 1995] Smith, J.Q., "Handling multiple sources of variation using influence diagrams", *European Journal of Operations Research*, **86**, 1995.

[Smith and Wood, 1989] Smith, D.J. and Wood, K.B., *Engineering Quality Software*, Elsevier, 2nd ed, 1989.

[Sperner, 1928] Sperner, E., "Ein Satz über Untermengen einer endlichen Menge", *Mathematisch Zeitschrift*, **27**, 544–548, 1928.

[SRD, 1983] *A review by SRD of 'Precursors to potential severe core damage accidents 1969–1979. A status report'*, Safety and Reliability Directorate, June 1983.

[Stephens, 1986] Stephens, M.A., "Tests based on EDF statistics", in *Goodness-of-fit Techniques*, eds R.B. D'Agostino abd M.A. Stephens, Marcel Dekker, New York, 1986.

[Swain, 1989] Swain, A.D., *Comparative Evaluations of Methods for Human Reliability Analysis*, GRS-71, ISBN 3-923875-21-5, GRS, Köln, 1989.

[Swain and Guttmann, 1983] Swain, A.D. and Guttmann, H.E., *Handbook of Human Reliability Analysis with Emphasis on Nuclear Power Plant Applications*, US Nuclear Regulatory Commission, August 1983.

[Tenga et al., 1995] Tenga, T.O., Adams, M.E., Pliskin, J.S., Safran, D.G., Siegel, J.E., Weinstein, M.C. and Graham, J.D., "Five hundred life-saving interventions and their cost-effectiveness", *Risk Analysis*, **15**, 369–390, 1995.

[Thayer et al., 1978] Thayer, T.A., Lipow, M. and Nelson, E.C. *Software Reliability: a Study of a Large Project Reality*, North-Holland, Amsterdam, 1978.

[Thompson, 1986] Thompson, M., "To hell with the turkeys", in *Values at Risk*, ed MacLean, D, pp. 113–135. Rowman and Allanheld, Totowa, NJ, 1986.

[Thompson and Perry, 1992] Thompson, P.A. and Perry, J.G. (eds), *Engineering Construction Risks*, Thomas Telford, London, 1992.

[TNO, 1983] TNO, "LPG: a study, a comprehensive analysis on the risks inherent in the storage, transshipment, transport and use of LPG and motor spirit", TNO, Apeldoorn, Netherlands, vols. 1–3, May 1983.

[Tomic, 1993] Tomic, B., "Multi-purpose in-plant data system" International Atomic Energy Authority (IAEA), June 1993.

[TRW, 1976] TRW Defence and Space Systems Group, *Software Reliability Study*, Redondo Beach, Calif., rep. 76-2260.1-9-5, 1976.

[Tsiatis, 1975] Tsiatis, A., "A nonidentifiabilty aspect of the problem of competing risks", *Proceedings of the National Academy of Sciences of the USA*, **72**, no. 1, 20–22, 1975.

[UN, 1992] United Nations Rio Convention (1992), United Nations Conference on Environment and Development: Rio Declaration on Environment and Development, June 14, 1992, reprinted in *International Legal Materials*, **31**, 874–879, 1992.

[Union of Concerned Scientists, 1977] Union of Concerned Scientists, *The Risks of Nuclear Power Reactors: a Review of the NRC Reactor Safety Study WASH-1400*. Cambridge, Mass., 1977.

[USAF, 1991] United States Air Force, *Military Handbook; Reliability Prediction of Electronic Equipment*, MIL-HDBK-217F, Griffiss Air Force Base, Rome, New York, 1991.

[US Navy, 1987] United States Navy, *Sneak Circuit Analysis*, NAVSO P-3634, August 1987.

[van der Weide and Bedford, 1998] van der Weide, J.A.M. and Bedford, T., "Competing risks and eternal life", in *Safety and Reliability (Proceedings of ESREL'98)*, eds S. Lydersen, G.K. Hansen and H.A. Sandtorv, vol. 2, pp. 1359–1364, Balkema, Rotterdam, 1998.

[van Elst et al., 1998] van Elst, N., Bedford, T., Jorissen R. and Klaassen, D., "A generic risk model for the closing procedure of moveable water barriers", *Safety and Reliability (Proceedings of ESREL'98)*, eds S. Lydersen, G.K. Hansen and H.A. Sandtorv, vol. 2, pp. 435–442, Balkema, Rotterdam, 1998.

[van Lambalgen, 1987] van Lambalgen, M., *Random Sequences* PhD dissertation, University of Amsterdam 1987.

[Versteeg, 1988] Versteeg, M., "External safety policy in the Netherlands: an approach to risk management", *Journal of Hazardous Materials*, **17**, 215–221.

[Vesely, 1977] Vesely, W.E., "Estimating common cause failure probabilities in reliability and risk analyses; Marshall–Olkin specialization", IL-0454, 1977.

[Vesely et al., 1981] Vesely, W.E., Goldberg, F.F., Roberts, N.H. and Haasi, D.F., *The Fault Tree Handbook*, US Nuclear Regulatory Commission, NUREG 0492, 1981.

[von Mises, 1919] von Mises, R., "Grundlagen der Wahrscheinlichkeitsrechnung", *Mathematische Zeitschrift*, **5**, 52–99, 1919.

[von Mises, 1981] von Mises, R., Wahrscheinlichkeit, Statistike und Wahrheit,

Springer, Berlin, 1936. (English translation: *Probability, Statistics and Truth*), Dover, New York, 1981.

[von Neumann and Morgenstern, 1947] von Neumann, J. and Morgenstern, O., *Theory of Games and Economic Behavior*, 2nd ed, Princeton University Press, Princeton, (First published by Wiley, New York, 1994) 1947.

[Vrijling and van Gelder, 1997] Vrijling, J.K. and van Gelder, P.H.A.J.M., "Societal risk and the concept of risk aversion", *Proceedings of ESREL'97*, ed C. Guedes Soares, pp. 45–52, Pergamon, 1997.

[Vrijling *et al.*, 1995] Vrijling, J.K. et al, 'A framework for risk evaluation', *Journal of Hazardous Materials*, **43**, 245–261, 1995.

[Whetton, 1993] Whetton, C.P., "Sneak analysis of process systems", *Transactions of the Institute of Chemical Engineers*, **71**, Part B, August 1993.

[Wiggins, 1985] Wiggins, J., *ESA Safety Optimization Study*, Hernandez Engineering, HEI-685/1026, Houston, Texas, 1985.

[Winkler, 1981] Winkler, R.L., "Combining probability distributions from dependent information sources", *Management Science*, **27**, 479–488, 1981.

[Winkler, 1996] Winkler, R.L., "Uncertainty in probabilistic risk assessment", *Reliability Engineering and System Safety*, **54**, 127–132, 1996.

[Woo, 1999] Woo, G., *The Mathematics of Natural Catastrophies*, Imperial College Press, London, 1999.

[Wright and Ayton, 1994] Wright, G. and Ayton, P. (eds), *Subjective Probability*, Wiley, New York, 1994.

[Xiaozhong and Cooke, 1991] Xiaozhong, W. and Cooke, R.M., "Optimal inspection sequence in fault diagnosis", *Reliability and System Safety*, **37**, 207–210, 1991.

[Zadeh, 1965] Zadeh, L., "Fuzzy sets", *Information and Control*, **8**, 338–353, 1965.

Index

absolute probability judgement, 235
accident sequence, 99
act, 24
active components, 105
additive value model, 271
age dependent censoring, 175
ALARA, 356
ALARP, 8, 362
aleatory uncertainty, 33
alpha-factor model, 151
ambiguity, 19, 35
analytic hierarchy process, 283
APJ, 220
Apollo, 4
arrow reversal, 294
attribute hierarchy, 270
attributes, 269
availability, 49

background measure, 202
Barlow–Proschan importance, 136
base rate participation, 351
Bayes' Theorem, 64
Bayesian, 61
    belief networks, 286
    inference, 63
belief nets, 286
Bernoulli distribution, 41
beta distribution, 53, 69, 87
beta-factor model, 146
binary decision diagram, 123
binomial distribution, 41, 58, 69, 76, 78
binomial failure rate model, 148
Boolean logic, 100

calibration, 200, 251
Cauchy–Schwarz, 59
censored data, 88
central limit theorem, 58
Challenger, 4
chance node, 289, 294
Chernobyl, 7
chi-squared test, 79, 80

child node, 289
classical statistical inference, 75
cobweb plots, 342
coherent, 102
coloring, 162
command failure, 105
common cause, 118, 142
competing risk, 166
    dependent, 184
    independent exponential, 172
    random clipping, 175
    random signs, 175
concatenation, 162
conditional independence, 288
conditional independence property, 236
conditional preferential independence, 274
conditionality principle, 62
confidence interval, 78
conjugate distribution, 69
consequence, 25, 300
consequence analysis, 12
consistency, 70, 74
convergence in distribution, 58
convolution, 52
copula, 330
    diagonal band, 331
    minimum information, 331
correlation, 44
cost–benefit analysis, 281, 363
countably additive, 40
countermeasures, 300
counting process, 55
covariance, 44
critical, 136
    item list, 136
    path, 302
cumulative distribution function, 41
cut set, 110
cut set representation, 114

deaths per million, 350
decision node, 289
decision theory, 63, 259

decision tree, 262
decreasing failure rate, 46
degree of belief, 24
Delphi method, 194, 235
delta yearly probability of death, 353
density, 41
dependence, 227
dependent failures, 140
descriptive decision theory, 259
deterministic node, 289
directed acyclic graph, 286
dispersion models, 208
disutility, 360
dual tree, 115
duality, 112

elicitation variables, 318
empirical Bayes prior, 73
empirical distribution function, 77
entropy, 72, 346
epistemic uncertainty, 33
equilibrium availability, 49
error seeding, 247
errors
  of commission, 225
  of omission, 225
event, 22
event tree, 99
exchangeable, 35
expectation, 42
expected reward, 205
expert opinion, 191
exponential, 42
exponential distribution, 47, 67, 76, 78
external boundaries, 103
extreme value theory, 94

failure,
  critical, 153, 157
  degraded, 153, 157
  entrained functional unavailability, 158
  incipient, 153
  total, 158
failure effect, 105
failure likelihood index, 231
failure mechanism, 105
failure mode, 105
failure rate, 19, 45, 46, 62
  naked, 155
  observed, 155
fault tree, 99
field, 39
finitely additive, 39
Fisher information, 75
FMEA, 99
force of mortality, 45
frequency, 10
frequentist, 61, 75
Fussell–Vesely importance, 136
fuzzy, 20

gamma distribution, 52, 67

GEMS, 222
generic data, 64
God, 18
Goel–Okumoto model, 250
graphical methods, 85
group risk, 356
Gumbel distribution, 96

hazard, 10
hazard rate, 45
human cognitive reliability, 219
human life, value of, 363
human reliability, 218
hyperparameters, 74
hypothesis testing, 79

identifiability, 168
implicant, 124
importance, 135
inclusion–exclusion principle, 40
increasing failure rate, 46
independence preservation property, 198
independent, 40, 43, 59
influence diagrams, 236, 286
information, 73, 347
informativeness, 200
integration error, 253
internal boundaries, 104
intrinsic range, 200
involuntary, 351

Jelinski–Moranda model, 247
joint density, 43
joint distribution, 43

Kaplan–Meier estimator, 91
knowledge-based, 221
Kolmogorov–Smirnov test, 80
Kullback–Liebler divergence, 73

Laplace's test, 80
Latin hypercube sampling, 328
Lewis report, 6
likelihood principle, 62
linear opinion pools, 199
Littlewood's model, 248
Littlewood–Verral model, 249
log-linear NHPP, 76
lognormal distribution, 54
loss of life expectancy, 351

maintenance
  corrective, 156
  preventive, 156
marginal value functions, 271
marginalization property, 197
Markov chain, 287
Markov property, 293
Marshall–Olkin model, 143
maximum entropy prior, 72
maximum likelihood, 63, 75, 93
mean time to failure, 42

measure, 39
median rank, 87
memoryless property, 47
minimal cut set, 110
minimal p-cut, 130
minimal path set, 112
minimally informative probability, 320
mixture, 48
MOCUS, 121
model parameters, 318
model uncertainty, 19, 34
monotone, 45
Monte Carlo simulation, 327
moralized graph, 292
mortality rate, 354
multi-attribute decision theory, 269
multi-attribute utility theory, 232
multi-attribute value theory, 270
multiple greek letter model, 151
mutual preferential independence, 271
mutually independent, 40

network, 302
nominal group technique, 235
normal distribution, 42, 58
normative decision theory, 259

observable quantities, 317
observation, 28
operating mode
  alternating, 159
  continuous, 159
  standby, 159

pairwise comparison, 232
parallel system, 117
parameter uncertainty, 34
parent node, 289
Pareto distribution, 69
partial belief, 24
partial correlation, 334
passive components, 105
path set, 112
path set representation, 115
performance based weighting, 199
performance shaping factors, 223
Peterson bounds, 181
Poisson distribution, 51, 58
Poisson process, 52, 55, 145
  homogeneous, 56
  nonhomogeneous, 56
pooling, 161
posterior distribution, 64
power law NHPP, 77
predictive mean, 69
predictive variance, 69
preventive maintenance, 83
primary fault, 105
prime implicant, 124
prior
  improper, 71
  non-informative, 72

prior distribution, 64
probabilistic inversion, 316
probability, 10, 20, 21
  classical, 22
  frequency, 23
  qualitative, 27
  subjective, 24
probability-conditional, 40
product moment correlation, 44
product-limit estimator, 91
proper scoring rule, 205

quantile function, 45

random variable, 41
rank correlation, 45
rank distribution, 87
rare event approximation, 118
rational consensus, 71, 199
rational preference, 24
Reactor Safety Study, 6
relative information, 201
reliability, 41
repair, 50
ROCOF, 56
rule-based, 221

scenario, 10
scheduling
  calendar based, 156
  condition based, 156
  emergency, 157
  opportunity based, 157
scoring rule, 205
secondary fault, 105
sensitivity analysis, 327
series, 48
series system, 117
Seveso, 8
σ-field, 39
significance level, 79
single point failure, 135
skill-based, 221
SLIM, 220
sneak analysis, 241
software reliability, 240
splicing, 162
stationarity, 58
stationary increments, 57
statistical inference, 61
statistical likelihood, 200
stochastic processes, 55
strong set-wise function property, 197
subsurvival function, 168
  conditional, 169
Success Likelihood Index, 231
Success Likelihood Index Methodology, 230
sufficiency principle, 62
sufficient statistic, 62
superposition, 145, 161
survivor, 41

target variables, 318
task analysis, 220
temporal boundaries, 104
test statistic, 79
THERP, 224
Three Mile Island, 7
time dependence, 228
time reliability correlations, 232
time window censoring, 179
top event, 100
total probability, 65
total risk, 359
total time on test, 173
transitivity, 261
truth conditions, 18

u-plot, 251
uncertainty, 17, 179
  aleatory, 317
  model parameter, 317

uncertainty analysis, 316, 326
utility, 24, 261, 268

value
  of human life, 363
  of information, 264
  of perfect information, 267
value function, 261
value node, 289
variance, 42
vine, 334
volitional uncertainty, 18, 35
voluntary, 351

Weibull distribution, 83, 85
weighting factors model, 271

Printed in the United States
By Bookmasters